Montenbruck · Pfleger

Astronomy on the Personal Computer

Oliver Montenbruck Thomas Pfleger

Astronomy on the Personal Computer

Translated by Storm Dunlop
With a Foreword by Richard M. West

Fourth, Completely Revised Edition
With 46 Figures and CD-Rom

Springer

Dr. rer. nat. Oliver Montenbruck
DLR-GSOC, D-82230 Weßling
e-mail: oliver.montenbruck@dlr.de

Dipl.-Ing. Thomas Pfleger
Schützenstraße 25, D-53773 Hennef
e-mail: th.pfleger@t-online.de

Translator
Dr. Storm Dunlop
140 Stocks Lane
East Wittering, Chichester
West Sussex PO20 8NT, United Kingdom

Title of the original German edition:
O. Montenbruck, T. Pfleger: *Astronomie mit dem Personal Computer*. Dritte Auflage

Cover picture: Comet Hale-Bopp, taken on March 29, 1997 in Potzlow near Seehausen, Germany. To the upper left one sees the stellar cluster $h + \chi$ in Perseus and to the lower right the Andromeda nebula (M 31). The exposure time was 10 min.
From the picture archive of Sterne und Weltraum; Photo: Thomas Helms.

Library of Congress Cataloging-in-Publication Data applied for.
Die Deutsche Bibliothek - CIP-Einheitsaufnahme: *Astronomy on the personal computer* [Medienkombination] / Oliver Montenbruck ; Thomas Pfleger.
Dt. Ausg. u.d.T.: Astronomie mit dem Personal-Computer

Additional material to this book can be downloaded from http://extras.springer.com

DOI 10.1007/978-3-642-03436-7
Buch - 4., completely rev. ed. - 2000; CD-ROM - 4., completely rev. ed. - 2000

MyCopy version of the original edition 2000

Catalogue on the CD-ROM
PPM Star Catalogue, Volume I & II
© Astronomisches Rechen-Institut Heidelberg, Germany 1991
PPM Star Catalogue, Volume III & IV
© Astronomisches Rechen-Institut Heidelberg, Germany 1993

Typesetting: camera-ready copy from the authors
Cover design: Erich Kirchner, Heidelberg

Printed on acid-free paper 55/3141/ts 5 4 3 2 1 0
www.springer.com/mycopy

Foreword to the Fourth Edition

Astronomy is a venerable science with roots in the distant past. At the same time, it is one of the most modern and dynamic areas of basic research, involved in the study of truly basic questions that are of fundamental significance for mankind. From being a mainly observational-descriptive activity over the past millennia, it has recently evolved into a high-technology venture aimed at a profound understanding of our surroundings in space and, most importantly, our own place therein. It is obviously a very broad science with natural links in a great many directions, including physics and chemistry, as well as meteorology and mineralogy, from optics, mechanics and electronics to advanced mathematics and special computational techniques and, not least, philosophy.

Recent years have seen tremendous advances in this science, above all in observational and computational techniques. With larger telescopes and better instruments, professional observers are reaching for the unknown, beyond the current horizons. With faster and more reliable communication links, in particular the all-pervading world wide web, entirely new international research opportunities have materialized. What is no less impressive, however, is the enormous speed and determination with which amateur astronomers have reacted to the new possibilities. Thanks to improved and affordable observational facilities, in particular the wide availability of reasonably priced CCD cameras, they are now entering many fields that were until recently the exclusive domains of their professional colleagues with force.

The present book is another clear sign of this welcome development. Since the appearance of the first German edition, just 10 years ago, it has become a basic text for the serious amateur. The first English edition, published just one year later, made it accessible to a world wide community of astronomy fans, and one that has continued to grow ever since. It is no secret either that amateur astronomers are not the only ones to have profited from its clear and thoughtful presentation. Quite a few professionals – I am one of them – gladly admit that they have learned important details from its various informative chapters.

Why is it then that this book is so important and so basic in its field? Many answers may be given. Every reader will probably have his or her favourite reasons, be it the possibility to predict lunar phases and solar eclipses many years into the future, or the accurate computation of asteroid positions by means of astrometric standard stars in the observed field. However, I believe that there are other important, more general considerations.

The subjects covered and the methods described in this book constitute a very fundamental and absolutely indispensable part of astronomy. Without profound knowledge of the exact timings of celestial phenomena, without an understanding of the dynamics of the solar system and without the ability to calculate accurate positions of planets, comets and stars from any given place at any given time in the (near) future, there would be no observational astronomy as we know it today.

The readers will find here clear and detailed explanations about how such calculations are made, as well as the software means to do so with a high degree of accuracy. Of particular value is the possibility to read, to change and adapt the source codes to suit specific personal needs and preferences, now further facilitated by the use of the C++ language. This is a feature not offered by the increasingly popular 'planetarium-type' software packages that are directed towards a less demanding sector. The extensive object catalogues on CD-ROM provide a wonderful database for use with these programmes. The availability of a dedicated webpage with updates and related links is an added bonus.

Indeed, as the reader/user steps onto 'professional' territory by means of this book, he or she will find many new and exciting horizons opening. These encompass more than the valuable ability to serve the technical needs of astronomical observations – just consider the fun of working with well-conceived software on the home computer, and the satisfaction of better understanding the 'mechanics' of nature. Above all, however, it will help the serious amateur to achieve professional-quality results that will directly contribute to the advancement of astronomy.

The authors are to be congratulated for this fine service to the astronomical community. I have no doubt that this marvellous book will long continue to be a major reference source for all astronomy clubs and individuals who regularly observe the skies.

Garching
November 1999

Richard M. West (ESO)

Preface to the Fourth Edition

When we look back over the last ten years, it is largely reassuring that the astronomical and mathematical procedures in our book *Astronomy on the Personal Computer* have lost little of their validity and relevance at the end of this millennium. What is impressive, however, is the frenzied, or rather explosive, development of computer techniques, such that a computer has already won a firm place in many households alongside the telephone and the television set.

So it is not surprising then, that for this new edition we have concentrated on making a fundamental revision and modernization of all the programs. In view of the continuing decline in the use of the Pascal programming language, we have been happy to accede to the frequently expressed desire for our programs to be available in C++. If we exclude system-specific languages such as VisualBasic, then C and C++ constitute the most popular and widely used programming languages. As well as their availability for the most diverse computer systems, their significance also derives from their excellent support of applications close to the system level. This is particularly the case for C++, which, with its object-oriented programming techniques, is especially suitable for the development of graphical user interfaces.

In adapting the previous Pascal programs, we have not restricted ourselves to a simple port to the new language, but have instead undertaken a comprehensive, new implementation, taking the multiple possibilities offered by C++ into account. We hope that in this way the reader will be provided with an up-to-date and powerful suite of programs and libraries, without sacrificing the comprehensibility that had proved itself in the previous editions.

Changes in content result from (among other things) the consistent introduction of vector and matrix notation, which permits a compact and up-to-date representation of all coordinate transformations, and which is handled in the C++ programs through corresponding classes and operators. Wherever necessary, modifications were made to conform to current conventions and values. This involves both improved rotation elements for the major planets, and also the reckoning of geographical longitude on Earth. As in geography and satellite geodesy, this is now measured positively towards the east in astronomy. This has caused, for example, differences of sign in calculating local sidereal time or the station coefficients for stellar occultations. These are correspondingly incorporated into the individual algorithms and programs.

The chapter on the determination of orbits from three observations has been extended to include a description of the full Gaussian method of orbital determination, so that cases with multiple solutions may now be handled without any problems. An addition is the `Phases` program, which identifies possible solar or lunar eclipses, and which also calculates phases of the Moon. Many improvements have been introduced to make the programs more user-friendly. In calculating rise and set times, for example, it is now possible to choose between civil, nautical, and astronomical twilight. Similarly, in calculating stellar occultations, events that are difficult to observe are flagged, using a comprehensive set of criteria. Wherever necessary, input and output files may be specified in the command line for all programs.

With the accompanying CD-ROM, it has been possible, for the first time, to provide the executable programs for PCs running Windows and Linux in uncompressed form. Readers without a C++ compiler or programming experience are thus able to use the CD-ROM with its data and programs immediately for their own calculations. The addition of a compiler is necessary only for those readers who want to adapt or expand the programs for their own personal requirements. To this end, advice on the corresponding installation and implementation is given in the Appendix.

As an additional bonus, the CD-ROM contains the *Position und Proper Motion Catalogue* by S. Röser and U. Bastian (published by Spektrum Akademie Verlag), with detailed information on approximately 470 000 stars, as well as the *Asteroid Orbital Elements* database by E. Bowell with orbital information for more than 100 000 minor planets. The corresponding files were kindly made available by the Astronomische Rechen-Institut Heidelberg and the Lowell Observatory, to whom – as to the authors themselves – we must here express our grateful thanks. To simplify the use of these extremely comprehensive catalogues, various additional programs are included on the CD-ROM, which may be used to manipulate the data for entry into `Foto` and `Numint` programs.

We would also like to advise readers here about the newly established website `http://www.springeronline.com/3-540-67221-4/`, on which we, together with the publisher, provide useful information relating to the book. Here, for example, you will find the source code for the Pascal programs from the previous edition of *Astronomy on the Personal Computer*, and references to useful Internet resources. If necessary, corrections and software updates will also be posted here.

We wish to convey our particular thanks to Dr Ch. Caron, C.-D. Bachem, and B. Reichel-Mayer of Springer-Verlag in Heidelberg for their excellent cooperation and support during the preparation of this book. Similarly, we want to thank Dr L. Weidinger, whose support first made it possible to port our programs into a Linux environment.

December 2003 *O. Montenbruck and T. Pfleger*

Preface to the Second Edition

Since the publication of the first edition of *Astronomy on the Personal Computer*, we have received numerous comments and suggestions. Together with the publishers' interest in a new edition, this prompted us to revise the book and to incorporate a wide range of improvements to the text.

The first important addition is a chapter on the calculation of perturbations. This shows how gravitational perturbations by the major planets may be incorporated into the calculation of ephemerides for minor planets and comets. The `NUMINT` program described here enables more accurate positions to be calculated. This will be of considerable assistance in searching for what are often extremely faint objects. This tool for calculating perturbations complements the chapters on the calculation of ephemerides, the determination of orbits, and astrometry.

The second additional chapter discusses the calculation of physical ephemerides for the major planets and the Sun, and thus fills a gap in the earlier edition's coverage. Amateurs now have at their disposal the means of both predicting and subsequently reducing interesting planetary observations.

Other changes mainly concern the calculation of rising and setting times for the planets, and of the local circumstances that apply to solar eclipses.

Finally, there is now a single version of the program diskette that is supplied with the book. Since the firm of Application Systems Heidelberg have introduced their Pure Pascal compiler for Atari ST/TT computers that is compatible with Borland's Turbo Pascal, there is no need for a special Atari version of the program diskette. The enclosed diskette may, therefore, be used without modifications with Turbo Pascal on IBM-compatible machines or with Pure Pascal on Atari computers. Details about the appropriate installation procedures may be found in the `AAREADME.DOC` file on the diskette.

We should like to thank Springer-Verlag for their helpful co-operation, and also Application Systems Heidelberg for their technical support.

Munich, July 1994 *O. Montenbruck and T. Pfleger*

Preface to the First Edition

Nowadays anyone who deals with astronomical computations, either as a hobby or as part of their job, inevitably turns to using a computer. This is particularly true now that personal computers have become firmly established as ubiquitous aids to living. Calculations that could not even be contemplated a short time ago are now available to a whole range of users, and at no farther remove than their desks. Not only has the technical capacity of computers grown, but so has the need for powerful – i.e., fast and accurate – programs. So the wish to avoid conventional, astronomical yearbooks as much as possible is quite understandable. We were therefore delighted to take up our publisher's suggestion that we explain the fundamental principles of spherical astronomy, ephemeris calculations, and celestial mechanics in the form of this book.

Astronomy on the Personal Computer offers readers who develop their own programs a comprehensive library of Pascal procedures for solving a whole range of individual steps that frequently occur in problems. This includes routines for common coordinate transformations, for time and calendar calculations, and for handling the two-body problem. Specific procedures allow the exact positions of the Sun, the Moon, and the planets to be calculated, taking mutual perturbations into account. Thanks to the widespread use of Pascal as a computing language, and by avoiding computer-specific commands, the programs may be employed on a wide range of modern computers from the PC to the largest mainframes. The large number of routines discussed should at least mean that few readers will have to 're-invent the wheel', and that they will therefore be free to concentrate on their own particular interests.

Each chapter of this book deals with a fairly restricted theme and ends with a complete main program. From simple questions, such as the determination of rising and setting times or the calculation of the positions of the planets, more complex themes are developed, such as the calculation of solar eclipses and stellar occultations. The programs for the astrometric reduction of photographs of star fields and for orbit determination enable users to derive orbital elements of comets or minor planets for themselves. Even readers without programming experience will be able to use the appropriate applications.

Sufficient details are given of the astronomical and mathematical grounds on which solutions of specific problems are based for readers to understand the programs presented. This knowledge will enable them to adapt any of the programs to their individual needs. This close link between theory and practice also enables us to explain what are sometimes quite complex aspects in a much easier fashion than the descriptions found in classical textbooks. To sum up, we hope that we have given readers a fundamental grounding in using computers for astronomy.

We should like to thank S. Dunlop for producing such an excellent translation and Dr. G. Wolschin and C.-D. Bachem of Springer-Verlag for their cordial co-operation and interest during the process of publishing this book. Our thanks are also due to all our friends and colleagues, who, with their ideas and advice, and their help in correcting the manuscript and in testing the programs, have played an important part in the success of this book.

Munich, August 1990 *O. Montenbruck and T. Pfleger*

Contents

1. Introduction

Recent years have seen a continuous increase in the power of small computers and a simultaneous decrease in their price. As a result many people interested in astronomy have such equipment at their disposal. This prompts the idea of using these computers for astronomical computation. What positive advantages are there in using one's own computer, when one can obtain all the most important data required for observing, fully as accurately, in one of the many yearly handbooks?

1.1 Some Examples

Let us first consider the calculation of the rising and setting times of the Sun and the Moon. Moving from one observation site to another differences in time soon occur, and these cannot be neglected. Yet in a handbook the rising and setting times are generally given for just a few places. If one wants the times at a given position, one is generally advised to use an interpolation routine or else read the corrections from nomograms. In this case it is significantly more convenient and more accurate if the computer can give the required data without further work.

The major portion of an almanac consists of pages and pages of tables of the positions of the Sun, the Moon, and the planets. Of all the information published, frequently only a small fraction is of specific interest. Using appropriate programs, just the values actually required can be calculated and simultaneously expressed in the desired coordinate system.

Even more important is the possibility of being able to calculate the orbit of a celestial body oneself. With new discoveries, actual ephemerides are frequently not available or only obtainable after some delay. A single set of orbital elements is sufficient for the motion of a comet to be followed with an adequate degree of accuracy.

Solar eclipses are impressive celestial phenomena, which often prompt amateur astronomers to undertake long journeys to observe them. Even though official sources may provide detailed information on a given eclipse, it is often desirable to plan a trip long before these data are published. If one requires information for a particular observing site that is not covered in the almanacs, a special program for computing local circumstances will be extremely useful.

Anyone who wants to observe stellar occultations by the Moon requires contact times for their observing site. When handbooks contain such predictions, these are generally restricted to a few selected major cities. Anyone wanting to observe from another site is once again reduced to resorting to approximations. The `Occult` program given in this book enables both the occurrence of stellar occultations and precise data about them to be calculated for any arbitrary selection of stars.

Many amateur astronomers are keen on astrophotography. Using reference stars with known coordinates, the position of a minor planet or a comet can be determined from photographs that one has taken oneself. If at least three exposures, taken at intervals, are available then the orbital elements may be calculated in the process of determining an orbit. In the programs `Photo`, `Gauss` and `Comet` we provide the necessary tools for such highly complicated calculations that are required. Observers of minor planets will appreciate having the `Numint` program to use for computing precise ephemerides that take gravitational perturbations by the major planets into account, and that may be used even if the available orbital elements refer to an epoch that is several years ago.

1.2 Astronomy and Computing

Any introduction to the calculation of astronomical phenomena using computer programs would be incomplete without a discussion of the accuracy that can be attained. It is important, however, to first establish that the value of such programs primarily depends on the mathematical and physical description of the problem, as well as on the quality of the algorithms that are employed. It is often incorrectly assumed that the use of double-precision arithmetic will ensure the accuracy of a specific program. Yet even an accurately calculated program must return inaccurate results if the methods of calculation are not accurate in themselves.

An example of this is the calculation of ephemerides using the laws governing the two-body problem. Here the orbit of a planet is represented in a simplified manner by an ellipse. But apart from the gravitational force exerted by the Sun, which is responsible for this form of orbit, other forces are produced by the remaining planets. These lead to periodic and long-term perturbations, which may amount to about one arc-minute for the inner planets, and up to a degree for the outer planets. A theoretical Keplerian orbit is therefore only an approximation of the actual circumstances. If we require greater accuracy, then the use of more complicated descriptions of planetary orbits that model the mutual perturbations between the planets is unavoidable. Using high-precision arithmetic in the computer would not help in the slightest!

In developing our programs, we have tried to obtain an accuracy that is approximately the same as that found in astronomical yearbooks. The errors in the fundamental routines for determining the coordinates of the Sun, the Moon, and the planets amount to about $1''$-$3''$. This accuracy is sufficient for the calculation of solar eclipses or stellar occultations, and should therefore be sufficient for most other purposes.

Anyone who wants to carry out accurate calculations consistently, however, must first be clear about the exact definitions of the coordinate systems used. Unfortunately, experience shows that values in different coordinate systems are cheerfully compared with one another. Then the cry goes up that the program is not accurate in its calculations, because different values are given in a yearbook. Frequently the discrepancy can be explained by the coordinates having a slightly different basis, which was not recognized by the user, who thus suspected an 'error'.

We have therefore noted, at all relevant points, important corrections such as precession and nutation, or aberration and light time. The difference between Universal Time and Ephemeris Time, which repeatedly causes difficulties, is frequently discussed. Readers should not allow the frequent mention of these effects to confuse them. It is rather our intention that this form of discussion and its immediate application to appropriate programs will enable them to obtain a better feel of the magnitude and the practical significance of the individual corrections. The occasional repetitions appear to be necessary, because each chapter – and with it the description of a very specific topic – may also be read on its own. But as the programs are frequently constructed around the same routines, readers should also take the opportunity of reading other relevant chapters.

The power of a program not only depends on the mathematical description of the problem, but also on the use of appropriate computational methods. These are discussed at the time, in connection with the various applications. An example of this is the evaluation of angular functions in calculating the periodic perturbations in the orbit of the Moon and of the planets. By making use of the addition theorem and recursive relations for the trigonometric functions, the computing times required by the programs `Planpos` and `Luna` are significantly reduced. If a particularly large number of positions, separated by short intervals, of a celestial body are required for some specific purpose, it is advisable to use Chebyshev polynomial approximations. It is then possible to obtain, without a loss of accuracy, valid expressions – covering a limited period of time – that can be particularly easily and quickly evaluated. Use of this technique is made in the programs `Eclipse`, `EclTimer` and `Occult`. Other aspects of mathematical computation are discussed in association with solving equations by Newton's procedure, *regula falsi*, and quadratic interpolation. In addition, a simple, but numerically stable algorithm for determining least-squares-fits is described in connection with the astrometric reduction of photographic plates.

1.3 Programming Languages and Techniques

Alongside the phenomenally rapid changes in hardware in recent years, software techniques have also undergone major developments. Applications programmes, which are constantly becoming more and more powerful and comprehensive, generally have a graphical user interface, and here Microsoft Windows has become the recognized standard. But a lot has changed from the programmer's point of view. Because it has been either desirable or essential to develop such graphical and event-driven applications that respond to the user's actions, object-oriented methods of software development have become the norm. In addition, business data-processing has introduced various new trends. Software needs to be network-aware, and to be conceived and developed as largely independent of hardware platforms and specific operating systems. One result of this is the spread of object-oriented programming languages such as C++ and, more recently, Java (which is very similar to C++), the advantages of which are seen primarily in the field of data-processing over firms' intranets and via the Internet.

When it comes to the development of technical and scientific applications, as before, the structured languages such as Fortran and Ada are employed, although both have been significantly modernized as Fortran 90 and Ada 95. Apart from these languages, however, even in this field the object-oriented C++ is gaining new adherents, which must primarily be because of its universal applicability, the excellent performance of the compiled programs, and its general availability. C++ allows both structured and object-oriented programming, and is therefore often described as being a hybrid language. It is true, of course, that astronomical and numerical algorithms are able to make only restricted use of the advantages of object-oriented programming. If, however, such core algorithms are to be used in programs with a graphical user interface, then C++ proves to be a flexible and powerful language, that allows algorithms, user input and graphics to be equally efficiently implemented. This is why we have chosen C++ as the programming language for the current, new edition of this book.

It must be said here, however, that when compared with other languages, C++ does have many deficiencies that make the development of complex scientific programs more difficult. These include, for instance, the poor support for single- and multi-dimensional arrays, the lack of local procedures, the inadequate module concept, as well as non-existent index checking, and unchecked type conversion.

As a consequence of the problems that are dealt with, and the methods that are available in the programming language, elements of both structured and object-oriented programming are equally employed in the programs given here. As in the earlier Pascal implementation the most important aims in doing so are a transparent and understandable conversion of the basic algorithms as well as the provision of modular and reusable components.

In creating astronomical programs, we repeatedly encounter certain basic subsidiary problems that are particularly well suited for this approach. Examples are the evaluation of mathematical functions, the conversion of coordinates between various systems, and the determination of precise positions for the planets, the Sun, or the Moon. It would be extremely laborious to have to repeat these tasks every time. If the required functions are always readily available in the form of a library of powerful and reliable functions, then one can concentrate on the specific features required from the program being developed, and thus achieve one's goal more rapidly and easily.

Many of the operators, functions, and classes that we set out in this book thus form the basis for a powerful, astronomical program library. It includes separate modules for the following:

- input and output routines (`APC_IO.cpp`),
- mathematical and astronomical constants (`APC_Const.cpp`),
- technical and scientific classes and functions (`APC_Math.cpp`, `APC_VecMat3D.cpp`),
- classes and functions for solving specific mathematical problems (`APC_Cheb.cpp`, `APC_DE.cpp`),
- functions and auxiliary types for time and calendrical calculations (`APC_Time.cpp`),
- various functions for spherical astronomy (`APC_Spheric.cpp`),
- special functions for precession and nutation (`APC_PrecNut.cpp`),
- functions for calculating elliptical, parabolic and hyperbolic Keplerian orbits (`APC_Kepler.cpp`),
- functions for calculating precise positions for the Sun, Moon, and planets taking the various perturbations into account (`APC_Sun.cpp`, `APC_Moon.cpp`, `APC_Planets.cpp`) as well as
- functions for calculating physical ephemerides of the planets (`APC_Phys.cpp`).

These individual components of the library will be developed and explained in stages in the corresponding chapters of this book. By using the appropriate header files (`APC_*.h`), the necessary modules may be introduced into application programs in a simple fashion.

In employing classes and objects we have restricted ourselves to specific applications, because overenthusiastic use of these language elements can also make programs more involved and difficult to read. Immediate advantages are obtained, for example, by defining individual classes `Vec3D` and `Mat3D` for vectors and matrices together with the possiblity for overloading operators. C++ allows the language's operators (such as +, - oder *, for example) to be overloaded so that they can also handle user-defined data types. Instead of involved function calls, the addition of vectors or the multiplication of matrices can be carried out in a much more elegant way by employing these operators. The use of specifically defined classes thus permits broad equivalence between instructions and equations, and contributes to a greater clarity and legibility in the source code.

In addition, we felt that it was sensible to define the auxiliary classes `Time`, `DateTime` and `Angle` to simplify the formatted output of times and angles. They support the use of the shift operator (<<) and thus allow a simple and C++ compliant method of writing this output operation.

Examples of a profitable use of object-oriented techniques in the implementation of numerical procedures are the class `SolverDE` for the numerical integration of differential equations and the class `SolverLSQ` for solving linear estimation problems. Although the individual steps in the calculations do not differ in the implementations of classical programming languages, the concept of classes provides a high level of abstraction, and gives optimal encapsulation of the necessary data. Tasks such as the preparation of workspace are no longer the user's responsibility, but are independently carried out by the constructor of the corresponding objects.

2. Coordinate Systems

Astronomy uses a whole series of different coordinate systems to specify the positions of stars and planets. Here we must distinguish between heliocentric coordinates, which are based on the Sun, and geocentric coordinates, which are centred on the Earth. Ecliptic coordinates refer the position of a point to the plane of the Earth's orbit, while equatorial coordinates are measured relative to the position of the Earth's or celestial equators. The slow shift of this reference plane because of precession means that the equinox of the coordinate system that is being used also has to be taken into account. Although it would obviously be possible to decide, once and for all, to use a single, fixed coordinate system, every system has specific advantages that make it particularly suitable for certain purposes.

Because there are various coordinate systems, it is often necessary to convert coordinates given in one system to those of another. The `Coco` ('Coordinate Conversion') program is designed to do just this. It provides accurate conversions between ecliptic and equatorial coordinates, and between geocentric and heliocentric coordinates, for various epochs. The individual transformations, which will be described shortly, are written as separate functions. They may therefore be used elsewhere, and are an important basis for later programs.

2.1 Making a Start

The C++ programming language has at its disposal powerful standard libraries that already contain many mathematical functions. Nevertheless, it proves very useful for subsequent work if we start by defining our own basic library modules. The first of these contains the definitions of important constants that will be repeatedly required subsequently. The corresponding module `APC_Const` consists solely of a header file `APC_Const.h`, which may be passed to other modules by the use of an `#include` instruction. Apart from mathematical constants such as the quantity π and the factors required for the conversion of degrees and angular measure, this contains astronomical and physical constants, such as the length of the astronomical unit, and the speed of light.

```
const double pi      = 3.14159265358979324;
const double pi2     = 2.0*pi;
const double Rad     = pi / 180.0;
const double Deg     = 180.0 / pi;
const double Arcs    = 3600.0*180.0/pi;
...
```

```
const double AU      = 149597870.0;      // [km]
const double c_light = 173.14;           // [AU/d]
```

The module APC_Math provides (among others) the two functions

```
//------------------------------------------------------------------------------
// Frac: Gives the fractional part of a number
//------------------------------------------------------------------------------
double Frac ( double x )
{
  return x-floor(x);
}
```

and

```
//------------------------------------------------------------------------------
// Modulo: calculates x mod y
//------------------------------------------------------------------------------
double Modulo ( double x, double y )
{
  return y*Frac(x/y);
}
```

which enable us to calculate the fractional and integral parts of a number. These functions are, for example, repeatedly required, when one wants to express the monotonically increasing longitude of a celestial body as lying between 0° bis 360°.

In addition, APC_Math contains two functions Ddd and DMS, which support the sexagesimal expression of times and angles. In specifying angles in degrees, fractions of a degree are normally given in minutes and seconds of arc. For time, this corresponds to the normal use of minutes and seconds. Conversion of the two forms is normally required once (at most) when a value is input or ouput. Provided all values are positive, this, in principle, causes no difficulty. It is, however, important to take precautions to ensure that the sign is correctly handled with negative angles. The action of the functions Ddd and DMS is best explained by a few examples:

```
     Dd         D  M  S

 15.50000      15 30 00.0
 -8.15278      -8 09 10.0
  0.01667       0  1  0.0
 -0.08334       0 -5  0.0
```

In converting a negative number Dd into degrees, minutes and seconds, only the leading (non-zero) figure of the three numbers D, M and S is negative. It should also be noted that in both functions D and M, which by the very nature of things take integer values only, are defined as int variables, whereas S is a double variable.

```
//------------------------------------------------------------------------------
// Ddd: Conversion of degrees, minutes and seconds of arc to decimal
//      representation of an angle
```

```
//   D,M,S      Degrees, minutes and seconds of arc
//   <return>   Angle in decimal representation
//------------------------------------------------------------------------------
double Ddd( int D, int M, double S )
{
  double sign;
  if ( (D<0) || (M<0) || (S<0) ) sign = -1.0; else sign = 1.0;
  return  sign * ( fabs((double)D)+fabs((double)M)/60.0+fabs(S)/3600.0 );
}

//------------------------------------------------------------------------------
// DMS: Finds degrees, minutes and seconds of arc for a given angle
//   Dd         Angle in degrees in decimal representation
//   D,M,S      Degrees, minutes and seconds of arc
//------------------------------------------------------------------------------
void DMS ( double Dd, int& D, int& M, double& S )
{
  double x;
  x = fabs(Dd);    D = int(x);
  x = (x-D)*60.0; M = int(x);   S = (x-M)*60.0;
  if (Dd<0.0) { if (D!=0) D*=-1; else if (M!=0) M*=-1; else S*=-1.0; }
}
```

Because the conversion of angles into degrees, minutes and seconds is almost exclusively required for output, it is worthwhile combining both functions in a suitable manner. To do so, we first make use of the possibility offered in C++ of defining a suitable class:

```
// Format tag for Angle output (used by Angle class output operator)
enum AngleFormat {
  Dd,     // decimal representation
  DMM,    // degrees and whole minutes of arc
  DMMm,   // degrees and minutes of arc in decimal representation
  DMMSS,  // degrees, minutes of arc and whole seconds of arc
  DMMSSs  // degrees, minutes, and seconds of arc in decimal representation
};

//
// Auxiliary class for sexagesimal angle output
//
class Angle
{
  public:
    // Constructor
    Angle (double alpha, AngleFormat Format=Dd);
    // Modifiers
    void Set (AngleFormat Format=Dd);
    // Angle output
    friend ostream& operator << (ostream& os, const Angle& alpha);
  private:
    double       m_angle;
    AngleFormat  m_Format;
};
```

Objects of the `Angle` class may be created by the equally named constructor by specifying the angle and a format descriptor. The corresponding data are stored in the class attributes and used for later output by employing the shift operator <<. Before doing so the angle is converted into degrees, minutes and seconds as required. Finally the resulting values are output, individually formatted. The definition of the shift operator implements the common manipulators for the number of places, sign, etc., so that the output may be easily laid out within broad limits. Because of the differences between the cases for the individual formats, the implementation of the class is quite comprehensive and will not, therefore, be discussed in detail here. Its application is, however, extremely simple, as the following short program shows:

```
#include <iostream>
#include <iomanip>
#include "APC_Math.h"

void main() {

  using namespace std;
  double x=12.3456;

  cout << setprecision(2) << setw(12) << Angle(x,Dd) << endl;
  cout << setprecision(2) << setw(12) << Angle(x,DMM) << endl;
  cout << setprecision(2) << setw(12) << Angle(x,DMMm) << endl;
  cout << setprecision(2) << setw(12) << Angle(x,DMMSS) << endl;
  cout << setprecision(2) << setw(12) << Angle(x,DMMSSs) << endl;

}
```

The output for the angle $12^\circ\!.3456$ is represented in five different forms:

```
      12.35
      12 21
   12 20.74
   12 20 44
12 20 44.16
```

Any sign would be printed left justified.

We should also introduce a third module here, the `APC_VecMat3D` library, which provides two classes for three-dimensional vectors and matrices as well as numerous related operators and functions. Three-dimensional vectors serve to describe the position of a point $\boldsymbol{r}$ in space, where, in addition to the Cartesian coordinates $\boldsymbol{r} = (x, y, z)$, the polar coordinates r, ϑ and φ may also be employed (Fig. 2.1). The relationship between the two forms is given by

$$\begin{array}{rclcrcl} x &=& r\cos\vartheta\cos\varphi & \qquad & r &=& \sqrt{x^2+y^2+z^2} \\ y &=& r\cos\vartheta\sin\varphi & & \tan\varphi &=& y/x \\ z &=& r\sin\vartheta & & \tan\vartheta &=& z/\sqrt{x^2+y^2} \quad . \end{array} \tag{2.1}$$

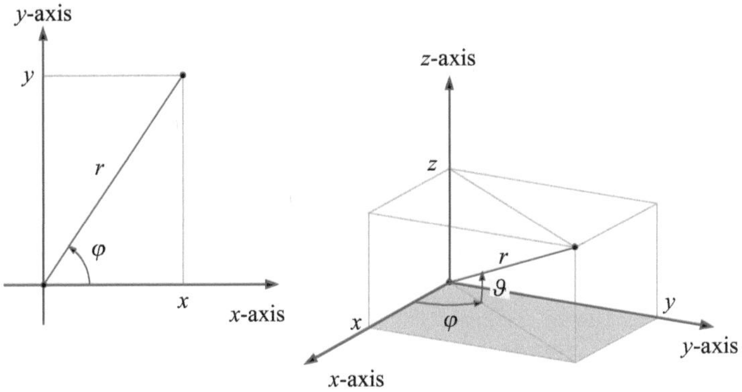

Fig. 2.1. Plane and spatial polar coordinates

The class Vec3D allows both representations to be employed alongside one another in a simple fashion. To this end, the corresponding constructors Vec3D and the access operators [] are provided:

```
//
// Vec3D: three dimensional vectors
//
class Vec3D
{
  public:
    // Constructors
    Vec3D ();  // default constructor initializes to zero vector
    Vec3D (double X, double Y, double Z);
    Vec3D (const Polar& polar);
    // component access (read only)
    double operator [] (xyz_index Index) const;
    // retrieves polar angles or norm of vector
    double operator [] (pol_index Index);
    // simple vector output
    friend ostream& operator << (ostream& os, const Vec3D& Vec);
    // further operators and functions
    ...
  private:
    // Members
    double m_Vec[3];        // components of vector
    double m_phi;           // polar angle (azimuth)
    double m_theta;         // polar angle (altitude)
    double m_r;             // norm of vector
    bool   m_bPolarValid;   // flag for validity of polar coordinates
    // On-demand calculation of polar components
    void CalcPolarAngles ();
};
```

Here the data types index, pol_index and Polar are defined as

```
enum xyz_index { x=0, y=1, z=2 };
enum pol_index { phi=0, theta=1, r=2 };  // azimuth, altitude, radius
```

and

```
struct Polar {
  // Constructors
  Polar();
  Polar(double Az, double Elev, double R = 1.0);
  // Members
  double phi;    // azimuth of vector
  double theta;  // altitude of vector
  double r;      // norm of vector
};
```

Conversion of Cartesian coordinates is carried out using the private function

```
//
// Calculate polar components
//
void Vec3D::CalcPolarAngles ()
{
  // Length of projection in x-y-plane:
  const double rhoSqr = m_Vec[0] * m_Vec[0] + m_Vec[1] * m_Vec[1];
  // Norm of vector
  m_r = sqrt ( rhoSqr + m_Vec[2] * m_Vec[2] );
  // Azimuth of vector
  if ( (m_Vec[0]==0.0) && (m_Vec[1]==0.0) )
    m_phi = 0.0;
  else
    m_phi = atan2 (m_Vec[1], m_Vec[0]);
  if ( m_phi < 0.0 )  m_phi += 2.0*pi;
  // Altitude of vector
  const double rho = sqrt ( rhoSqr );
  if ( (m_Vec[2]==0.0) && (rho==0.0) )
    m_theta = 0.0;
  else
    m_theta = atan2(m_Vec[2], rho);
}
```

as soon as polar coordinates are accessed for the first time with the `[]` operator. Finally the status variable `m_bPolarValid` is set to true, so that the conversion is at most carried out once. It should be noted that angles are required, and will be calculated, in radians and are to be converted on input and output. `APC_Const` provides the constants $\texttt{Deg} = 180°/\pi$ and $\texttt{Rad} = \pi/180°$ for this purpose.

As well as the basic constructors and access operators, the `Vec3D` class includes numerous operators and functions, that allow vector calculations to be carried out easily. For example, vectors may be added by using the + operator, while the * operator allows the multiplication of a vector with a scalar (see Table 2.1).

The class `Vec3D` is complimented by the three-dimensional matrix class `Mat3D`, which is defined as follows:

Table 2.1. Vector- und matrix operations in APC_VecMat3D

Name	Arg1 (a)	Arg2 (b)	Value (c)	Notation	Explanation
-	Vec3D		Vec3D	$\boldsymbol{c} = -\boldsymbol{a}$	Unary minus
	Mat3D		Mat3D	$\boldsymbol{C} = -\boldsymbol{A}$	
+	Vec3D	Vec3D	Vec3D	$\boldsymbol{c} = \boldsymbol{a} + \boldsymbol{b}$	Vector addition
	Mat3D	Mat3D	Mat3D	$\boldsymbol{C} = \boldsymbol{A} + \boldsymbol{B}$	Matrix addition
*	double	Vec3D	Vec3D	$\boldsymbol{c} = a\boldsymbol{b}$	Scalar multiplication
	Vec3D	double	Vec3D	$\boldsymbol{c} = \boldsymbol{a}b$	
	double	Mat3D	Mat3D	$\boldsymbol{C} = a\boldsymbol{B}$	
	Mat3D	double	Mat3D	$\boldsymbol{C} = \boldsymbol{A}b$	
/	Vec3D	double	Vec3D	$\boldsymbol{c} = \boldsymbol{a}/b$	Scalar division
	Mat3D	double	Mat3D	$\boldsymbol{C} = \boldsymbol{A}/b$	
*	Mat3D	Vec3D	Vec3D	$\boldsymbol{c} = \boldsymbol{A}\boldsymbol{b}$	Matrix/vector product
	Vec3D	Mat3D	Vec3D	$\boldsymbol{c} = \boldsymbol{a}\boldsymbol{B}$	
Norm	Vec3D		double	$c = \lvert\boldsymbol{a}\rvert$	Euclidean norm
Dot	Vec3D	Vec3D	double	$c = \boldsymbol{a}^T\boldsymbol{b}$	Dot product
Cross	Vec3D	Vec3D	Vec3D	$\boldsymbol{c} = \boldsymbol{a} \times \boldsymbol{b}$	Cross product
Id3D			Mat3D	$\mathbf{1}$	Identity matrix
R_x	double		Mat3D	$\boldsymbol{R}_x(a)$	Elementary rotation
R_y	double		Mat3D	$\boldsymbol{R}_y(a)$	matrix
R_z	double		Mat3D	$\boldsymbol{R}_z(a)$	

```
//
// Mat3D: 3 dimensional transformation matrices
//

class Mat3D
{
  public:
    // Constructors (empty matrix; matrix from column vectors)
    Mat3D ();
    Mat3D ( const Vec3D& e_1, const Vec3D& e_2, const Vec3D& e_3 );
    // component access
    friend Vec3D Col(const Mat3D& Mat, xyz_index Index);
    friend Vec3D Row(const Mat3D& Mat, xyz_index Index);
    // simple matrix output
    friend ostream& operator << (ostream& os, const Mat3D& Mat);
    // further operators and functions
    ...

  private:
    double m_Mat[3][3];   // Elements of the matrix
};
```

Whereas – as mentioned – objects in the vector class describe spatial coordinates, in what follows, objects in the matrix class primarily serve to describe the rotation of coordinate systems in space. Simple rotations around the x-, y-, or z-axes, may be obtained using the R_x, R_y and R_z functions, which, for a given rotation angle

ϕ, calculate the following matrices:

$$\boldsymbol{R}_x(\phi) = \begin{pmatrix} 1 & 0 & 0 \\ 0 & +\cos\phi & +\sin\phi \\ 0 & -\sin\phi & +\cos\phi \end{pmatrix} \tag{2.2}$$

$$\boldsymbol{R}_y(\phi) = \begin{pmatrix} +\cos\phi & 0 & -\sin\phi \\ 0 & 1 & 0 \\ +\sin\phi & 0 & +\cos\phi \end{pmatrix} \tag{2.3}$$

$$\boldsymbol{R}_z(\phi) = \begin{pmatrix} +\cos\phi & +\sin\phi & 0 \\ -\sin\phi & +\cos\phi & 0 \\ 0 & 0 & 1 \end{pmatrix} \quad . \tag{2.4}$$

The signs here are chosen so that a positive angle corresponds to a positive rotation (i.e., an anticlockwise rotation) of the reference frame around the rotational axis.

2.2 Calendar and Julian Dates

When calculating ephemerides, the difference in time between two given dates is generally required. A running day number, as has been commonly used in astronomy for a very long time, is therefore extremely convenient. The Julian Date gives the total number of days that have elapsed since 4713 B.C. January 1. It is named after Julius Scaliger, the father of Joseph Justus Scaliger, who was the first to use it for chronological purposes. As the count begins in biblical times, the Julian Date has now reached a very high value. At noon on 1980 July 23 it amounted to 2 444 444.0 days, for example. When we recall that a second is approximately equal to 0.00001 day, then we find that we require a twelve-figure Julian Date to express an accurate time. The first two numbers hardly alter over a period of three centuries, however, and therefore the Modified Julian Date

$$\mathrm{MJD} = \mathrm{JD} - 2400000.5$$

is employed alongside the true Julian Date. The MJD is the total number of days that has elapsed since 1858 November 17 00:00 hours. Like the ordinary civil date, the MJD therefore changes at midnight, and not at noon like the Julian Date.

```
//------------------------------------------------------------------------------
//
// Mjd: Modified Julian Date from calendar date and time
//   Year,Month,Day  Calendar date components
//   Hour,Min,Sec    Time components (optional)
//   <return>        Modified Julian Date
//
//------------------------------------------------------------------------------
```

```
double Mjd ( int Year, int Month, int Day,
             int Hour=0, int Min=0, double Sec=0.0 )
{
  // Variables
  long    MjdMidnight;
  double  FracOfDay;
  int     b;

  if (Month<=2) { Month+=12; --Year;}
  if ( (10000L*Year+100L*Month+Day) <= 15821004L )
    b = -2 + ((Year+4716)/4) - 1179;      // Julian calendar
  else
    b = (Year/400)-(Year/100)+(Year/4);  // Gregorian calendar

  MjdMidnight = 365L*Year - 679004L + b + int(30.6001*(Month+1)) + Day;
  FracOfDay   = Ddd(Hour,Min,Sec) / 24.0;

  return MjdMidnight + FracOfDay;
}
```

`Mjd` takes into account the fact that, because of the Gregorian calendar reform, 1582 October 4 (JD 2299159.5) was followed by 1582 October 15 (JD 2299160.5).[1] Until that date the Julian calendar is used, in which every fourth year included February 29. After that date, leap days are omitted in every century year that is not evenly divisible by 400. As a result, the average length of the year according to the Gregorian calendar is

$$365 + 1/4 - 1/100 + 1/400 = 365.2425 \text{ days},$$

slightly shorter than the 365.25 days of the Julian year.

`Caldat` is a function to calculate the normal calendar date from the Modified Julian Date. Because it is closely linked to the subject of this section it will be given here.

```
//------------------------------------------------------------------------
// CalDat: Calendar date and time from Modified Julian Date
//   Mjd              Modified Julian Date
//   Year,Month,Day   Calendar date components
//   Hour             Decimal hours
//------------------------------------------------------------------------
void CalDat ( double Mjd,
              int& Year, int& Month, int& Day, double & Hour )
{
  long    a,b,c,d,e,f;
```

[1] Caution needs to be exercised, because only Italy, Spain and Portugal adopted the new calendar on this date. Most other Catholic states in Continental Europe changed at various dates in 1583 and 1584, with Protestant states and cities following in the following century, most by 1700–1701. Great Britain, its colonies and dominions changed on 1752 September 3 (which was followed by September 14). A full listing is given in the *Explanatory Supplement to the Astronomical Ephemeris and the American Ephemeris and Nautical Almanac*, published jointly by the U. S. Government Printing Office, Washington, and Her Majesty's Stationery Office, London, in 1974, pp.412–416.

```
  double  FracOfDay;
  // Convert Julian day number to calendar date
  a = long(Mjd+2400001.0);
  if ( a < 2299161 ) {  // Julian calendar
    b = 0;  c = a + 1524;
  }
  else {                // Gregorian calendar
    b = long((a-1867216.25)/36524.25);
    c = a +  b - (b/4) + 1525;
  }
  d     = long ( (c-122.1)/365.25 );
  e     = 365*d + d/4;
  f     = long ( (c-e)/30.6001 );
  Day   = c - e - int(30.6001*f);
  Month = f - 1 - 12*(f/14);
  Year  = d - 4715 - ((7+Month)/10);
  FracOfDay = Mjd - floor(Mjd);
  Hour = 24.0*FracOfDay;
}
//---------------------------------------------------------------------------
// CalDat: Calendar date and time from Modified Julian Date
//   Mjd              Modified Julian Date
//   Year,Month,Day   Calendar date components
//   Hour,Min,Sec     Time components
//---------------------------------------------------------------------------
void CalDat ( double Mjd,
              int& Year, int& Month, int& Day,
              int& Hour, int& Min, double& Sec ) {
  double Hours;
  CalDat (Mjd, Year, Month, Day, Hours);
  DMS (Hours, Hour, Min, Sec);
}
```

The two overloaded versions of `CalDat` differ in the way they handle the fraction of a day since midnight, which is given either in decimal hours or in hours, minutes, and seconds. As with the sexagesimal expression of angles, the latter form is generally required only for the output of times. The `APC_Time` module therefore provides, alongside `Mjd` and `CalDat`, two special classes `Time` and `DateTime`, with the corresponding constructors

```
  Time ( double Hour, TimeFormat Format=HHMMSS );
  DateTime ( double Mjd, TimeFormat Format=None );
```

which provide an even better means of carrying out this task. The enumeration type

```
enum TimeFormat {
  None,   // don't output time (date only)
  DDd,    // output time as fractional part of a day
  HHh,    // output time as hours with one decimal place
  HHMM,   // output time as hours and minutes (rounded to the next minute)
  HHMMSS  // output time as hours, minutes and seconds (rounded to next s)
};
```

here serves to establish the format and the required rounding for the output of `Time` and `DateTime` objects, using the shift operator (<<). Here is an example:

```
double MJD = Mjd ( 1961,01,14, 03,30,10.0 ); // 14 Jan. 1961, 3:30:10
double JD  = MJD + 2400000.5;                // Julian Date
cout << setprecision(1)
     << "Date:" << DateTime(MJD,HHMMSS)      // Output
     << setprecision(3)
     << "MJD: " << setw(12) << MJD
     << "JD:  " << setw(12) << JD;
```

2.3 Ecliptic and Equatorial Coordinates

Ecliptic and equatorial coordinates differ in the reference plane from which they are measured. In the former case, the (x, y) plane is the Earth's orbital plane (the *ecliptic*), and in the latter, the plane perpendicular to the Earth's axis, i.e., parallel to the plane of the Earth's equator (Fig. 2.2). The x-x'-axis is common to both systems, and is defined as being the direction of the vernal equinox or *First Point of Aries*, designated by (Υ). This direction is perpendicular to the north pole of the ecliptic (the z-axis) and to the North Celestial Pole (the z'-axis). The angle ε between the ecliptic and the equator amounts to approximately 23°.5.

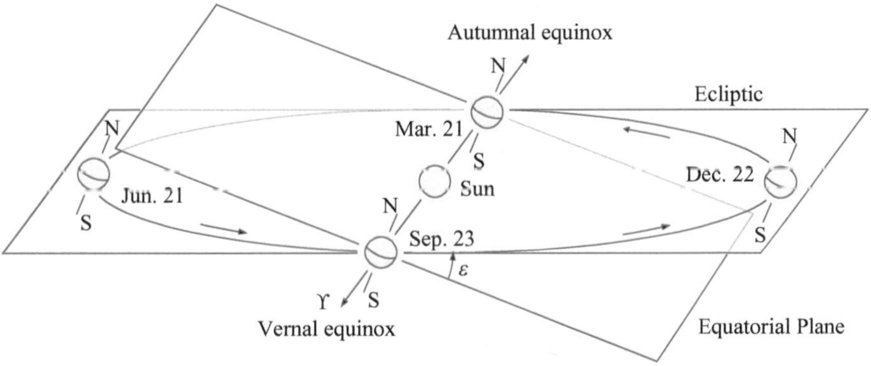

Fig. 2.2. Ecliptic and equator

What are the relationships between the ecliptic coordinates (x, y, z) and the equatorial coordinates (x', y', z') of a point? As $x = x'$, only the transformations $(y, z) \leftrightarrow (y', z')$ have to be considered. A point on the y-axis ($z = 0$) has equatorial coordinates $y' = +y\cos\varepsilon$ and $z' = +y\sin\varepsilon$. If the point lies on the z-axis, then its equatorial coordinates are $y' = -z\sin(\varepsilon)$ and $z' = +z\cos\varepsilon$. In the equatorial system, therefore, a given point (x, y, z) has the components

$$\begin{aligned} x' &= +x \\ y' &= +y\cos\varepsilon - z\sin\varepsilon \\ z' &= +y\sin\varepsilon + z\cos\varepsilon \quad . \end{aligned} \tag{2.5}$$

The corresponding inverse relationships are

$$\begin{aligned} x &= +x' \\ y &= +y'\cos\varepsilon + z'\sin\varepsilon \\ z &= -y'\sin\varepsilon + z'\cos\varepsilon \quad . \end{aligned} \tag{2.6}$$

The polar coordinates corresponding to (x, y, z) are ecliptic longitude l, ecliptic latitude b and distance r (see Fig. 2.3).

$$\begin{aligned} x &= r\cos b\cos l & x' &= r\cos\delta\cos\alpha \\ y &= r\cos b\sin l & y' &= r\cos\delta\sin\alpha \\ z &= r\sin b & z' &= r\sin\delta \quad . \end{aligned} \tag{2.7}$$

The equatorial coordinates corresponding to these are right ascension α, declination δ, and again distance r, which has the same value in both systems.

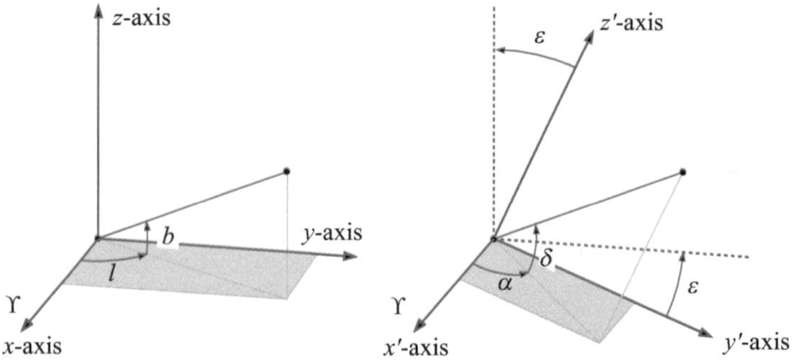

Fig. 2.3. Ecliptic and equatorial coordinates

The transformations between ecliptic and equatorial coordinates may be described in a particularly compact and clear manner, when expressed in vector and matrix notation. Using the elementary rotational matrices introduced earlier, we have

$$\boldsymbol{r} = \boldsymbol{R}_x(+\varepsilon)\,\boldsymbol{r}' \qquad \text{and} \qquad \boldsymbol{r}' = \boldsymbol{R}_x^T(+\varepsilon)\,\boldsymbol{r} = \boldsymbol{R}_x(-\varepsilon)\,\boldsymbol{r} \quad , \tag{2.8}$$

where $\boldsymbol{R}_x^T$ indicates the transposition (reflection) of the matrix $\boldsymbol{R}_x$. Similarly, we obtain the transformed coordinates of a vector of type Vec3D by multiplying the corresponding rotational matrix. Here is another short sample program:

```
double eps = Rad*23.5;                    // Obliquity of ecliptic in [rad]
Vec3D  e   = Vec3D(1.0,2.0,3.0);          // Ecliptic coordinates
Vec3D  a   = R_x(-eps)*e;                 // Equatorial coordinates
cout << "l   " << Deg*e[phi]       << " b   " << Deg*e[theta] << endl
     << "RA  " << Deg*a[phi]/15.0 << " Dec " << Deg*a[theta] << endl;
```

The value of right ascension will be given – as usual – in hours ($1^{\mathrm{h}} \equiv 15°$), instead of in degrees.

So far, nothing has been said about the exact value of the obliquity of the ecliptic ε. ε is not constant, but is decreasing by about 47″ per century. This primarily arises from slow alterations in the Earth's orbit as a result of perturbations caused by the other planets. If we wish to avoid errors in coordinate transformations, it is particularly important whether the ecliptic for 1950 or for 2000 is used. An expression of the form 'equinox 2000' is therefore used, to specify that coordinates refer to the equinox and the ecliptic (or the equator) of the year 2000, for example. The exact value for the obliquity of the ecliptic as a function of time is

$$\varepsilon = 23\overset{\circ}{.}43929111 - 46\overset{\prime\prime}{.}8150\,T - 0\overset{\prime\prime}{.}00059\,T^2 + 0\overset{\prime\prime}{.}001813\,T^3 \quad , \tag{2.9}$$

where T is the number of Julian centuries between the epoch and 2000 January 1 (12^{h}). In contrast to a century in the Gregorian calendar, which averages 36 524.25 days, a Julian century is always exactly 36 525 days. Important Julian epochs are J1900 (1900 January 0.5, JD 2415020.0) and J2000 (2000 January 1.5, JD 2451545.0). There is exactly one Julian century between these dates. T may be calculated for a given epoch with the Julian Date JD from

$$T = (\mathrm{JD} - 2451545)/36525 \quad .$$

To a close approximation, the value of T may be obtained from the year y by $T \approx (y-2000)/100$. For a given epoch T, the following function provides directly the matrix for the transformation of equatorial into ecliptic coordinates.

```
//------------------------------------------------------------------------
// Equ2EclMatrix: Transformation of equatorial to ecliptical coordinates
//   T          Time in Julian centuries since J2000
//   <return>   Transformation matrix
//------------------------------------------------------------------------
Mat3D Equ2EclMatrix (double T) {
  const double
    eps = ( 23.43929111-(46.8150+(0.00059-0.001813*T)*T)*T/3600.0 ) * Rad;
  return R_x(eps);
}
```

The opposite transformation is correspondingly described by the transformation matrix that follows.

```
//------------------------------------------------------------------------
// Ecl2EquMatrix: Transformation of ecliptical to equatorial coordinates
//   T          Time in Julian centuries since J2000
//   <return>   Transformation matrix
//------------------------------------------------------------------------
Mat3D Ecl2EquMatrix (double T)
{
  return Transp(Equ2EclMatrix(T));
}
```

Using `Ecl2EquMatrix`, we may write the above example as follows (let the equinox now be J1950 = JD2433282.50):

```
double MJD = 33282.0;                          // Epoch
double T   = (MJD-MJD_J2000)/36525.0;
Mat3D  U   = Ecl2EquMatrix(T);                 // Transformation matrix
Vec3D  e   = Vec3D(1.0,2.0,3.0);               // Ecliptic coordinates
Vec3D  a   = U*e;                              // Equatorial coordinates
cout << "l   " << Deg*e[phi]      << " b   " << Deg*e[theta] << endl
     << "RA  " << Deg*a[phi]/15.0 << " Dec " << Deg*a[theta] << endl;
```

2.4 Precession

The shift in the position of the Earth's axis and of the ecliptic caused by forces exerted by the Sun, Moon and planets not only causes a slight change in the angle ε between the equator and the ecliptic, but also a shift of the vernal equinox by about $1\overset{\circ}{.}5$ per century ($1'$ per year). For precise calculations, therefore, the equinox of the coordinate system used must be stated. The equinoxes most frequently used are

- the equinox of date,
- equinox J2000 and
- equinox B1950.

'Equinox of date' means that the values used are those for the equator, ecliptic, and vernal equinox for the actual date under consideration. Such daily alteration of the coordinate system is sensible if one requires the coordinates of a planet, for example, for use in conjunction with the setting circles on an equatorially mounted telescope, or on a transit circle. Because of the shift in the Earth's axis, the orientation of the polar axis of a telescope will also alter. On the other hand, if one wants to study the actual spatial motion of a planet, then it is better to use a fixed equinox, such as that for Julian Epoch J2000 (2000 January 1.5 = JD 2451545.0), which was generally introduced in 1984. Before that, the older equinox B1950 had been used for a long time, and was employed for many stellar catalogues and atlases (such as the *SAO Star Catalog* and *Atlas Coeli*). The prefix B indicates that it is not half-way between the epochs J1900 and J2000 (which would, of course, be J1950 = JD 2433282.5), but the beginning of the Besselian year 1950 (1950 Jan. 0.923 = JD 2433282.423).

Transformation of coordinates from one equinox to another requires similar steps to those required in converting from ecliptic to equatorial coordinates. Whereas those two coordinate systems could be brought into coincidence by a simple rotation about the x-axis, we now have to deal with a total of three rotations: first about the z-axis, second about the x-axis, and finally about the z-axis again.

If we first consider the ecliptic at time T_0 and time $T_0 + T$, then the angle between the two planes is

$$\begin{aligned}\pi &= (47\overset{''}{.}0029 - 0\overset{''}{.}06603 \cdot T_0 + 0\overset{''}{.}000598 \cdot T_0^2) \cdot T \\ &\quad +(-0\overset{''}{.}03302 + 0\overset{''}{.}000598 \cdot T_0) \cdot T^2 + 0\overset{''}{.}000060 \cdot T^3\end{aligned} \tag{2.10}$$

(see Fig. 2.4). If two coordinate systems (x', y', z') and (x'', y'', z'') are established, such that the x'-y'-plane coincides with the ecliptic for time T_0, and the x''-y''-plane

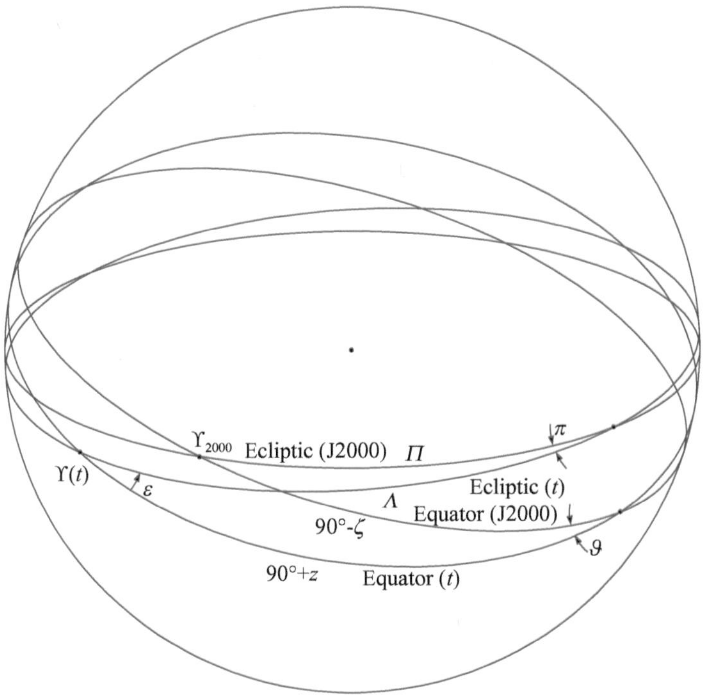

Fig. 2.4. The effects of precession on the ecliptic, equator, and vernal equinox

with that for time $T_0 + T$, then we have:

$$\begin{aligned} x'' &= x' \\ y'' &= +\cos\pi \cdot y' + \sin\pi \cdot z' \\ z'' &= -\sin\pi \cdot y' + \cos\pi \cdot z' \quad . \end{aligned}$$

This assumes that the x'-axis and the x''-axis coincide with the line of intersection of the two planes. Let (x_0, y_0, z_0) be the ecliptic coordinates corresponding to the epoch T_0 and (x, y, z) the coordinates at epoch $T_0 + T$. We then have

$$\begin{aligned} x' &= +\cos\Pi \cdot x_0 + \sin\Pi \cdot y_0 \\ y' &= -\sin\Pi \cdot x_0 + \cos\Pi \cdot y_0 \\ z' &= z_0 \quad , \end{aligned}$$

and

$$\begin{aligned} x &= +\cos\Lambda \cdot x'' - \sin\Lambda \cdot y'' \\ y &= +\sin\Lambda \cdot x'' + \cos\Lambda \cdot y'' \\ z &= z'' \quad , \end{aligned}$$

where

$$\begin{aligned}
\Pi &= (174^\circ\!.876383889 + 3289''\!.4789 T_0 + 0''\!.60622 T_0^2) \\
&\qquad +(-869''\!.8089 - 0''\!.50491 T_0)T + 0''\!.03536 T^2 \\
p &= (5029''\!.0966 + 2''\!.22226 T_0 - 0''\!.000042 T_0^2)T \\
&\qquad +(1''\!.11113 - 0''\!.000042 T_0)T^2 - 0''\!.000006 T^3 \\
\Lambda &= \Pi + p \quad .
\end{aligned} \tag{2.11}$$

Π and Λ are the angles between the x'/x''-axis and the vernal equinox Υ_0 at epoch T_0 (x_0-axis), and the vernal equinox Υ at epoch $T_0 + T$ (x-axis), respectively. p denotes precession in longitude, because the change in the vernal equinox mainly takes place in ecliptic longitude. Substituting for the various relationships, we obtain the equation

$$\boldsymbol{r} = \boldsymbol{P}\,\boldsymbol{r}_0 \quad \text{with} \quad \boldsymbol{P} = \boldsymbol{R}_z(-\Pi - p)\,\boldsymbol{R}_x(\pi)\,\boldsymbol{R}_z(\Pi) \quad . \tag{2.12}$$

Explicitly, the elements of $\boldsymbol{P}$ are given by

$$\begin{aligned}
p_{11} &= +\cos\Lambda\cos\Pi + \sin\Lambda\cos\pi\sin\Pi \\
p_{21} &= +\sin\Lambda\cos\Pi - \cos\Lambda\cos\pi\sin\Pi \\
p_{31} &= +\sin\pi\sin\Pi \\
p_{12} &= +\cos\Lambda\sin\Pi - \sin\Lambda\cos\pi\cos\Pi \\
p_{22} &= +\sin\Lambda\sin\Pi + \cos\Lambda\cos\pi\cos\Pi \\
p_{32} &= -\sin\pi\cos\Pi \\
p_{13} &= -\sin\Lambda\sin\pi \\
p_{23} &= +\cos\Lambda\sin\pi \\
p_{33} &= +\cos\pi \quad .
\end{aligned} \tag{2.13}$$

The angles in equatorial coordinates corresponding to the three angles π, Π and Λ are $90^\circ - \zeta$, ϑ and $90^\circ + z$:

$$\begin{aligned}
\zeta &= (2306''\!.2181 + 1''\!.39656 T_0 - 0''\!.000139 T_0^2)T \\
&\qquad +(0''\!.30188 - 0''\!.000345 T_0)T^2 + 0''\!.017998 T^3 \\
\vartheta &= (2004''\!.3109 - 0''\!.85330 T_0 - 0''\!.000217 T_0^2)T \\
&\qquad +(-0''\!.42665 - 0''\!.000217 T_0)T^2 - 0''\!.041833 T^3 \\
z &= \zeta + (0''\!.79280 + 0''\!.000411 T_0)T^2 + 0''\!.000205 T^3 \quad .
\end{aligned} \tag{2.14}$$

This gives corresponding formulae where

$$\boldsymbol{P} = \boldsymbol{R}_z(-z)\,\boldsymbol{R}_y(\vartheta)\,\boldsymbol{R}_z(-\zeta) \tag{2.15}$$

or, by component,

$$
\begin{aligned}
p_{11} &= -\sin z \sin\zeta + \cos z \cos\vartheta \cos\zeta \\
p_{21} &= +\cos z \sin\zeta + \sin z \cos\vartheta \cos\zeta \\
p_{31} &= +\sin\vartheta \cos\zeta \\
p_{12} &= -\sin z \cos\zeta - \cos z \cos\vartheta \sin\zeta \\
p_{22} &= +\cos z \cos\zeta - \sin z \cos\vartheta \sin\zeta \\
p_{32} &= -\sin\vartheta \sin\zeta \\
p_{13} &= -\cos z \sin\vartheta \\
p_{23} &= -\sin z \sin\vartheta \\
p_{33} &= +\cos\vartheta \quad .
\end{aligned} \tag{2.16}
$$

The most onerous part of calculating precession is determining the values for the various angles and for p_{ij}. However, the transformation matrices depend only on the two epochs T_0 and $T_0 + T$, and may be reused if a whole series of different positions are to be converted from one fixed epoch to another.

```
//------------------------------------------------------------------------------
// PrecMatrix_Ecl: Precession of ecliptic coordinates
//   T1          Epoch given
//   T2          Epoch to precess to
//   <return>    Precession transformation matrix
// Note: T1 and T2 in Julian centuries since J2000
//------------------------------------------------------------------------------
Mat3D PrecMatrix_Ecl (double T1, double T2)
{
  const double dT = T2-T1;
  double Pi, pi, p_a;

  Pi  = 174.876383889*Rad +
        ( ((3289.4789+0.60622*T1)*T1) +
              ((-869.8089-0.50491*T1) + 0.03536*dT)*dT )/Arcs;
  pi  = ( (47.0029-(0.06603-0.000598*T1)*T1)+
              ((-0.03302+0.000598*T1)+0.000060*dT)*dT )*dT/Arcs;
  p_a = ( (5029.0966+(2.22226-0.000042*T1)*T1)+
              ((1.11113-0.000042*T1)-0.000006*dT)*dT )*dT/Arcs;
  return R_z(-(Pi+p_a)) * R_x(pi) * R_z(Pi);
}

//------------------------------------------------------------------------------
// PrecMatrix_Equ: Precession of equatorial coordinates
//   T1          Epoch given
//   T2          Epoch to precess to
//   <return>    Precession transformation matrix
// Note: T1 and T2 in Julian centuries since J2000
//------------------------------------------------------------------------------
Mat3D PrecMatrix_Equ (double T1, double T2)
{
```

```
  const double dT = T2-T1;
  double zeta,z,theta;
  zeta  =  ( (2306.2181+(1.39656-0.000139*T1)*T1)+
               ((0.30188-0.000344*T1)+0.017998*dT)*dT )*dT/Arcs;
  z     =  zeta + ( (0.79280+0.000411*T1)+0.000205*dT)*dT*dT/Arcs;
  theta =  ( (2004.3109-(0.85330+0.000217*T1)*T1)-
               ((0.42665+0.000217*T1)+0.041833*dT)*dT )*dT/Arcs;
  return R_z(-z) * R_y(theta) * R_z(-zeta);
}
```

To give an idea of the effects of precession, and also to allow some practice in using the various routines, a somewhat longer example will now be given. A set of coordinates for equinox B1950 is to be converted to a corresponding set for equinox J2000. This may be carried out in two ways, according to whether precession or the conversion between ecliptic and equatorial coordinates is applied first.

```
// Declarations
double T_B1950 = -0.500002108;                // Epoch B1950 (JD2433282.423)
double l   = 200.0*Rad;                       // Longitude
double b   =  10.0*Rad;                       // Latitude
Vec3D  r_0 = Vec3D(Polar(l,b));               // Ecliptic coordinates B1950
Vec3D  r;

// Method 1
r = PrecMatrixEcl(T_B1950,T_J2000) * r_0;   // Ecliptic J2000
cout << "l,b (J2000)      " << Deg*r[phi]      << Deg*r[theta] << endl;
r = Ecl2EquMatrix(T_J2000) * r;             // Equator J2000
cout << "RA,Dec (J2000)   " << Deg*r[phi]/15.0 << Deg*r[theta] << endl;

// Method 2
r = Ecl2EquMatrix(T_B1950) * r_0;           // Equator B1950
cout << "RA,Dec (B1950)   " << Deg*r[phi]/15.0 << Deg*r[theta] << endl;
r = PrecMatrixEqu(T_B1950,T_J2000) * r;     // Equator J2000
cout << "RA,Dec (J2000)   " << Deg*r[phi]/15.0 << Deg*r[theta] << endl;
```

2.5 Geocentric Coordinates and the Orbit of the Sun

All that is now required to complete the program is a method for converting heliocentric coordinates (referred to the centre of the Sun) into geocentric ones (referred to the centre of the Earth), and the converse. Conversion of the origin of the coordinates is described by the equations

$$\boldsymbol{r}_{\rm p} = \boldsymbol{r}_{\odot{\rm p}} + \boldsymbol{r}_{\odot} \quad \text{and} \quad \boldsymbol{r}_{\odot{\rm p}} = \boldsymbol{r}_{\rm p} - \boldsymbol{r}_{\odot} \tag{2.17}$$

which are derived from the Sun-Earth-Planet triangle illustrated in Fig. 2.5. Here $\boldsymbol{r}_{\odot{\rm p}}$ and $\boldsymbol{r}_{\rm p}$ are the heliocentric and geocentric position vectors of a point P, and $\boldsymbol{r}_{\odot}$ is the geocentric position of the Sun. Written in the form of individual components, the transformation equations are

$$\begin{aligned} x_{\rm p} &= x_{\odot{\rm p}} + x_{\odot} \\ y_{\rm p} &= y_{\odot{\rm p}} + y_{\odot} \\ z_{\rm p} &= z_{\odot{\rm p}} + z_{\odot} \end{aligned} \quad \text{and} \quad \begin{aligned} x_{\odot{\rm p}} &= x_{\rm p} - x_{\odot} \\ y_{\odot{\rm p}} &= y_{\rm p} - y_{\odot} \\ z_{\odot{\rm p}} &= z_{\rm p} - z_{\odot} \end{aligned} \quad ,$$

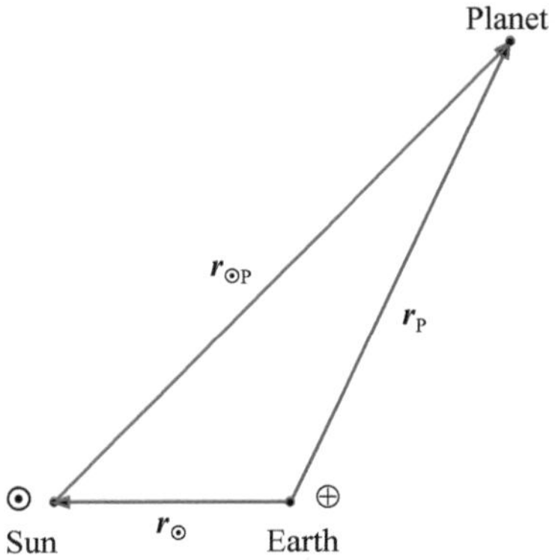

Fig. 2.5. The Earth-Sun-Planet triangle

where it makes no difference whether ecliptic or equatorial coordinates are employed to express the various vectors.

As any error in the position of the Sun inevitably affects the accuracy of the conversions, it is worthwhile taking some trouble in calculating the coordinates of the Sun. The `SunPos` function, introduced here, attains an accuracy of about $1''$, which should be more than sufficient for most purposes. For a time

$$T = (\mathrm{JD} - 2451545)/36525$$

it provides the Cartesian ecliptic coordinates

$$\boldsymbol{r}_\odot = \begin{pmatrix} x_\odot \\ y_\odot \\ z_\odot \end{pmatrix} = \begin{pmatrix} R\cos B\cos L \\ R\cos B\sin L \\ R\sin B \end{pmatrix} \quad . \tag{2.18}$$

Like the associated ecliptic longitude L and latitude B these coordinates initially refer to the equinox of date, but may be converted to another equinox by means of the functions already discussed.

```
//------------------------------------------------------------------------------
//
// SunPos: Computes the Sun's ecliptical position using analytical series
//   T          Time in Julian centuries since J2000
//   <return>   Geocentric position of the Sun (in [AU]), referred to the
//              ecliptic and equinox of date
//
//------------------------------------------------------------------------------
Vec3D SunPos (double T) {

  // Variables
  double  M2,M3,M4,M5,M6;             // Mean anomalies
  double  D, A, U;                    // Mean arguments of lunar orbit
  Pert    Ven, Mar, Jup, Sat;         // Perturbations
  double  dl, dr, db;                 // Corrections in longitude ["],
                                      // radius [AU] and latitude ["]
  double  l,b,r;                      // Ecliptic coordinates
```

```
// Mean anomalies of planets and mean arguments of lunar orbit [rad]
M2 = pi2 * Frac ( 0.1387306 + 162.5485917*T );
M3 = pi2 * Frac ( 0.9931266 +  99.9973604*T );
M4 = pi2 * Frac ( 0.0543250 +  53.1666028*T );
M5 = pi2 * Frac ( 0.0551750 +   8.4293972*T );
M6 = pi2 * Frac ( 0.8816500 +   3.3938722*T );
D  = pi2 * Frac ( 0.8274 + 1236.8531*T );
A  = pi2 * Frac ( 0.3749 + 1325.5524*T );
U  = pi2 * Frac ( 0.2591 + 1342.2278*T );

// Keplerian terms and perturbations by Venus
Ven.Init ( T, M3,0,7, M2,-6,0 );
Ven.Term ( 1, 0,0,-0.22,6892.76,-16707.37, -0.54, 0.00, 0.00);
Ven.Term ( 1, 0,1,-0.06, -17.35,    42.04, -0.15, 0.00, 0.00);
Ven.Term ( 1, 0,2,-0.01,  -0.05,     0.13, -0.02, 0.00, 0.00);
Ven.Term ( 2, 0,0, 0.00,  71.98,  -139.57,  0.00, 0.00, 0.00);
Ven.Term ( 2, 0,1, 0.00,  -0.36,     0.70,  0.00, 0.00, 0.00);
Ven.Term ( 3, 0,0, 0.00,   1.04,    -1.75,  0.00, 0.00, 0.00);
Ven.Term ( 0,-1,0, 0.03,  -0.07,    -0.16, -0.07, 0.02,-0.02);
Ven.Term ( 1,-1,0, 2.35,  -4.23,    -4.75, -2.64, 0.00, 0.00);
Ven.Term ( 1,-2,0,-0.10,   0.06,     0.12,  0.20, 0.02, 0.00);
Ven.Term ( 2,-1,0,-0.06,  -0.03,     0.20, -0.01, 0.01,-0.09);
Ven.Term ( 2,-2,0,-4.70,   2.90,     8.28, 13.42, 0.01,-0.01);
Ven.Term ( 3,-2,0, 1.80,  -1.74,    -1.44, -1.57, 0.04,-0.06);
Ven.Term ( 3,-3,0,-0.67,   0.03,     0.11,  2.43, 0.01, 0.00);
Ven.Term ( 4,-2,0, 0.03,  -0.03,     0.10,  0.09, 0.01,-0.01);
Ven.Term ( 4,-3,0, 1.51,  -0.40,    -0.88, -3.36, 0.18,-0.10);
Ven.Term ( 4,-4,0,-0.19,  -0.09,    -0.38,  0.77, 0.00, 0.00);
Ven.Term ( 5,-3,0, 0.76,  -0.68,     0.30,  0.37, 0.01, 0.00);
Ven.Term ( 5,-4,0,-0.14,  -0.04,    -0.11,  0.43,-0.03, 0.00);
Ven.Term ( 5,-5,0,-0.05,  -0.07,    -0.31,  0.21, 0.00, 0.00);
Ven.Term ( 6,-4,0, 0.15,  -0.04,    -0.06, -0.21, 0.01, 0.00);
Ven.Term ( 6,-5,0,-0.03,  -0.03,    -0.09,  0.09,-0.01, 0.00);
Ven.Term ( 6,-6,0, 0.00,  -0.04,    -0.18,  0.02, 0.00, 0.00);
Ven.Term ( 7,-5,0,-0.12,  -0.03,    -0.08,  0.31,-0.02,-0.01);
dl = Ven.dl(); dr = Ven.dr();  db = Ven.db();

// Perturbations by Mars
Mar.Init ( T, M3,1,5, M4,-8,-1 );
Mar.Term ( 1,-1,0,-0.22,   0.17,    -0.21, -0.27, 0.00, 0.00);
Mar.Term ( 1,-2,0,-1.66,   0.62,     0.16,  0.28, 0.00, 0.00);
Mar.Term ( 2,-2,0, 1.96,   0.57,    -1.32,  4.55, 0.00, 0.01);
Mar.Term ( 2,-3,0, 0.40,   0.15,    -0.17,  0.46, 0.00, 0.00);
Mar.Term ( 2,-4,0, 0.53,   0.26,     0.09, -0.22, 0.00, 0.00);
Mar.Term ( 3,-3,0, 0.05,   0.12,    -0.35,  0.15, 0.00, 0.00);
Mar.Term ( 3,-4,0,-0.13,  -0.48,     1.06, -0.29, 0.01, 0.00);
Mar.Term ( 3,-5,0,-0.04,  -0.20,     0.20, -0.04, 0.00, 0.00);
Mar.Term ( 4,-4,0, 0.00,  -0.03,     0.10,  0.04, 0.00, 0.00);
Mar.Term ( 4,-5,0, 0.05,  -0.07,     0.20,  0.14, 0.00, 0.00);
Mar.Term ( 4,-6,0,-0.10,   0.11,    -0.23, -0.22, 0.00, 0.00);
Mar.Term ( 5,-7,0,-0.05,   0.00,     0.01, -0.14, 0.00, 0.00);
Mar.Term ( 5,-8,0, 0.05,   0.01,    -0.02,  0.10, 0.00, 0.00);
dl += Mar.dl(); dr += Mar.dr();  db += Mar.db();
```

```
  // Perturbations by Jupiter
  Jup.Init ( T, M3,-1,3, M5,-4,-1 );
  Jup.Term (-1,-1,0, 0.01,   0.07,    0.18, -0.02, 0.00,-0.02);
  Jup.Term ( 0,-1,0,-0.31,   2.58,    0.52,  0.34, 0.02, 0.00);
  Jup.Term ( 1,-1,0,-7.21,  -0.06,    0.13,-16.27, 0.00,-0.02);
  Jup.Term ( 1,-2,0,-0.54,  -1.52,    3.09, -1.12, 0.01,-0.17);
  Jup.Term ( 1,-3,0,-0.03,  -0.21,    0.38, -0.06, 0.00,-0.02);
  Jup.Term ( 2,-1,0,-0.16,   0.05,   -0.18, -0.31, 0.01, 0.00);
  Jup.Term ( 2,-2,0, 0.14,  -2.73,    9.23,  0.48, 0.00, 0.00);
  Jup.Term ( 2,-3,0, 0.07,  -0.55,    1.83,  0.25, 0.01, 0.00);
  Jup.Term ( 2,-4,0, 0.02,  -0.08,    0.25,  0.06, 0.00, 0.00);
  Jup.Term ( 3,-2,0, 0.01,  -0.07,    0.16,  0.04, 0.00, 0.00);
  Jup.Term ( 3,-3,0,-0.16,  -0.03,    0.08, -0.64, 0.00, 0.00);
  Jup.Term ( 3,-4,0,-0.04,  -0.01,    0.03, -0.17, 0.00, 0.00);
  dl += Jup.dl(); dr += Jup.dr();  db += Jup.db();

  // Perturbations by Saturn
  Sat.Init ( T, M3,0,2, M6,-2,-1 );
  Sat.Term ( 0,-1,0, 0.00,   0.32,    0.01,  0.00, 0.00, 0.00);
  Sat.Term ( 1,-1,0,-0.08,  -0.41,    0.97, -0.18, 0.00,-0.01);
  Sat.Term ( 1,-2,0, 0.04,   0.10,   -0.23,  0.10, 0.00, 0.00);
  Sat.Term ( 2,-2,0, 0.04,   0.10,   -0.35,  0.13, 0.00, 0.00);
  dl += Sat.dl(); dr += Sat.dr();  db += Sat.db();

  // Difference of Earth-Moon-barycentre and centre of the Earth
  dl += +  6.45*sin(D) - 0.42*sin(D-A) + 0.18*sin(D+A)
        +  0.17*sin(D-M3) - 0.06*sin(D+M3);
  dr += + 30.76*cos(D) - 3.06*cos(D-A) + 0.85*cos(D+A)
        -  0.58*cos(D+M3) + 0.57*cos(D-M3);
  db += + 0.576*sin(U);

  // Long-periodic perturbations
  dl += + 6.40 * sin ( pi2*(0.6983 + 0.0561*T) )
        + 1.87 * sin ( pi2*(0.5764 + 0.4174*T) )
        + 0.27 * sin ( pi2*(0.4189 + 0.3306*T) )
        + 0.20 * sin ( pi2*(0.3581 + 2.4814*T) );

  // Ecliptic coordinates ([rad],[AU])
  l = pi2 * Frac ( 0.7859453 + M3/pi2 +
                 ( (6191.2+1.1*T)*T + dl ) / 1296.0e3 );
  r = 1.0001398 - 0.0000007 * T + dr * 1.0e-6;
  b = db / Arcs;

  return Vec3D ( Polar(l,b,r) );  // Position vector
}
```

The extent of the SunPos function is explained by the fact that the relative motion of the Sun and the Earth cannot be described to the required accuracy by simple elliptical orbits. Apart from the perturbations of the other planets – particularly those of Venus and Jupiter – there is also the monthly oscillation of the centre of the Earth around the barycentre of the Earth-Moon system. Strictly speaking, not the Earth itself, but rather the position of the barycentre, is moving in the plane of the ecliptic. The Moon's orbit around the Earth is therefore reflected in a small periodic perturbation of the geocentric orbit of the Sun. The details of the function and its basis will not be discussed here. Both will be treated thoroughly in Chap. 6, where appropriate routines for all the planets will be given.

2.6 The COCO Program

The functions described so far will now be combined into a complete program. In the following, all the relevant routines as well as the main program are given. The functions that are already familiar are combined into the corresponding library modules, whose matching header files are incorporated at the beginning of the program.

The core functions of the program are incorporated into a class `Position`, which, as well as the input and output of positions, has three methods for selecting the origin (heliocentric or geocentric), the equinox, and the reference frame (ecliptic or equatorial). The main program is therefore restricted to the declaration of an object of this class and the user-selected call to its methods. Although the class `Position` has no immediate application outside the `Coco` program, the object-oriented approach here, fosters a compact and clean programming.

```
//------------------------------------------------------------------------------
// File:    Coco.cpp
// Purpose: Coordinate transformations
// (c) 1999 Oliver Montenbruck, Thomas Pfleger
//------------------------------------------------------------------------------

#include <iomanip>
#include <iostream>

#include "APC_Const.h"
#include "APC_Math.h"
#include "APC_PrecNut.h"
#include "APC_Spheric.h"
#include "APC_Sun.h"
#include "APC_Time.h"
#include "APC_VecMat3D.h"

using namespace std;

// Definition of class "Position"
enum enOrigin   { Heliocentric, Geocentric };  // Origin of coordinates
enum enRefSys   { Ecliptic, Equator };         // Reference system
class Position {
  public:
    void Input(); // Query user for parameters
    void SetOrigin(enOrigin Origin);
    void SetRefSys(enRefSys RefSys);
    void SetEquinox(double T_Equinox);
    void Print();
```

```
  private:
    Vec3D      m_R;           // Coordinate vector
    enOrigin   m_Origin;      // Origin of coordinate system
    enRefSys   m_RefSys;      // Reference system
    double     m_TEquinox;    // Equinox (centuries since J2000)
    double     m_MjdEpoch;    // Epoch (Modif. Julian Date)
};

// Data input and output
void Position::Input() { ... }
void Position::Print() { ... }

// Change of origin
void Position::SetOrigin(enOrigin Origin)
{
  double T_Epoch;
  Vec3D  R_Sun;
  if (Origin!=m_Origin) {
    // Geocentric coordinates of the Sun at epoch w.r.t.
    // given reference system and equinox
    T_Epoch = (m_MjdEpoch-MJD_J2000)/36525.0;
    if (m_RefSys==Ecliptic)
      R_Sun = PrecMatrix_Ecl(T_Epoch,m_TEquinox) * SunPos(T_Epoch);
    else
      R_Sun = Ecl2EquMatrix(m_TEquinox) *
              PrecMatrix_Ecl(T_Epoch,m_TEquinox) * SunPos(T_Epoch);
    // Change origin
    if (m_Origin==Heliocentric) {
      m_R += R_Sun; m_Origin = Geocentric;
    }
    else {
      m_R -= R_Sun; m_Origin = Heliocentric;
    };
  };
}
// Change of reference system
void Position::SetRefSys(enRefSys RefSys)
{
  if (RefSys!=m_RefSys) {
    if (m_RefSys==Equator) {
      m_R = Equ2EclMatrix(m_TEquinox) * m_R;  m_RefSys = Ecliptic;
    }
    else {
      m_R = Ecl2EquMatrix(m_TEquinox) * m_R;  m_RefSys = Equator;
    }
  };
}
// Change of equinox
void Position::SetEquinox(double T_Equinox)
{
  if (T_Equinox!=m_TEquinox) {
    if (m_RefSys==Equator)
      m_R = PrecMatrix_Equ(m_TEquinox,T_Equinox) * m_R;
```

```
    else
      m_R = PrecMatrix_Ecl(m_TEquinox,T_Equinox) * m_R;
    m_TEquinox = T_Equinox;
  };
}

//------------------------------------------------------------------------------
//
// Main program
//
//------------------------------------------------------------------------------
void main() {
  // Variables
  Position  Pos;
  char      c;
  bool      End = false;
  double    Year;

  // Header
  cout << endl
       << "                COCO: coordinate conversions       " << endl
       << "        (c) 1999 Oliver Montenbruck, Thomas Pfleger" << endl
       << endl;

  // Initialization
  Pos.Input();
  Pos.Print();
  // Command loop

  do {
    // Command input
    cout << "Enter command (?=Help) ... ";
    cin >> c; cin.ignore(81,'\n'); c=tolower(c);
    // Actions
    switch (c) {
      case 'x':
        End = true;  break;
      case 'a':
        Pos.SetRefSys(Equator); Pos.Print();  break;
      case 'e':
        Pos.SetRefSys(Ecliptic); Pos.Print();  break;
      case 'g':
        Pos.SetOrigin(Geocentric); Pos.Print();  break;
      case 'h':
        Pos.SetOrigin(Heliocentric); Pos.Print();  break;
      case 'n':
        Pos.Input(); Pos.Print();  break;
      case 'p':
        cout << "New equinox (yyyy.y)   ...  ";
        cin >> Year; cin.ignore(81,'\n');
        Pos.SetEquinox( (Year-2000.0)/100.0 );
        Pos.Print();
        break;
```

```
      default:
        // Display help text
        cout << endl
             << "Available commands" << endl
             << "  e=ecliptic,    a=equatorial,    p=precession, " << endl
             << "  g=geocentric, h=heliocentric, n=new input,   " << endl
             << "  x=exit                                        " << endl
             << endl;
        break;
    }
  }
  while (!End);
  cout << endl;
}
```

The following example should illustrate the use and possibilities offered by Coco. The coordinates to be converted may be entered in either the equatorial or the ecliptic form. One can also choose between Cartesian coordinates or polar coordinates. This choice is offered by Coco at the beginning. All entries are indicated in italic. We select equatorial coordinates in polar representation and enter the coordinates of the vernal equinox of 1950.0. The distance is arbitrarily chosen as 1 AU. It must always be positive.

```
                COCO: coordinate conversions
        (c) 1999 Oliver Montenbruck, Thomas Pfleger

New input:

  Reference system (e=ecliptic,a=equator) ... a
  Format (c=cartesian,p=polar)            ... p
  Coordinates (RA [h m s] Dec [o ' "] R)  ... 0 0 0.0 0 0 0.0 1.0
  Equinox (yyyy.y)                        ... 1950.0
  Origin (h=heliocentric,g=geocentric)    ... g
  Epoch (yyyy mm dd hh.h)                 ... 1989 1 1 0.0
```

Coco acknowledges the input by displaying these data in Cartesian and polar coordinates:

```
Geocentric equatorial coordinates
(Equinox J1950.0,  Epoch 1989/01/01 00.0)

  (x,y,z) =  (  1.00000000,  0.00000000,  0.00000000)

         h  m  s               o  '  "
  RA =  0 00 00.00    Dec = + 0 00 00.0    R =   1.00000000

Enter command (?=Help) ... ?

Available commands
  e=ecliptic,   a=equatorial,   p=precession,
  g=geocentric, h=heliocentric, n=new input,
  x=exit
```

```
Enter command (?=Help) ...
```

Subsequently, the program expects the input of a command for converting the coordinates. Possible inputs are:

- **a** Conversion to equatorial coordinates,
- **e** Conversion to ecliptic coordinates,
- **p** Precession (Choice of equinox),
- **g** Conversion to geocentric coordinates,
- **h** Conversion to heliocentric coordinates,
- **n** New input,
- **?** Help,
- **x** Exit (end of program) .

We first want to convert the given position to epoch 2000, and therefore choose option p to calculate the precession. After entering the year, we obtain the position of the vernal equinox for 1950 referred to the new epoch 2000. The precession amounts to slightly more than half a degree.

```
Enter command (?=Help) ... p
New equinox (yyyy.y)   ... 2000.0

Geocentric equatorial coordinates
(Equinox J2000.0,  Epoch 1989/01/01 00.0)

  (x,y,z) =  (  0.99992571,  0.01117889,  0.00485898)

             h  m  s                  o  '  "
  RA =   0 02 33.73     Dec = + 0 16 42.2     R =    1.00000000
```

We now need the given equatorial coordinates to be converted into the ecliptic system. We therefore choose option e and obtain the converted coordinates – again in Cartesian and polar forms. Instead of right ascension RA and declination Dec we now have ecliptic longitude L and latitude B. As in the previous step, the distance remains unchanged.

```
Enter command (?=Help) ... e

Geocentric ecliptic coordinates
(Equinox J2000.0,  Epoch 1989/01/01 00.0)

  (x,y,z) =  (  0.99992571,  0.01218922,  0.00001132)

             o  '  "              o  '  "
  L =    0 41 54.27     B = + 0 00 02.3     R =    1.00000000
```

It is now possible to convert the given coordinates from geocentric to heliocentric coordinates for the initially specified epoch using option h. As in the previous step we chose ecliptic coordinates, we now obtain heliocentric ecliptic coordinates.

```
Enter command (?=Help) ... h

Heliocentric ecliptic coordinates
(Equinox J2000.0,  Epoch 1989/01/01 00.0)

  (x,y,z) =  (  0.81725247,  0.97838164,  0.00003597)

            o  '  "              o  '  "
  L =  50 07 39.50    B = + 0 00 05.8    R =    1.27480674
```

The conversion into ecliptic coordinates may be reversed by selecting option a, which converts the current (ecliptic) coordinates into equatorial coordinates.

```
Enter command (?=Help) ... a

Heliocentric equatorial coordinates
(Equinox J2000.0,  Epoch 1989/01/01 00.0)

  (x,y,z) =  (  0.81725247,  0.89763329,  0.38921086)

            h  m  s                o  '  "
  RA =  3 10 44.07    Dec = +17 46 36.5    R =    1.27480674
```

Let us now recalculate the coordinates for the original equinox of 1950.

```
Enter command (?=Help) ... p
New equinox (yyyy.y)   ... 1950.0

Heliocentric equatorial coordinates
(Equinox J1950.0,  Epoch 1989/01/01 00.0)

  (x,y,z) =  (  0.82911747,  0.88843066,  0.38521087)

            h  m  s                o  '  "
  RA =  3 07 54.68    Dec = +17 35 17.2    R =    1.27480674
```

Conversion of these coordinates into geocentric coordinates should now yield the original coordinates of the vernal equinox, because we have applied the inverse transformation to the three conversions that we invoked. Unavoidable rounding errors within the computer may, however, prevent exactly the same coordinates as those entered at the beginning of this example from being obtained.

```
Enter command (?=Help) ... g

Geocentric equatorial coordinates
(Equinox J1950.0,  Epoch 1989/01/01 00.0)

  (x,y,z) =  (  1.00000000, -0.00000000,  0.00000000)

            h  m  s                o  '  "
  RA = 24 00 00.00    Dec = + 0 00 00.0    R =    1.00000000
```

Entering x then terminates the program.

```
Enter command (?=Help) ... x
```

The values shown here may differ slightly depending on the particular C++ compiler being used.

3. Calculation of Rising and Setting Times

3.1 The Observer's Horizon System

The ecliptic and equatorial coordinates discussed so far are specified in terms of the mean plane of the Earth's orbit and of the position of the Earth's axis. Neither of these systems is particularly suitable, however, for an observer situated on the surface of the Earth. As such an observer (without being aware of it) takes part in the Earth's daily rotation, it appears as if the Sun, Moon, and stars follow large arcs across the sky from East to West during the course of a day, reaching their highest point above the horizon when they are on the meridian.

The stars' apparent paths depend on the geographical latitude of the point of observation. In the Southern Hemisphere most stars do not reach their greatest altitude in the South, but in the North. As a result, orienting themselves on the sky is more difficult for observers who are used to the appearance of the sky at moderate northern latitudes. All the well-known constellations appear to be standing on their heads.

Two points on the celestial sphere are particularly significant: the zenith (the point directly above the observer), and the North Celestial Pole (Fig. 3.1). The latter is taken to be the point on the celestial sphere towards which the Earth's axis is pointing, and about which, in consequence, all the stars appear to be moving in concentric circles. The altitude of this point above the horizon corresponds to the observer's geographic latitude φ. A great circle drawn through the North Celestial Pole and the zenith, known as the *meridian*, intersects the horizon at the exact North and South points.

The coordinates used in the horizontal system are azimuth (A) and altitude (h), which may, for instance, be determined by the use of a theodolite (see Fig. 3.1). The altitude is simply the angular height above the horizon. The azimuth, determined by rotating the theodolite around its vertical axis, is defined as the angle between the South point and the required position. On this basis, the azimuth of the West point is $A = 90°$ and, correspondingly, for the East point $A = 270°$ (or $A = -90°$). Unfortunately, alongside the definition just given, there is a second one that is used, in particular, in navigation. Here, azimuth is taken to originate at the North point. According to this method, the East point therefore has an azimuth of 90°. In case of any doubt, it is essential to check which of these two definitions of azimuth is being used!

Because of the rotation of the Earth, the altitude and azimuth of an object with a given right ascension and declination alter continuously. If a telescope is held at

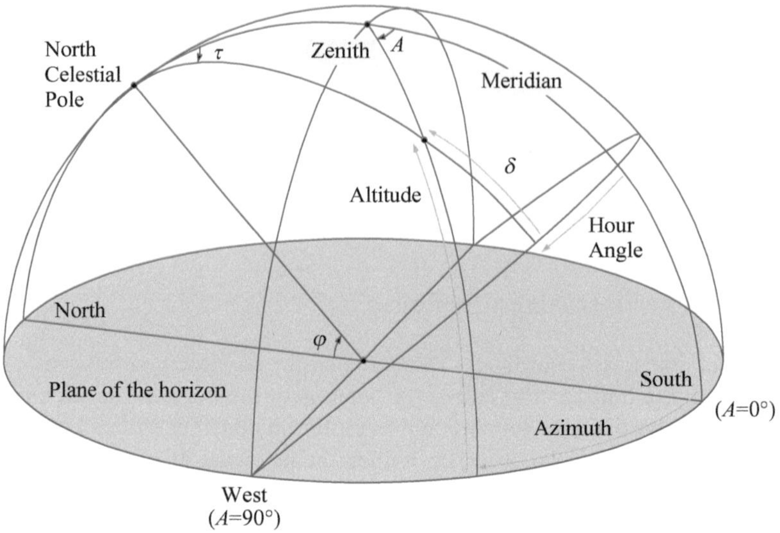

Fig. 3.1. The horizontal system

a fixed position in the horizontal system, with time, the stars that cross the field of view are those that have the same declination, and which differ only in their right ascension. The determination of the declination of a star from altitude and azimuth is therefore independent of time. For right ascension, on the other hand, only the difference between the object and stars on the meridian may be determined.

To obtain reciprocal conversions between horizontal and equatorial coordinates, the concept of *hour angle* τ is used. This is the difference between the right ascension of the star being observed and the right ascension of stars on the meridian. Hour angle τ, like right ascension α, is generally measured in units of time ($1^{\mathrm{h}} \equiv 15°$), and therefore corresponds approximately[1] to the amount of time that has passed since the star crossed the meridian. In the system employed here, A and τ have the same mathematical sign. If we have the Cartesian coordinates

$$\boldsymbol{r} = \begin{pmatrix} x \\ y \\ z \end{pmatrix} = \begin{pmatrix} \cos h \cos A \\ \cos h \sin A \\ \sin h \end{pmatrix} \qquad \hat{\boldsymbol{r}} = \begin{pmatrix} \hat{x} \\ \hat{y} \\ \hat{z} \end{pmatrix} = \begin{pmatrix} \cos\delta\cos\tau \\ \cos\delta\sin\tau \\ \sin\delta \end{pmatrix} \tag{3.1}$$

the transformations

$$\begin{aligned} x &= +\hat{x}\sin\varphi - \hat{z}\cos\varphi & \qquad \hat{x} &= +x\sin\varphi + z\cos\varphi \\ y &= +\hat{y} & \hat{y} &= +y \\ z &= +\hat{x}\cos\varphi + \hat{z}\sin\varphi & \hat{z} &= -x\cos\varphi + z\sin\varphi \end{aligned} \tag{3.2}$$

apply, which may in brief be written as $\boldsymbol{r} = \boldsymbol{R}_y(\pi/2 - \varphi)\hat{\boldsymbol{r}}$. Here φ is the geographical latitude of the observer, and ^ indicates equatorial coordinates. Substitution

[1]One rotation of the Earth takes only $23^{\mathrm{h}}56^{\mathrm{m}}$.

then yields a system of three equations

$$\begin{aligned}\cos h \cos A &= +\cos\delta \cos\tau \sin\varphi - \sin\delta \cos\varphi \\ \cos h \sin A &= +\cos\delta \sin\tau \\ \sin h &= +\cos\delta \cos\tau \cos\varphi + \sin\delta \sin\varphi\end{aligned} \qquad (3.3)$$

for computing azimuth and altitude from hour angle and declination. Conversely, the hour angle and declination may be determined from

$$\begin{aligned}\cos\delta \cos\tau &= +\cos h \cos A \sin\varphi + \sin h \cos\varphi \\ \cos\delta \sin\tau &= +\cos h \sin A \\ \sin\delta &= -\cos h \cos A \cos\varphi + \sin h \sin\varphi\end{aligned} \qquad (3.4)$$

for a given azimuth and altitude. The transformation between the equatorial system and the horizontal system may be accomplished by the functions `Equ2Hor` and `Hor2Equ`, which are given here for the sake of completeness.

```
//------------------------------------------------------------------------------
// Equ2Hor: Transformation of equatorial coordinates to the horizon system
//   Dec        Declination [rad]
//   tau        Hour angle [rad]
//   lat        Geographical latitude of the observer [rad]
//   h,Az       Altitude and azimuth [rad]
//------------------------------------------------------------------------------
void Equ2Hor ( double Dec, double tau, double lat,
               double& h, double& Az )
{
  Vec3D e_equ, e_hor;
  e_equ = Vec3D(Polar(tau,Dec));        // unit vector in horizontal system
  e_hor = R_y(pi/2.0-lat) * e_equ;      // unit vector in equatorial system
  Az = e_hor[phi];                      // polar angles
  h  = e_hor[theta];
}
//------------------------------------------------------------------------------
// Hor2Equ: Transformation of horizon system coordinates
//          to equatorial coordinates
//   h,Az       Altitude and azimuth [rad]
//   lat        Geographical latitude of the observer [rad]
//   Dec        Declination [rad]
//   tau        Hour angle [rad]
//------------------------------------------------------------------------------
void Hor2Equ ( double h, double Az, double lat,
               double& Dec, double& tau )
{
  Vec3D e_equ, e_hor;
  e_hor = Vec3D(Polar(Az,h));           // unit vector in horizontal system
  e_equ = R_y(-(pi/2.0-lat)) * e_hor;   // unit vector in equatorial system
  tau = e_equ[phi];                     // polar angles
  Dec = e_equ[theta];
}
```

For determining the times of rising and setting, which is the aim of this chapter, one equation only is required, namely

$$\sin h = \cos\varphi \cos\delta \cos\tau + \sin\varphi \sin\delta \quad . \qquad (3.5)$$

It allows the altitude h to be calculated from given values of geographical latitude φ, declination δ, and hour angle τ.

Before we can carry out this conversion, however, some preparation is required. First, we lack the means of determining the coordinates (α, δ) of the Sun and the Moon. Then we need to clarify how we can calculate the hour angle from the known right ascension at a particular time. Finally, we have to take into account a whole series of corrections that affect the altitude of the observed horizon. Only after these steps will we return to the equation just mentioned.

3.2 Sun and Moon

Calculation of rising and setting does not make demands for a high degree of accuracy in the coordinates of the Sun and the Moon. The `MiniSun` and `MiniMoon` functions therefore contain only the most important terms describing the respective orbits. They are greatly reduced versions of `SunPos` and `MoonPos`, which will be described in detail in Chap. 6 and Chap. 8. Conversion of ecliptic longitude and latitude into equatorial coordinates is also included, therefore both functions may be used without further operations to obtain the right ascension and declination of the object in question.

```
//------------------------------------------------------------------------
// MiniMoon: Computes the Moon's RA and declination using a low precision
//           analytical series
//   T         Time in Julian centuries since J2000
//   RA        Right Ascension of the Moon in [rad]
//   Dec       Declination of the Moon in [rad]
//------------------------------------------------------------------------
void MiniMoon (double T, double& RA, double& Dec)
{
  const double eps = 23.43929111*Rad;
  double L_0, l,ls, F, D, dL, S, h, N, l_Moon, b_Moon;
  Vec3D  e_Moon;
  // Mean elements of lunar orbit
  L_0 = Frac (0.606433 + 1336.855225*T);        // mean longitude [rev]
  l  = pi2*Frac ( 0.374897 + 1325.552410*T );  // Moon's mean anomaly
  ls = pi2*Frac ( 0.993133 +   99.997361*T );  // Sun's mean anomaly
  D  = pi2*Frac ( 0.827361 + 1236.853086*T );  // Diff. long. Moon-Sun
  F  = pi2*Frac ( 0.259086 + 1342.227825*T );  // Dist. from ascending node
  // Perturbations in longitude and latitude
  dL = +22640*sin(l) - 4586*sin(l-2*D) + 2370*sin(2*D) +  769*sin(2*l)
       -668*sin(ls) - 412*sin(2*F) - 212*sin(2*l-2*D) - 206*sin(l+ls-2*D)
       +192*sin(l+2*D) - 165*sin(ls-2*D) - 125*sin(D) - 110*sin(l+ls)
       +148*sin(l-ls) - 55*sin(2*F-2*D);
  S  = F + (dL+412*sin(2*F)+541*sin(ls)) / Arcs;
  h  = F-2*D;
  N  = -526*sin(h) + 44*sin(l+h) - 31*sin(-l+h) - 23*sin(ls+h)
       + 11*sin(-ls+h) - 25*sin(-2*l+F) + 21*sin(-l+F);
  // Ecliptic longitude and latitude
  l_Moon = pi2 * Frac( L_0 + dL/1296.0e3 ); // [rad]
```

```
  b_Moon = ( 18520.0*sin(S) + N ) / Arcs;    // [rad]
  // Equatorial coordinates
  e_Moon = R_x(-eps) * Vec3D(Polar(l_Moon,b_Moon));
  RA   = e_Moon[phi];
  Dec = e_Moon[theta];
}

//------------------------------------------------------------------------------
// MiniSun: Computes the Sun's RA and declination using a low precision
//          analytical series
//   T         Time in Julian centuries since J2000
//   RA        Right Ascension of the Sun in [rad]
//   Dec       Declination of the Sun in [rad]
//------------------------------------------------------------------------------
void MiniSun (double T, double& RA, double& Dec)
{
  const double eps = 23.43929111*Rad;
  double L,M;
  Vec3D  e_Sun;
  // Mean anomaly and ecliptic longitude
  M  = pi2 * Frac ( 0.993133 + 99.997361*T);
  L  = pi2 * Frac ( 0.7859453 + M/pi2 +
                    (6893.0*sin(M)+72.0*sin(2.0*M)+6191.2*T) / 1296.0e3);
  // Equatorial coordinates
  e_Sun = R_x(-eps) * Vec3D(Polar(L,0.0));
  RA   = e_Sun[phi];
  Dec = e_Sun[theta];
}
```

3.3 Sidereal Time and Hour Angle

The right ascensions of the Sun and Moon are not sufficient to calculate their altitudes above the horizon. We also need to know the right ascension of stars on the meridian and, from that, determine the hour angle. Precisely which stars these are depends on the observer's location and the exact time. As the Earth rotates around its axis once a day, an observer sees all right ascensions between 0^h and 24^h transit the meridian. Over an hour, the right ascension of the stars that culminate alters by about 1^h.

The regular motion of the stars is very suitable for forming the basis for another system of time reckoning, which is known as *sidereal time*. For any given place, it is defined as the right ascension of the stars that are on the meridian at that particular instant. Because of this definition, sidereal time may be determined directly by observation of the sky.

The need for sidereal time, in addition to solar time, arises from the slight difference between the length of a solar day and one rotation of the Earth. Clocks, which we use continually to tell the time, are arranged to divide the day into 24 hours. Here, one day is the alternation of light and dark, which is determined by the Sun, and 24 hours are the amount of time, on average, between one meridian passage of the Sun and the next. If, however, we measure the corresponding interval for a star, we find that this takes only $23^h56^m4\overset{s}{.}091$. This shorter period, known as a sidereal day in distinction to a solar day, is exactly the duration of one rotation of the Earth. The cause of the difference of about 4^m arises from the Earth's year-long orbit around the Sun. Because of this motion, the Sun's right ascension changes by $360° \equiv 24^h$ per year, which is about 4^m per day. So the time between two meridian transits of the Sun is greater than that for a star by this amount.

For any given Universal Time UT, the sidereal time at Greenwich may be determined from

$$\begin{aligned} \Theta_0 &= 24110\overset{s}{.}54841 + 8640184\overset{s}{.}812866 \cdot T_0 + 1.0027379093 \cdot \mathrm{UT} \\ &\quad + 0\overset{s}{.}093104 \cdot T^2 - 0\overset{s}{.}0000062 \cdot T^3 \end{aligned} \quad (3.6)$$

where

$$T_0 = \frac{\mathrm{JD}_0 - 2451545}{36525} \quad \text{and} \quad T = \frac{\mathrm{JD} - 2451545}{36525} \quad .$$

JD and JD_0 are the Julian Date of the time of observation, and the Julian Date at 0^h UT on the date of observation.

```
//------------------------------------------------------------------------------
// GMST: Greenwich mean sidereal time
//   MJD        Time as Modified Julian Date
//   <return>   GMST in [rad]
//------------------------------------------------------------------------------
double GMST (double MJD)
{
  const double Secs = 86400.0;          // Seconds per day
  double MJD_0, UT, T_0, T, gmst;
  MJD_0 = floor ( MJD );
  UT    = Secs*(MJD-MJD_0);      // [s]
  T_0   = (MJD_0-51544.5)/36525.0;
  T     = (MJD  -51544.5)/36525.0;
  gmst  = 24110.54841 + 8640184.812866*T_0 + 1.0027379093*UT
          + (0.093104-6.2e-6*T)*T*T;        // [sec]
  return  (pi2/Secs)*Modulo(gmst,Secs);    // [Rad]
}
```

For a position with a geographical longitude λ, the local sidereal time differs by $\lambda/15°$ hours from the sidereal time at Greenwich:

$$\Theta = \Theta_0 + \lambda \cdot 1^h/15° \quad . \quad (3.7)$$

Here, longitude λ is reckoned *positive towards the east* (for Munich in Germany, for example, $\lambda = +11\overset{\circ}{.}6$). The hour angle τ of a star of right ascension α is therefore given by

$$\tau = \Theta - \alpha \quad . \quad (3.8)$$

As an example for the use of sidereal time in the conversion of right ascension and hour angle, the following sample program computes the azimuth and altitude of the star Deneb (α Cyg) for a given place and time:

```
double lambda = +11.6*Rad;                     // Observing site Munich
double phi    = +48.1*Rad;
double RA     = 15.0*Ddd(20,41,12.8)*Rad       // Equ. coord. Deneb [rad]
double Dec    =      Ddd(45,15,25.8)*Rad
double ModJD  = Mjd ( 1993,8,1, 21,0,0.0 );    // Aug. 1 1993, 21:00 UT

double tau,h,Az;

tau = GMST(ModJD) + lambda - RA;               // Hour angle [rad]
Equ2Hor ( Dec,tau,phi, h,Az );                 // Azimuth and altitude

cout << "Deneb:" << endl                       // Output
     << setprecision(2)
     << "Altitude   =" << setw(7) << Deg*h  << " deg"
     << "Azimuth    =" << setw(7) << Deg*Az << " deg";
```

3.4 Universal Time and Ephemeris Time

We have already frequently used the concept of 'Time', without considering it in more detail. In astronomy, however, we encounter a whole series of ways of reckoning time that are employed alongside one another. In general, they may be divided into two classes, the most important members of which are *dynamical time* (TT, TDB) and *Universal Time* (UT). The different objectives that are the reasons for these different times need to be explained in moderate detail.

The concept behind dynamical time is that it is a time that allows astronomical processes to be described in terms of physics. One second of dynamical time is defined by the period of a specific transition occurring in the caesium atom, and is correspondingly measured nowadays by atomic clocks. Since 1984, dynamical time has replaced *Ephemeris Time* (ET). The basis on which the latter had previously been established was that of tables of the motion of the Sun, the Moon, and the planets, which had been calculated according to classical mechanics. Ephemeris Time could be determined by comparing the observed positions of these celestial bodies with the ephemerides calculated earlier. The grounds for the introduction of dynamical time lay in relativity theory, which established that there is no truly universally applicable time, but that its rate also depends on the position and motion of the reference system in which time is being measured. This necessitates, in particular, differentiating between Terrestrial (Dynamical) Time (TT or TDT), which refers to the centre of the Earth, and Barycentric Dynamical Time (TDB), which relates to the centre of gravity of the Solar System. For our purposes, however, the three forms of time ET, TT, and TDB may be regarded as equivalent.

In contrast to Ephemeris Time and dynamical time, *Universal Time* (UT) is a *non-uniform* timescale. UT is the best current realization of a solar time. It was established to try to ensure that over several thousand years one day has an average length of 24 hours. But the result of this is that the length of one second of Universal Time is not constant, because the actual mean length of a day depends on the rotation of the Earth and the apparent motion of the Sun (i.e., the length of the year). Unfortunately, it is not possible to determine Universal Time by a suitable conversion from dynamical time, because the rotation of the Earth cannot be predicted accurately. Every change in the Earth's rotation alters the length of the day, and must therefore be taken into account in UT. Universal Time is therefore defined as a function of sidereal time, which directly reflects the rotation of the Earth. For any particular day, 0^{h} UT is defined as the instant at which Greenwich Mean Sidereal Time (GMST) has the value

$$\begin{aligned}\mathrm{GMST}(0^{\mathrm{h}}\mathrm{UT}) &= 24110^{\mathrm{s}}.54841 + 8640184^{\mathrm{s}}.812866 \cdot T_0 \\ &\quad + 0^{\mathrm{s}}.093104 \cdot T_0^2 - 0^{\mathrm{s}}.0000062 \cdot T_0^3\end{aligned}$$

where

$$T_0 = \frac{\mathrm{JD}(0^{\mathrm{h}}\mathrm{UT}) - 2451545}{36525} \quad .$$

We have already encountered this equation in Sect. 3.3 where we used it to calculate sidereal time for a given instant of Universal Time. We have now, however, established that sidereal time is the only observable quantity from which we can derive Universal Time.

The difference between Universal Time and Ephemeris Time or Terrestrial Time may be determined only retrospectively. Table 3.1 summarizes the value of ΔT=ET-UT (ET=TT=TDB) over the course of the last century. At present ΔT is increasing by about 0.5 to 1.0 seconds per year.

Table 3.1. Value of ΔT=ET−UT (in s) for 1900-2000

Year	ET-UT	Year	ET-UT	Year	ET-UT	Year	ET-UT
1900	-2.72						
1905	3.86	1930	24.02	1955	31.07	1980	50.54
1910	10.46	1935	23.93	1960	33.15	1985	54.34
1915	17.20	1940	24.33	1965	35.73	1990	56.86
1920	21.16	1945	26.77	1970	40.18	1995	60.82
1925	23.62	1950	29.15	1975	45.48	2000	63.83

Clock time, which we use for everyday purposes, is derived from *Coordinated Universal Time* (UTC). UTC is obtained from atomic clocks, and therefore has the same rate as Terrestrial Time. By the use of leap seconds, which may be inserted once or twice a year, care is taken to ensure that UTC never deviates more than 0.9 seconds from Universal Time, UT. UTC is the basis for time measurement over the whole of the Earth. Every point is part of a specific time zone, within which the official time differs by whole hours (or half hours) from UTC. In practice, the time zones have been arranged to agree with geographical features and, in particular, with national borders, to avoid having several time zones across a single country. The time zone employed in a particular area is established by international agreement and may be determined from a chart of the time zones.

At this point we may well enquire which time we should use when we want to calculate the rising and setting times of the Sun and the Moon. The preceding definitions give us the following general rule of thumb for the use of Universal Time and Ephemeris Time:

- Universal Time (UT) is used to calculate sidereal time.
- Ephemeris Time (ET) or dynamical time (TT, TDB) are used to calculate solar, lunar, or planetary ephemerides.

Let us take as an example the calculation of the altitude of the Moon, and consider the individual steps in the process. Let us take the time as being 1982 January 1, 0^{h} Central European Time, and the observing site as Munich ($\varphi = 48\overset{\circ}{.}1$, $\lambda = -11\overset{\circ}{.}6$). We can ignore the slight difference between UTC and UT, so we obtain the Universal Time from UT=CET-1^{h}. As discussed earlier, Universal Time has to be used in calculating sidereal time, while Ephemeris Time is used in calculating the coordinates of the Moon. We obtain the conversion as shown in this portion of a program:

```
double lambda = +11.6*Rad;                  // Geogr. coord. Munich
double phi    = +48.1*Rad;
double MjdCET = Mjd ( 1982,1,1, 0,0,0.0 );  // Jan. 1, 1982, 00:00 CET
double CET_UT = 1.0                          // Zone time difference [h]
double ET_UT  = 52.17;                       // in sec; for 1982

double MjdUT,MjdET, RA,Dec, sinH;

MjdUT  = MjdCET - CET_UT/24.0;              // MJD Universal Time
MjdET  = MjdUT  + ET_UT/86400.0;            // MJD Ephemeris Time

MiniMoon ( MjdET,RA,Dec);                   // Equatorial coordinates

tau  = GMST(MjdUT) + lambda - RA;           // Hour angle [rad]
sinH = sin(phi)*sin(Dec)                    // Sine of altitude
       + cos(phi)*cos(Dec)*cos(tau);        //   above the horizon
```

We can also ask what errors occur if we ignore the difference between Universal Time and Ephemeris Time. If the coordinates of the Moon are calculated with the simplification that we use UT instead of ET, the example just given becomes:

```
MjdUT  = MjdCET - CET_UT/24.0;              // MJD Universal Time

MiniMoon ( MjdUT,RA,Dec);                   // Equatorial coordinates

tau  = GMST(MjdUT) + lambda - RA;           // Hour angle [rad]
sinH = sin(phi)*sin(Dec)                    // Sine of altitude
       + cos(phi)*cos(Dec)*cos(tau);        //   above the horizon
```

Although the sidereal time is calculated correctly, the coordinates of the Moon are now calculated for a time that is ΔT=ET−UT too *early*. At present, ΔT is about one minute, an interval of time in which the Moon moves by about 30 arc-seconds. This error is even smaller than the error obtained by using the simplified `MiniMoon` function. The times of moonrise and moonset will be in error by only about 3^{s}. For the Sun, these values are even smaller. In the `Sunset` program, we can therefore ignore ΔT with a clear conscience. In later applications, such as the calculation of exact planetary positions or in predicting stellar occultations, it is, however, essential to distinguish carefully between the various forms of time. This is why we have discussed in detail the various methods of reckoning astronomical time.

3.5 Parallax and Refraction

So far, we have always given the right ascension and declination of a celestial body in geocentric equatorial coordinates, i.e., in a system with its origin at the centre of the Earth. We are not at the centre of the Earth, however, and observe from its surface. What differences does this cause between the coordinates that are calculated and those actually observed?

In observing fixed stars, in practice no difference is found between geocentric coordinates and the *topocentric* coordinates that are obtained by parallel translation of the zero point of the coordinates from the centre of the Earth to the position of the observer on the surface. The distance of the fixed stars is enormous in comparison with the distance between the observer and the centre of the Earth. But when we are dealing with nearer celestial bodies, such as the planets, the Sun, or the Moon, there are marked differences.

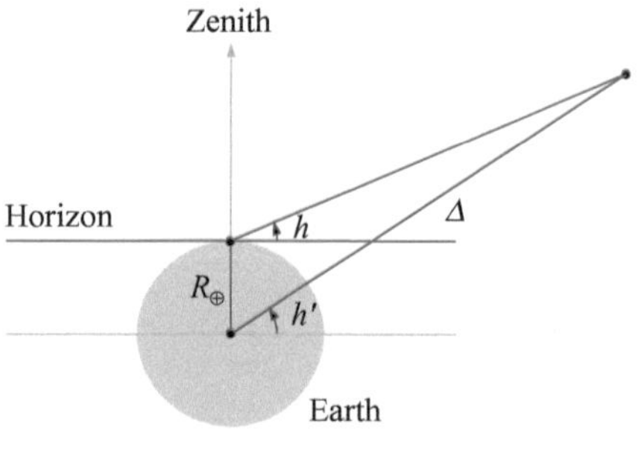

Fig. 3.2. Parallax

This difference is known as *parallax* and causes an object observed from the surface of the Earth to appear slightly lower than it would if it were observable from the centre of the Earth. The topocentric altitude h is less than the geocentric altitude h' (see Fig. 3.2). If the object is at the zenith, the parallax disappears. As the altitude decreases, however, the parallax becomes increasingly important, and becomes a maximum when the altitude $h = 0°$. This is known as the *horizontal parallax* and is calculated from

$$\pi = \arcsin(R_\oplus/\Delta) \quad , \tag{3.9}$$

where $R_\oplus \approx 6378$ km, the distance of the observer from the centre of the Earth, and Δ the *geo*centric distance of the object. If we substitute the distance of the Sun in (3.9) then we obtain a value of $\pi_\odot = 8''.8$. For the Moon, our nearest neighbour, on the other hand, the horizontal parallax amounts to the no longer insignificant amount of

$$\pi_{\text{Moon}} = \arcsin\left(\frac{6378\ \text{km}}{384400\ \text{km}}\right) \approx 57' \quad .$$

We can already see that for calculating rising and setting times we must use the topocentric altitude rather than the geocentric one. Before that, however, yet another effect must be described.

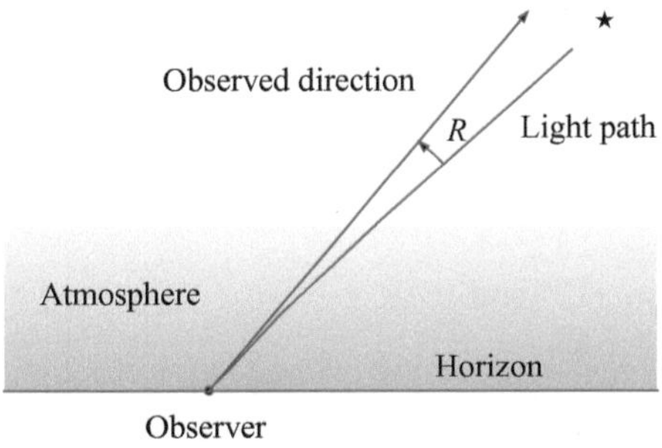

Fig. 3.3. Refraction

In passing from the vacuum of space into the optically denser atmosphere of the Earth, rays of light are deviated, in accordance with the law of refraction, towards the vertical (see Fig. 3.3). This is known as *atmospheric refraction*. To an observer on the ground, stars will appear to have been 'raised' slightly. A ray of light entering the atmosphere at a low angle has a longer path through layers of air of differing densities (and refractive indexes) than a ray entering vertically. Atmospheric refraction is therefore greater, the lower the altitude of the object. The maximum value of about 34 arc-minutes is attained at the horizon. Some further data are given in Table 3.2. To obtain the observed altitude, R should be added to the calculated topocentric altitude. Atmospheric refraction also depends on the density and temperature of the air. The figures quoted should therefore be taken as average values. Comprehensive tables of normal refraction and numerical approximation formulae may be found in appropriate reference works.

Table 3.2. Values of refraction near the horizon

h	10°	5°	2°	1°	0°
R	5′31″	10′15″	19′7″	25′36″	34′

We have seen that when dealing with rising and setting times, we must use topocentric altitudes. In addition, we also have to take refraction near the horizon into account. Because rising and setting times are always referred to the upper limb of the Sun or the Moon, we also need to make allowance for the apparent radius $s_\odot$ or $s_{\rm Moon}$. At the instant the body rises or sets, it therefore has the *geo*centric altitude

$$h_{\rm R/S} = 0^\circ + \pi - R_{h=0} - s \ . \tag{3.10}$$

Our task now consists of finding, for a given date, the time at which the body being observed reaches an altitude of $h = h_{\rm R/S}$. Because there is no sense in trying to obtain the times to an accuracy better than a few minutes, we can forgo accurate calculation of π and s and use average values. The values normally taken in calculating rising and setting times are therefore:

- Sunrise or sunset : $h_{\rm R/S} = -0^\circ 50'$,
- Moonrise or moonset : $h_{\rm R/S} = +0^\circ 08'$,
- for Stars or Planets : $h_{\rm R/S} = -0^\circ 34'$.

As our program also calculates the times of the beginning and end of twilight, the necessary definitions should also be discussed here. The types of twilight are

- *astronomical* twilight when $h_\odot = -18^\circ$,
- *nautical* twilight when $h_\odot = -12^\circ$ and
- *civil* twilight when $h_\odot = -6^\circ$.

No further corrections are applied to these values for parallax, refraction, or apparent semi-diameter. Twilight therefore begins or ends, by definition, when the *geocentric* altitude of the Sun attains one of the values quoted. What remains to be solved now, is the general problem of finding the instant of time at which a celestial body reaches a specified altitude.

3.6 Rising and Setting Times

Equation (3.5) allows us to determine the altitude h, given the geographical latitude φ, the declination δ, and the hour angle τ. The hour angle is obtained for a given local sidereal time from (3.8).

If, on the other hand, we have the altitude h, we can rearrange (3.5), and obtain a relationship for the hour angle:

$$\cos\tau = \frac{\sin h - \sin\varphi\sin\delta}{\cos\varphi\cos\delta} \ . \tag{3.11}$$

We can therefore calculate the hour angle of a star when it is observed at an altitude h above the horizon. The equation is not, however, valid for every possible value of h, but only for the altitudes that the star may actually attain. At a geographical latitude of $\varphi = 50^\circ$, a star of declination $\delta = 60^\circ$, for example, is circumpolar, and may be observed only between $h = 20^\circ$ and $h = 80^\circ$. If, for a predetermined altitude h, the value of $|\cos\tau|$ is consistently greater than 1, then the star will always be found either above or below that altitude.

For the rising and setting of stars, we substitute the altitude $h_{R/S} = -0°34'$ in (3.11), the hour angle thus obtained being expressed in units of time ($15° \equiv 1^h$). It gives the number of hours in sidereal time that a star that is just rising requires to reach culmination. A sidereal day is equal to $23^h56^m4\overset{s}{.}091$ solar time, giving a factor of $23^h56^m4\overset{s}{.}091/24^h = 0.9972696$ by which an interval stated in sidereal time may be converted into solar time. If we multiply the hour angle by this factor, we obtain what is known as the *semi-diurnal arc*. This is half the overall time that the star is visible above the horizon.

For a given right ascension $\alpha_\star$ of a star, the sidereal time at the instant of rising or setting is given by

$$\Theta_{R/S} = \begin{cases} \alpha_\star - \tau & \text{for rise} \\ \alpha_\star + \tau & \text{for set} \end{cases} .$$

If the Local Sidereal Time at 0^h is given by Θ_0, then the star rises

$$0.9973 \cdot (\Theta_0 - \Theta_R)$$

before midnight, and sets

$$0.9973 \cdot (\Theta_S - \Theta_0)$$

after midnight.

In contrast to the coordinates of the stars, those of the Sun and the Moon alter noticeably during the course of a day. To calculate the hour angle and the times of rising and setting from the equations just given, however, we require the right ascension and declination at the moment the bodies cross the horizon. This requires an iterative procedure. The rising and setting times are first calculated for the coordinates at an arbitrary time for that particular day (in general, $t = 12^h$ is chosen). With the approximation thus obtained, improved coordinates may be determined, and these then again used to obtain more accurate rising and setting times. In the case of the Moon, these steps are repeated until the times obtained differ by less than one minute. For the Sun and the planets, the first iteration gives a sufficiently accurate result.

Unfortunately, there are a few problem cases, particularly in determining the times of moonrise and moonset, where the simple iteration method is difficult to apply. The following points need to be borne in mind:

- As the Moon orbits the Earth from West to East, it remains longer in the sky than stars with the same declination. As a result, moonrise is, on average, about 50^m later each day. This means that there is one day in every month on which the Moon does not rise, and also another day on which the Moon does not set. To give an example: On 1988 February 8 at Munich, the Moon rose at 23^h30^m and set at 9^h41^m on the following day. Because of its motion relative to the stars, it then remained below the horizon until after midnight. So the next moonrise did not take place on February 9, but on February 10, at 0^h43^m.

- At high geographical latitudes, the Sun and the Moon frequently remain below the horizon for days. The rising and setting times are then primarily determined by the daily change in declination. Possible grazing contacts with the horizon are thus difficult to establish using the iteration method.

Because of this, the `Sunset` program is based on another principle. The rising and setting times are determined by inverse interpolation of a series of solar and lunar altitudes. This does require a greater computational effort, but leads to a significant simplification of the program's structure.

3.7 Quadratic Interpolation

If we calculate a table of altitudes at hourly intervals, it is then possible to represent the behaviour of the altitudes as a function of time through a simple interpolation function. Quadratic interpolation is suitable for our purposes: from three values we may calculate the coefficients of a parabola, which approximates the behaviour of the function between the given points. We work from three values of the function:

$$y_- = f(x{=}-1), \quad y_0 = f(x{=}0) \quad \text{and} \quad y_+ = f(x{=}+1) \quad .$$

We require the coefficients a, b and c of a parabola

$$y = a \cdot x^2 + b \cdot x + c \quad , \tag{3.12}$$

which runs through the points $(-1, y_-)$, $(0, y_0)$ and $(1, y_+)$. Substituting into the formula for the parabola, we have the three equations

$$\begin{aligned} y_- &= a - b + c \\ y_0 &= c \\ y_+ &= a + b + c \quad , \end{aligned}$$

from which the required coefficients of the parabola may be obtained:

$$\begin{aligned} a &= (y_+ + y_-)/2 - y_0 \\ b &= (y_+ - y_-)/2 \\ c &= y_0 \quad . \end{aligned} \tag{3.13}$$

For $b^2 \geq 4ac$, this parabola is zero at the points:

$$x_{1,2} = \frac{-b \pm \sqrt{b^2 - 4ac}}{2a} \tag{3.14}$$

and has an extreme value

$$\begin{aligned} x_\mathrm{E} &= -b/2a \\ y_\mathrm{E} &= a \cdot x_\mathrm{E}^2 + b \cdot x_\mathrm{E} + c \quad . \end{aligned} \tag{3.15}$$

The `Quad` function solves these equations and also chooses just the roots that lie between $x = -1$ and $x = +1$.

```
//------------------------------------------------------------------------------
// Quad: Quadratic interpolation
//       Performs root finding and search for extreme values based on three
//       equidistant values of a function.
//   y_minus    Value of function at x = -1
//   y_0        Value of function at x =  0
//   y_plus     Value of function at x =  1
//   xe         Abscissa of extremum (may be outside [-1, 1])
//   ye         Value of function at xe
//   root1      First root found
//   root2      Second root found
//   n_root     Number of roots found in [-1, 1]
//------------------------------------------------------------------------------
void Quad ( double y_minus, double y_0, double y_plus,
            double& xe, double& ye, double& root1,
            double& root2, int& n_root )
{
  double a,b,c, dis, dx;  n_root = 0;
  // Coefficients of interpolating parabola y=a*x^2+b*x+c
  a  = 0.5*(y_plus+y_minus) - y_0;
  b  = 0.5*(y_plus-y_minus);
  c  = y_0;
  // Find extreme value
  xe = -b/(2.0*a);
  ye = (a*xe+b) * xe + c;
  dis = b*b - 4.0*a*c; // Discriminant of y=a*x^2+b*x+c

  if (dis >= 0) // Parabola has roots  {
    dx = 0.5 * sqrt (dis) / fabs (a);
    root1 = xe - dx;
    root2 = xe + dx;
    if (fabs(root1) <= 1.0) ++n_root;
    if (fabs(root2) <= 1.0) ++n_root;
    if (root1       < -1.0) root1 = root2;
  }
}
```

3.8 The SUNSET Program

The program Sunset calculates the rising and setting times of the Sun and the Moon, as well as the beginning and ending of nautical twilight over a period of 10 days. Data to be entered are the starting date, the geographical coordinates of the observing site, and the difference between Local Time and Universal Time.

The search for the times of the various events is made by the method shown diagrammatically in Fig. 3.4. The SinAlt function calculates the sine of the solar or lunar altitude at hourly intervals. These values are interpolated in Quad and examined for zero points. If a root is found, and the Sun or Moon was below the horizon at the beginning of the day, then the event is a sunrise or moonrise. Otherwise, the Sun or Moon is setting. Finally, it may happen that two zero points

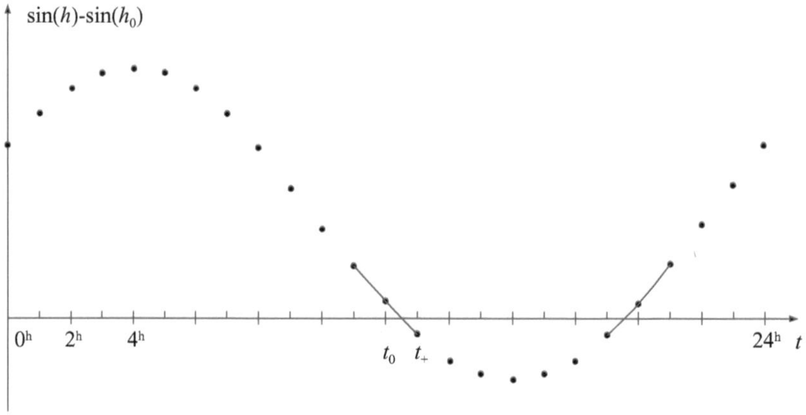

Fig. 3.4. Determining rising and setting times

are discovered in the interval being examined. To determine which of these two points is a rising and which a setting, we check to see if the altitude at the vertex is above or below the horizon. The process continues until the end of the day. It is then possible to determine whether the celestial body being considered is circumpolar, or whether it remains below the horizon for the entire day.

The times of civil, nautical, or astronomical twilight are calculated in a similar manner, in that a constant value of $\sin(-6°)$, $\sin(-12°)$ or $\sin(-18°)$ is subtracted from the sine of the altitude of the horizon.

```
//------------------------------------------------------------------------------
// File:     Sunset.cpp
// Purpose: Rising and setting times of Sun and Moon and twilight times
// (c) 1999 Oliver Montenbruck, Thomas Pfleger
//------------------------------------------------------------------------------
#include <cmath>
#include <iomanip>
#include <iostream>
#include "APC_Const.h"
#include "APC_Math.h"
#include "APC_Moon.h"
#include "APC_Spheric.h"
#include "APC_Sun.h"
#include "APC_Time.h"


using namespace std;
// Events to search for
enum enEvent {
  Moon,       // indicates moonrise or moonset
  Sun,        // indicates sunrise or sunset
  CivilTwi,   // indicates Civil twilight
  NautiTwi,   // indicates nautical twilight
  AstroTwi    // indicates astronomical twilight
};
```

```
//------------------------------------------------------------------------------
// SinAlt: Sine of the altitude of Sun or Moon
// Input:
//   Event     Indicates event to find
//   MJD0      0h at date to investigate (as Modified Julian Date)
//   Hour      Hour
//   lambda    Geographic east longitude in [rad]
//   Cphi      Cosine of geographic latitude
//   Sphi      Sine of geographic latitude
//   <return>  Sine of the altitude of Sun or Moon at instant of Event
//------------------------------------------------------------------------------
double SinAlt ( enEvent Event, double MJD0, double Hour,
                double lambda, double Cphi, double Sphi )
{
  double  MJD, T, RA, Dec, tau;
  MJD = MJD0 + Hour/24.0;
  T   = (MJD-51544.5)/36525.0;
  if ( Event == Moon ) MiniMoon ( T, RA, Dec);
  else MiniSun ( T, RA, Dec );
  tau = GMST(MJD) + lambda - RA;
  return ( Sphi*sin(Dec)+Cphi*cos(Dec)*cos(tau) );
}

//------------------------------------------------------------------------------
// GetInput: Prompts user for input
//   MJD       Time as Modified Julian Date
//   lambda    Geographic east longitude of the observer in [rad]
//   phi       Geographic latitude of the observer in [rad]
//   zone      Difference local time - universal time in [d]
//   twilight  Indicates civil, nautical or astronomical twilight
//------------------------------------------------------------------------------
void GetInput ( double& MJD, double& lambda,
                double& phi, double& zone, enEvent& twilight )
{
  int  year, month, day;
  char cTwilight;
  cout << " First date (yyyy mm dd)                  ... ";
  cin >> year >> month >> day; cin.ignore(81,'\n');
  cout << endl;
  cout << " Observing site:  east longitude [deg] ... ";
  cin  >> lambda; cin.ignore(81,'\n');
  cout << "                  latitude [deg]         ... ";
  cin  >> phi; cin.ignore(81,'\n');
  cout << "                  local time - UT [h]  ... ";
  cin  >> zone; cin.ignore(81,'\n');
  cout << " Twilight definition (c, n or a)         ... ";
  cin  >> cTwilight; cin.ignore(81,'\n');
  switch (tolower(cTwilight)) {
    case 'c': twilight = CivilTwi; break;
    case 'a': twilight = AstroTwi; break;
    case 'n':
    default:  twilight = NautiTwi;
  }
```

```
  lambda*=Rad;  phi*=Rad;
  zone/=24.0;
  MJD = Mjd(year,month,day) - zone;
}

//------------------------------------------------------------------------------
// FindEvents: Search for rise/set/twilight events of Sun or Moon
//   Event      Indicates event to search for
//   MJD0h      0h at desired date as Modified Julian Date
//   lambda     Geographic east longitude of the observer in [rad]
//   phi        Geographic latitude of the observer in [rad]
//   LT_Rise    Local time of rising or beginning of twilight
//   LT_Set     Local time of setting or end of twilight
//   rises      Event takes place
//   sets       Event takes place
//   above      Sun or Moon is circumpolar
//------------------------------------------------------------------------------
void FindEvents ( enEvent Event, double MJD0h, double lambda, double phi,
                  double& LT_Rise, double& LT_Set,
                  bool& rises, bool& sets, bool& above )
{
  static const double sinh0[5] = {
    sin(Rad*( +8.0/60.0)), // Moonrise                at h= +8'
    sin(Rad*(-50.0/60.0)), // Sunrise                 at h=-50'
    sin(Rad*(   - 6.0  )), // Civil twilight          at h=-6 deg
    sin(Rad*(   -12.0  )), // Nautical twilight       at h=-12deg
    sin(Rad*(   -18.0  )), // Astronomical twilight at h=-18deg
  };
  const double Cphi = cos(phi);
  const double Sphi = sin(phi);
  double hour = 1.0;
  double y_minus, y_0, y_plus;
  double xe, ye, root1, root2;
  int nRoot;

  // Initialize for search
  y_minus = SinAlt(Event, MJD0h, hour-1.0, lambda,Cphi,Sphi )-sinh0[Event];
  above = (y_minus>0.0);  rises = false;  sets  = false;
  // loop over search intervals from [0h-2h] to [22h-24h]
  do {
    y_0    = SinAlt( Event,MJD0h,hour    , lambda,Cphi,Sphi )-sinh0[Event];
    y_plus = SinAlt( Event,MJD0h,hour+1.0, lambda,Cphi,Sphi )-sinh0[Event];
    // find parabola through three values y_minus,y_0,y_plus
    Quad ( y_minus, y_0, y_plus, xe, ye, root1, root2, nRoot );
    if ( nRoot==1 ) {
      if ( y_minus < 0.0 )
        { LT_Rise = hour+root1;  rises = true; }
      else
        { LT_Set  = hour+root1;  sets  = true; }
    }
    if ( nRoot == 2 ) {
      if ( ye < 0.0 )
        { LT_Rise = hour+root2;  LT_Set = hour+root1; }
```

```
      else
        { LT_Rise = hour+root1;  LT_Set = hour+root2; }
      rises = true; sets = true;
    }
    y_minus = y_plus;      // prepare for next interval
    hour += 2.0;
  }
  while ( !( ( hour == 25.0 ) || ( rises && sets ) ) );
}

//------------------------------------------------------------------------------
//
// Main program
//
//------------------------------------------------------------------------------
void main() {

  // Variables
  bool     above, rise, sett;
  int      iEvent, day;
  double   lambda, zone, phi;
  double   date, start_date, LT_Rise, LT_Set;
  enEvent  Event, Twilight;

  // Title
  cout << endl
      << "      SUNSET: solar and lunar rising and setting times  " << endl
      << "        (c) 1999 Oliver Montenbruck, Thomas Pfleger     " << endl
      << endl;
  // Prompt user for input
  GetInput (start_date, lambda, phi, zone, Twilight );
  // Header
  cout << endl
      << "    Date"
      << "             Moon                 Sun                Twilight " << endl
      << "         "
      << "           rise/set            rise/set         beginning/end"
      << endl << endl;

  // loop over 10 subsequent days
  for (day=0; day<10; day++) {
    // current date
    date = start_date + day;
    cout << " " << DateTime(date+zone) << "  ";
    // loop over cases (Moon, Sun, Twilight)
    for (iEvent=0; iEvent<=2; iEvent++)  {
      // After events for Moon and Sun: specify kind of twilight event
      Event = (iEvent<2) ? (enEvent) iEvent : Twilight;
      // Now try to find times of events
      FindEvents ( Event, date, lambda, phi,
                   LT_Rise, LT_Set, rise, sett, above );
      // Output
      if ( rise || sett )  {
```

```
        if ( rise )
          cout << "   " << Time(LT_Rise,HHMM) << " ";
        else
          cout << "   ----- ";
        if ( sett )
          cout << "   " << Time(LT_Set,HHMM) << " ";
        else
          cout << "   ----- ";
      }
      else
        if ( above ) {
          if ( Event >= CivilTwi )
            cout << "    always bright ";
          else
            cout << "   always visible ";
        }
        else {
          if ( Event >= CivilTwi )
            cout << "     always dark  ";
          else
            cout << "  always invisible";
        }
    }
    cout << endl;
  }

  // Trailer
  cout << endl;
  cout << " all times in local time (=UT"
       << showpos << 24.0*zone << noshowpos << "h)"  << endl;
}
```

We will demonstrate the use of `Sunset` with some examples, which will show some of the factors that arise in calculating rising and setting times.

`Sunset` calculates the desired times for a period of ten days, beginning with the date entered. The geographical longitude is entered with positive values for those places east of the Greenwich meridian. Finally the difference between Local Time and UT is entered in hours. A value of 1, for example, would be used for Central European Time (CET), or 2 for Central European Summer Time (CEST). For time zones in North America, negative values are entered: -5 for Eastern Standard Time (EST), for example, or -8 for Pacific Standard Time (PST). Furthermore, the twilight definition may be specified to choose between civil twilight ('c'), nautical twilight ('n') and astronomical twilight ('a').

Let us, for example, calculate the rising and setting times of the Sun and Moon for Munich ($\lambda = 11\overset{\circ}{.}6$ East, $\varphi = 48\overset{\circ}{.}1$ North), beginning with 2000 March 23. We require the output for Central European Time (1^{h} different from Universal Time). First the starting date must be entered in the form: Year, Month, Day. Then `Sunset` asks for the geographical coordinates and the difference between LT and UT. (In the following, all data entered are shown in italic.) From the data that we have entered, `Sunset` determines the rising and setting times of the Sun and Moon, as well as the beginning and end of twilight. As the value for the altitude of the Sun that is used in the program is set at $-12°$, the output here and in the following example is for *nautical* twilight.

```
      SUNSET: solar and lunar rising and setting times
        (c) 1999 Oliver Montenbruck, Thomas Pfleger

 First date (yyyy mm dd)                     ... 2000 03 23

 Observing site:  east longitude [deg] ... +11.6
                  latitude [deg]       ... +48.1
                  local time - UT [h]  ... +1
 Twilight definition (c, n or a)        ... n

    Date           Moon               Sun             Twilight
                 rise/set           rise/set        beginning/end

 2000/03/23      22:12    08:01    06:10    18:31    05:02    19:39
 2000/03/24      23:17    08:28    06:08    18:32    05:00    19:41
 2000/03/25      -----    08:58    06:06    18:34    04:58    19:42
 2000/03/26      00:18    09:33    06:04    18:35    04:56    19:44
 2000/03/27      01:16    10:13    06:02    18:37    04:53    19:46
 2000/03/28      02:08    10:59    06:00    18:38    04:51    19:47
 2000/03/29      02:55    11:51    05:58    18:40    04:49    19:49
 2000/03/30      03:37    12:48    05:56    18:41    04:47    19:50
 2000/03/31      04:13    13:51    05:54    18:43    04:45    19:52
 2000/04/01      04:44    14:56    05:52    18:44    04:42    19:54

 all times in local time (=UT+1h)
```

It will be seen that there is no entry in the column under 'Moon rise' for 2000 March 25. On that night the Moon rose after midnight, so it is recorded under March 26. In general, there is one day per month on which it neither rises or sets.

To illustrate another interesting effect, let us calculate the rising and setting times for a fictitious point in Europe at latitude 65°, choosing 1989 June 15 as the starting date and Central European Summer Time as the local time. SUNSET gives the following output:

```
      SUNSET: solar and lunar rising and setting times
        (c) 1999 Oliver Montenbruck, Thomas Pfleger

 First date (yyyy mm dd)                     ... 1989 06 15

 Observing site:  East longitude [deg] ... +10.0
                  latitude [deg]       ... +65.0
                  local time - UT [h]  ... +2.0
 Twilight definition (c, n or a)        ... n

    Date           Moon               Sun             Twilight
                 rise/set           rise/set        beginning/end

 1989/06/15      19:58    01:00    02:24    00:16     always bright
```

```
1989/06/16     22:26      23:53    02:23    00:18     always bright
1989/06/17    always invisible     02:22    00:19     always bright
1989/06/18    always invisible     02:21    00:20     always bright
1989/06/19    always invisible     02:20    00:21     always bright
1989/06/20    always invisible     02:20    00:22     always bright
1989/06/21     02:39      03:24    02:20    00:23     always bright
1989/06/22     01:35      06:21    02:20    00:23     always bright
1989/06/23     01:15      08:29    02:21    00:23     always bright
1989/06/24     01:01      10:25    02:22    00:22     always bright

all times in local time (=UT+2h)
```

In this case, after setting on June 16 at 23^h53^m, the Moon remains below the horizon and is therefore invisible until June 21 at 2^h39^m. Correspondingly, at high latitudes the Moon may remain above the horizon for days.

Naturally, similar events may occur with the Sun, which we call the polar night or polar daylight ('midnight Sun'). Using `Sunset` we can estimate how long polar night or polar daylight last at a specific place. To do this we vary the period over which calculations are carried out, until the Sun is first shown as 'always invisible'. Polar night begins on the corresponding day. The end of the polar night may be found in a precisely analogous way.

The result 'always bright' in the 'Twilight' column indicates that it never becomes fully dark. This may even occur at a latitude of only 50°, when we determine astronomical instead of nautical or civil twilight.

3.9 The PLANRISE Program

In the previous sections we became acquainted with the necessary tools for computing rising and setting times by the method of quadratic interpolation. The latter may also be used to handle exceptional cases of rising and setting time calculations, which may otherwise present severe difficulties. Nevertheless, we also want to describe a program that illustrates the iterative method of obtaining planetary rising and setting times. The commented source code of `Planrise` is provided on the enclosed CD.

The function of `Planrise` is based on the equations given in Sect. 3.6. For the computation of planetary positions we make use of the `PertPosition` function, which is described in Sect. 5.2. To compute the rising and setting times of a planet we first determine its (geocentric) equatorial coordinates α and δ for an arbitrary time on the given day. Equation (3.11) then yields the value $\cos\tau(\varphi, \delta)$ for the hour angle at the instant of rising or setting. If the right-hand side results in a value whose magnitude is larger than one, the planet never rises or sets on that day. From the declination and the geographical latitude, `Planrise` then determines whether the planet is continuously above or below the horizon.

Because the horizontal parallax and the apparent diameter have virtually no impact on the rising and setting times, a value of $h_{R/S} = -0°34'$ for the geometric altitude at the time of rising or setting is employed in the computation, which thus takes refraction only into account. The iterative equation for determining the rising and setting times is then given by

$$t_{i+1} = t_i - 0.9973(\Theta(t_i) - \alpha(t_i) \pm \tau(\varphi, \delta(t_i))) \tag{3.16}$$

where

$$\tau = \arccos \frac{\sin h_{\mathrm{A/U}} - \sin\varphi \sin\delta}{\cos\varphi\cos\delta} \quad . \tag{3.17}$$

Here the positive sign applies for the computation of rising times. $\Theta(t_i)$ and $\alpha(t_i)$ denote the local sidereal time and the planet's right ascension at the time t_i of the ith approximation. The conversion factor 0.9973 is used to convert between universal time and sidereal time, as mentioned earlier.

As with the computation of rising and setting times, we may also employ the relationship

$$t_{i+1} = t_i - 0.9973(\Theta(t_i) - \alpha(t_i)) \tag{3.18}$$

to determine the time of culmination of a planet.

To reduce computing times, `Planrise` makes use of various simplifications. Linear interpolation, for example, is used to obtain the planetary coordinates α and δ, as well as the local sidereal time Θ, from precomputed values at 0^{h} and 24^{h} on the same day.

As an example we will compute the rising and setting times of the planets on 1999 December 31 in Munich ($\lambda = 11.6°$East, $\varphi = 48.1°$North). All input data for the `Planrise` program are shown in italic.

```
  PLANRISE: planetary and solar rising and setting times
            (c) 1999 Oliver Montenbruck, Thomas Pfleger

 Date (yyyy mm dd)                             ... 1999 12 31

 Observing site:  east longitude [deg] ...  +11.6
                  latitude [deg]       ...  +48.1
                  local time - UT [h]  ...  +1.0

                   rise     culmination     set

 Sun              08:04        12:16       16:29
 Mercury          07:33        11:37       15:41
 Venus            04:52        09:30       14:08
 Mars             10:33        15:35       20:37
 Jupiter          12:29        19:10       01:55
 Saturn           13:09        20:10       03:14
 Uranus           10:02        14:45       19:28
 Neptune          09:25        13:57       18:29
 Pluto            05:11        10:22       15:32

 all times in local time (=UT+1h)
```

Subject to good weather a total of three planets (Mars, Jupiter, and Saturn) can thus be seen with the naked eye in the evening sky.

4. Cometary Orbits

Whereas the orbits of the major planets are often calculated and published years in advance, predictions of cometary orbits are generally restricted to short-term ephemerides and the determination of orbital elements. The five or six orbital elements describe the orbit in space and time in a significantly shorter form than a long listing of positions. In general, the orbital elements of newly discovered comets are published by the International Astronomical Union (IAU) in the *IAU Circulars*. In addition, there are various catalogues that give the corresponding data for periodic comets and minor planets.

From the orbital elements, and using a suitable program, it is possible to calculate the position of a comet relative to the Earth and the Sun at any time. The `Comet` program that is described in this chapter is not restricted to the calculation of elliptical orbits, such as those of the planets and minor planets. It is also suitable for calculating cometary orbits that are parabolic, near-parabolic, or hyperbolic.

The following sections discuss the motion of comets relative to the Sun and to the ecliptic, neglecting perturbations by the major planets. Taking them into account would require an enormous, additional amount of work. The various coordinate conversions that are required for conversion into geocentric equatorial coordinates have already been described in Chap. 2.

4.1 Form and Orientation of the Orbit

The motion of comets around the Sun is governed by the same laws that control the orbits of the major planets:

1. The orbit is an ellipse, a parabola, or a hyperbola (i.e., a conic section), with the Sun at the focus (or at one of the two foci in the case of an ellipse). This means, in particular, that the orbit lies in a fixed plane.

2. The line joining the Sun and the comet (the radius vector) sweeps out equal areas in equal periods of time. This is known as the *law of equal areas*.

These laws were discovered by Johannes Kepler, and were later explained by Newton's theory of universal gravitation. Nevertheless, Kepler's Laws are valid only when the influence of the major planets on comets may be neglected in comparison with the Sun's gravity.

The orbit lies in a single plane because of the fact that the Sun's gravitational attraction always acts along the line joining the Sun and the comet (this is known as a centripetal force). In a short interval Δt, the comet moves at a velocity $\dot{\boldsymbol{r}}$ over a distance $\dot{\boldsymbol{r}} \cdot \Delta t$ between point $\boldsymbol{r}$ and point $\boldsymbol{r} + \dot{\boldsymbol{r}} \cdot \Delta t$ (Fig. 4.1). Taken together with the Sun, these two points determine a plane in space. Because the force of the Sun acts only in this plane, however, subsequent motion of the comet continues to lie in this plane. The law of equal areas is also a direct consequence of centripetal force. Proof of this requires a moderate amount of mathematics, so it will not be shown here.

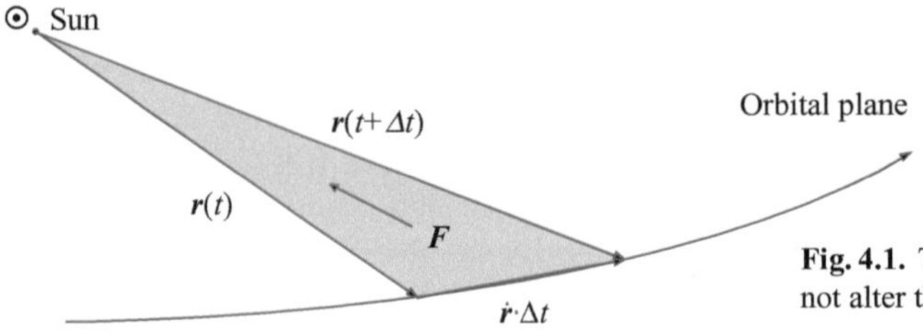

Fig. 4.1. The force on comets does not alter the plane of their orbits

The exact form of the orbit is determined by the decrease in the gravitational force with the square of the distance ($F \sim 1/r^2$). The three basic forms – the ellipse, parabola, and hyperbola – are shown in Fig. 4.2. While comets in elliptical orbits are periodic, making repeated returns, those in parabolic or hyperbolic orbits are close to the Sun only for a single, short period of time. The closest point on the orbit to the Sun is known as *perihelion*. The line joining perihelion to the Sun (known as the *line of apsides*) divides the orbit into two symmetrical halves.

A point on the orbit is defined by the distance from the Sun r, and the true anomaly[1] ν. The latter is the angle between the Sun-comet line and the Sun-perihelion direction. The relationship between r and ν is given by the equation of the conic section :

$$r = \frac{p}{1 + e \cdot \cos(\nu)} \quad . \tag{4.1}$$

The value p, which is the distance for $\nu = 90°$, is the semi-latus rectum (sometimes known as the *orbital parameter*) and it defines the size of the orbit. The *eccentricity* e, indicates how greatly the orbit deviates from a circle ($e = 0$). The higher the value, the more elongated the orbit, which finally becomes a parabola ($e = 1$), or a hyperbola ($e > 1$).

A rather more useful value than the orbital parameter is the semi-major axis, which is, however, defined only when $e \neq 1$:

$$a = \frac{p}{1 - e^2} \quad , \qquad p = a\,(1 - e^2) \quad . \tag{4.2}$$

For an elliptical orbit, a obviously equals the average of the least and greatest distances from the Sun:

$$r_{\min} + r_{\max} = \frac{p}{1 + e} + \frac{p}{1 - e} = \frac{2p}{1 - e^2} = 2a \quad .$$

[1] In celestial mechanics, 'anomaly' is used to mean an angle. In addition to the true anomaly, other parameters known as the mean anomaly and the eccentric anomaly are used.

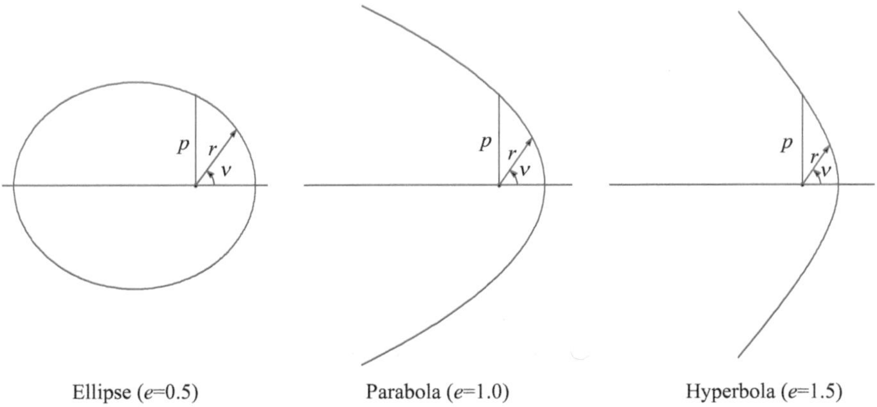

Fig. 4.2. Conic sections with eccentricities $e = 0.5$, $e = 1.0$ and $e = 1.5$ with the same orbital parameter p

It should be noted that the semi-major axis of a hyperbola – where $e > 1$ by definition – is negative. For parabolic orbits ($e \approx 1$), the perihelion distance $q = r_{\text{min}}$ is used instead of a:

$$q = \frac{p}{1+e} \quad . \tag{4.3}$$

In an orbit that is exactly a parabola, $q = p/2$.

4.2 Position in the Orbit

Although the equation for the conic section gives every possible position at which a comet may be found during its orbit, it does not establish at what time this will occur. To solve this problem, we use the law of equal areas. Using a closed elliptical orbit as an example, we will now discuss the major steps in this calculation.

An ellipse of eccentricity e may be obtained by reducing a single diameter of a circle by a factor of $\sqrt{1-e^2}$. The total area S of the ellipse with semi-major axis a is therefore smaller than that of a circle of radius a by this same factor, so we have:

$$S = (\pi a^2)\sqrt{1-e^2} \quad .$$

The line joining the Sun and the comet (the radius vector) sweeps out this area in the orbital period T, and the sector

$$S(t) = (\pi a^2)\sqrt{1-e^2}\left(\frac{t-t_0}{T}\right) \tag{4.4}$$

in the time t after perihelion passage t_0. If it is now possible to describe the area of this sector S in terms of the position in the orbit (r, ν), then we have established the required connection between the position in the orbit and time t. For a circular

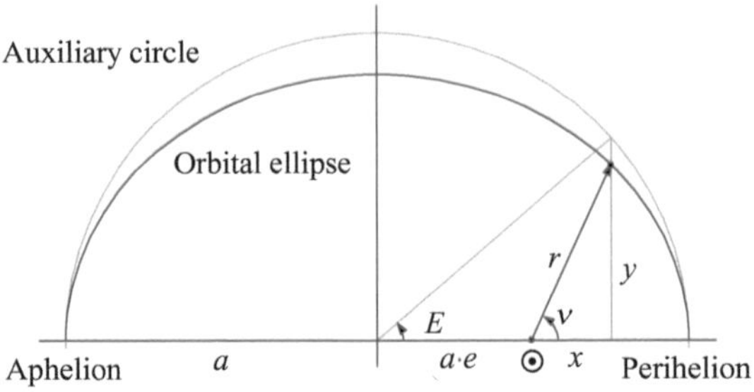

Fig. 4.3. The way in which eccentric anomaly E is defined

orbit this is extremely simple, because S is directly proportional to the angle ν and to the square of the semi-diameter. For an ellipse, however, we need to introduce an auxiliary parameter – the *eccentric anomaly*. Its geometrical significance may be seen from Fig. 4.3. If the x-axis is chosen as shown there, a point P on the ellipse has an x-coordinate of $x = a \cdot \cos E - a \cdot e$. The $a \cdot e$ term indicates the distance between the centre of the ellipse and the focus at which the Sun is situated. The y-coordinate is less than that of the point P′ on the circumscribed circle by the compression factor $\sqrt{1-e^2}$, so $y = \sqrt{1-e^2} \cdot a \cdot \sin E$. We then have

$$\begin{aligned} x &= r\cos\nu = a\,(\cos E - e) \\ y &= r\sin\nu = a\sqrt{1-e^2}\sin E \quad . \end{aligned} \tag{4.5}$$

To calculate S, two auxiliary areas may be considered, and these are illustrated in greater detail in Fig. 4.4:

Sector of the circle (P′OΠ) : $S_1 = \pi a^2 E/360°$

Sector of the Ellipse (POΠ) : $S_2 = \sqrt{1-e^2} \cdot S_1$

Triangle (O⊙P) : $S_3 = \frac{1}{2} y \cdot (a \cdot e) = \frac{1}{2} a^2 e \sqrt{1-e^2} \sin E$

Sector (P⊙Π) : $S = S_2 - S_3$.

By using the eccentric anomaly, the area S may be expressed in the form

$$S(t) = \frac{1}{2} a^2 \sqrt{1-e^2} \left(E \frac{\pi}{180°} - e \sin E \right) \quad . \tag{4.6}$$

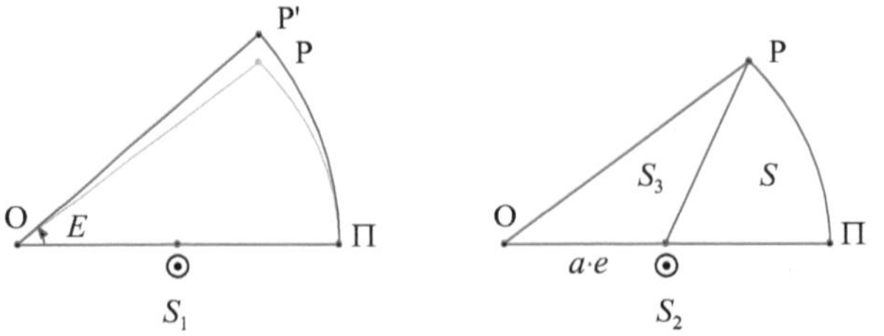

Fig. 4.4. The method of obtaining the area S of the sector

If this is compared with the formula for $S(t)$ given above, then from the relationship between the eccentric anomaly and the time, we obtain *Kepler's Equation*:

$$E(t) - \frac{180°}{\pi} e \sin E(t) = M(t) \tag{4.7}$$

where

$$M(t) = 360° \frac{(t - t_0)}{T} \quad . \tag{4.8}$$

To determine the position of the comet in its orbit at any given instant, we must solve this equation for E, and substitute the value thus obtained into Equation (4.5) to derive x and y. This process will be described in detail in the section about the mathematical treatment of Kepler's Equation.

Until now, the orbital period has been treated as an independent variable. In fact it depends on the semi-major axis as given by *Kepler's Third Law*:

$$\frac{a^3}{T^2} = \frac{GM_\odot}{4\pi^2} \qquad \text{where} \qquad GM_\odot = 1.32712 \cdot 10^{20} \mathrm{m}^3 \mathrm{s}^{-2} \quad . \tag{4.9}$$

The constant on the right-hand side of this equation contains the product $GM_\odot$: the gravitational constant times one solar mass. The units in which the value is given here – the metre (m) and second (s) – are rarely used in practice. The usual units are the day (1 d = 86400 s) and the astronomical unit (1 AU $\approx$ 149.59787 million km). The AU originally represented the semi-major axis of the Earth's orbit. Later the IAU decided that the astronomical unit should be defined such that the product of the gravitational constant and the solar mass should have a fixed value

$$\begin{aligned} GM_\odot &= k^2 \mathrm{AU}^3 \mathrm{d}^{-2} \qquad \text{where} \\ k &= 0.01720209895 \\ k^2 &\approx 2.959122083 \cdot 10^{-4} \quad . \end{aligned}$$

k is known as the Gaussian gravitational constant. Although at first sight this definition appears somewhat unusual, it means that one AU is defined irrespective of any possible long-term changes in the diameter of the Earth's orbit.

As well as the position of the comet in its orbit, we shall also require its velocity. The equations for determining this may be given here:

$$\begin{aligned} \dot{x} &= -\sqrt{\frac{GM_\odot}{a}} \frac{\sin E}{1 - e \cos E} \\ \dot{y} &= +\sqrt{\frac{GM_\odot(1 - e^2)}{a}} \frac{\cos E}{1 - e \cos E} \quad . \end{aligned} \tag{4.10}$$

For hyperbolic orbits we can determine equations similar to those for an elliptical orbit. Instead of the angular functions, we use the hyperbolic functions $\sinh x = (e^x - e^{-x})/2$ and $\cosh x = (e^x + e^{-x})/2$. We therefore have:

$$\begin{aligned} x &= r \cos \nu = |a| (e - \cosh H) \\ y &= r \sin \nu = |a| \sqrt{e^2 - 1} \sinh H \end{aligned} \tag{4.11}$$

and

$$\dot{x} = -\sqrt{\frac{GM_\odot}{|a|}}\frac{\sinh H}{e\cosh H - 1} \tag{4.12}$$
$$\dot{y} = +\sqrt{\frac{GM_\odot(e^2-1)}{|a|}}\frac{\cosh H}{e\cosh H - 1} \quad .$$

The parameter H, which corresponds to the eccentric anomaly, allows us to obtain a modified form of Kepler's Equation:

$$e\sinh H - H = M_\mathrm{h} = \sqrt{\frac{GM_\odot}{|a|^3}}\cdot(t-t_0) \quad . \tag{4.13}$$

With parabolic orbits, x and y are first expressed in terms of $\tan(\nu/2)$:

$$x = r\cos\nu = q\cdot\left(1-\tan^2\left(\frac{\nu}{2}\right)\right) \tag{4.14}$$
$$y = r\sin\nu = 2q\cdot\tan\left(\frac{\nu}{2}\right) \quad .$$

As mentioned earlier, q is the perihelion distance. The true anomaly ν is expressed in terms of time by *Barker's Equation*:

$$\tan\left(\frac{\nu}{2}\right) + \frac{1}{3}\tan^3\left(\frac{\nu}{2}\right) = \sqrt{\frac{GM_\odot}{2q^3}}(t-t_0) \quad . \tag{4.15}$$

This is a third-order equation of $\tan(\nu/2)$. In contrast to Kepler's Equation it may be solved directly for ν for a given time t. If we set

$$A = \frac{3}{2}\sqrt{\frac{GM_\odot}{2q^3}}(t-t_0) \qquad \text{and} \qquad B = \sqrt[3]{A+\sqrt{A^2+1}} \quad ,$$

then

$$\tan(\nu/2) = B - 1/B \quad , \qquad \nu = 2\arctan(B-1/B) \quad .$$

4.3 Mathematical Treatment of Kepler's Equation

Kepler's Equation cannot be solved analytically with respect to time, using the eccentric anomaly as a function of the mean anomaly. There are, however, appropriate mathematical procedures that enable this to be done.

We first calculate the mean anomaly M at a given time t. For low eccentricities this differs only slightly from the eccentric anomaly E. It may therefore be used as a first approximation E_0 for a subsequent iteration. If Kepler's Equation is slightly rearranged, we can see that to solve it, the root of the function

$$f(E) = E - \frac{180^\circ}{\pi}e\sin E - M(t)$$

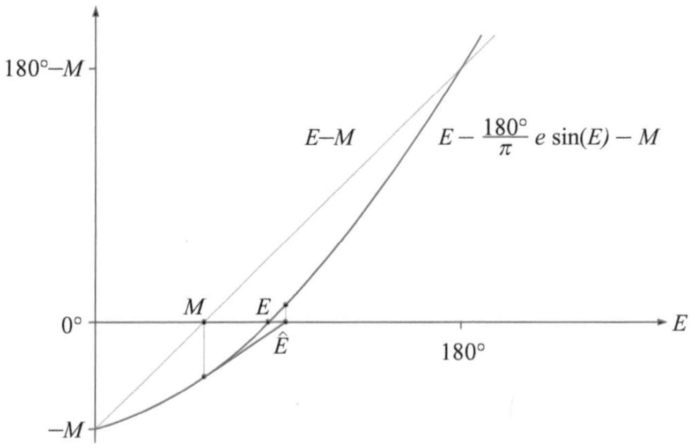

Fig. 4.5. Solving Kepler's Equation by Newton's method

has to be determined. One method of finding the point at which a function $f(x)$ is zero, is *Newton's method* (see Fig. 4.5). The tangent at a point x close to the zero point, and with a slope of $f'(x)$ is used as an approximation, and its intersection with the x-axis is calculated from:

$$\hat{x} = x - \frac{f(x)}{f'(x)} \quad .$$

In general one obtains an improved approximation $\hat{x}$ to the zero point of the function f. In the case of Kepler's Equation, the iterative expression has the form

$$\hat{E} = E - \frac{E - (180°/\pi) \cdot e \cdot \sin E - M(t)}{1 - e \cdot \cos E}$$

Here E must be given in degrees. If one is working in radians, then the factor $180°/\pi$ must be replaced by 1. For orbits that are close to a circle, three steps are typically sufficient to obtain a solution for E accurate to ten decimal places. Nevertheless, one should be careful not to apply Newton's method indiscriminately in cases of high eccentricity. It is suggested that readers should see for themselves what happens when the iteration is carried out with $e = 0.99$ and $E_0 = M = 5°$. Safety suggests that here another initial value should be used, namely $E_0 = \pi = 180°$. This initial value is used in the `EccAnom` function for eccentricities above $e = 0.8$. The iteration is continued until Kepler's equation is fulfilled with an accuracy that is about two decimals less than the relative accuracy of the `double` data type. The required value of the machine accuracy may be obtained from a call of `numeric_limits<double>::epsilon` in the `<limits>` module of the C++ Standard library.

```
const double eps_mach = numeric_limits<double>::epsilon();

//------------------------------------------------------------------------------
// EccAnom: computes the eccentric anomaly for elliptic orbits
//   M          Mean anomaly in [rad]
//   e          Eccentricity of the orbit [0,1[
//   <return>   Eccentric anomaly in [rad]
//------------------------------------------------------------------------------
double EccAnom (double M, double e)
{
  const int maxit = 15;
  const double eps = 100.0*eps_mach;
  int    i=0;
  double E, f;

  // Starting value
  M = Modulo(M, 2.0*pi);  if (e<0.8) E=M; else E=pi;
  // Iteration
  do {
    f = E - e*sin(E) - M;  E = E - f/(1.0-e*cos(E)); ++i;
    if (i==maxit) {cerr<<" Convergence problems in EccAnom"<<endl; break;}
  }
  while (fabs(f) > eps);
  return E;
}
```

It is now possible to determine the eccentric anomaly by calling this function just once. The remaining steps that are required to calculate an orbital position on the ellipse are included in the `Ellip` function:

```
//------------------------------------------------------------------------------
// Ellip: computes position and velocity vectors for elliptic orbits
//   GM        Product of gravitational constant and centre mass [AU^3*d^-2]
//   M         Mean anomaly in [rad]
//   a         Semi-major axis of the orbit in [AU]
//   e         Eccentricity of the orbit (<1)
//   r         Position w.r.t. orbital plane in [AU]
//   v         Velocity w.r.t. orbital plane in [AU/d]
//------------------------------------------------------------------------------
void Ellip ( double GM, double M, double a, double e,
             Vec3D& r, Vec3D& v )
{
  double k, E, cosE,sinE, fac, rho;

  k   = sqrt(GM/a);
  E   = EccAnom(M,e);
  cosE = cos(E);
  sinE = sin(E);
  fac = sqrt ( (1.0-e)*(1.0+e) );
  rho = 1.0 - e*cosE;
  r = Vec3D (a*(cosE-e),  a*fac*sinE,     0.0);
  v = Vec3D (-k*sinE/rho, k*fac*cosE/rho, 0.0);
}
```

Kepler's Equation may be solved in a similar manner for hyperbolic orbits. Here the iterative expression is:

$$\hat{H} = H - \frac{e \cdot \sinh H - H - M_{\mathrm{h}}(t)}{e \cdot \cosh H - 1} \quad .$$

A suitable initial value is

$$H_0 = \begin{cases} +\ln(+1.8 + 2M_{\mathrm{h}}/e) & \text{for } M_{\mathrm{h}} \geq 0 \\ -\ln(+1.8 - 2M_{\mathrm{h}}/e) & \text{for } M_{\mathrm{h}} < 0 \end{cases} \quad .$$

```
//------------------------------------------------------------------------------
// HypAnom: computes the eccentric anomaly for hyperbolic orbits
//   Mh        Mean anomaly in [rad]
//   e         Eccentricity of the orbit (>1)
//------------------------------------------------------------------------------
double HypAnom (double Mh, double e)
{
  const int maxit = 15;
  const double eps = 100.0*eps_mach;
  int    i=0;
  double H, f;
  // Starting value
  H = log (2.0*fabs(Mh)/e + 1.8);
  if (Mh < 0.0)  H = -H;
  // Iteration
  do {
    f = e*sinh(H) - H - Mh;
    H = H - f / ( e*cosh(H) - 1.0 );
    ++i;
    if (i==maxit) {
      cerr << " Convergence problems in HypAnom" << endl;
      break;
    }
  }
  while ( fabs(f) > eps*(1.0+fabs(H+Mh)) );
  return H;
}
```

The rectangular coordinates of the comet along its orbit may be calculated with the `Hyperb` function from the formulae given in the previous section. Unlike `Ellip`, the times of perihelion passage and the required instant are entered, instead of the mean anomaly M_{h}. The reason for the different choice of variables passed is that the `Ellip` function also has to be suitable for the calculation of the orbits of minor planets and planets. With these bodies, unlike the situation with comets, the time of perihelion passage is rarely employed as an orbital element. Instead, the mean anomaly M_{Ep} and the mean daily motion n at a specific epoch t_{Ep} are generally available. The mean anomaly $M = M_{\mathrm{Ep}} + n(t - t_{\mathrm{Ep}})$ at a given time t may then be used directly as input to `Ellip`. On the other hand, hyperbolic orbits are generally encountered only among cometary orbits, where it is customary for the time of perihelion passage to be given as one of the orbital elements.

```
//------------------------------------------------------------------------
// Hyperb: computes position and velocity vectors for elliptic orbits
//   GM        Product of gravitational constant and centre mass [AU^3*d^-2]
//   t0        Time of perihelion passage
//   t         Time for calculation
//   a         Semi-major axis of the orbit in [AU]
//   e         Eccentricity of the orbit (>1)
//   r         Position w.r.t. orbital plane in [AU]
//   v         Velocity w.r.t. orbital plane in [AU/d]
// Note: t0 and t in Julian centuries since J2000
//------------------------------------------------------------------------
void Hyperb ( double GM, double t0, double t, double a, double e,
              Vec3D& r, Vec3D& v )
{
  double k, Mh, H, coshH, sinhH, rho, fac;
  a  = fabs(a);
  k  = sqrt(GM/a);
  Mh = k*(t-t0)/a;
  H  = HypAnom(Mh,e);
  coshH = cosh(H);
  sinhH = sinh(H);
  fac = sqrt ( (e+1.0)*(e-1.0) );
  rho = e*coshH - 1.0;
  r = Vec3D (a*(e-coshH),  a*fac*sinhH,     0.0);
  v = Vec3D (-k*sinhH/rho, k*fac*coshH/rho, 0.0);
}
```

4.4 Near-Parabolic Orbits

As we indicated in the previous section, the calculation of elliptical or hyperbolic orbits with eccentricities close to one, and with small anomalies, is not always easy. We can avoid these difficulties from the start by making use of the fact that orbits with the characteristics described are very similar to a parabola. The method used here is that devised by Karl Stumpff.

From Kepler's equation

$$E(t) - e \sin E(t) = \sqrt{\frac{GM_{\odot}}{a^3}} \cdot (t - t_0) \quad ,$$

here expressed in radians ($180° = \pi$), and as

$$a = q/(1-e) \quad \text{and} \quad c_3 = (E - \sin E)/E^3 \quad ,$$

we have first

$$(1-e) \cdot E + e \cdot c_3(E) \cdot E^3 = \sqrt{GM_{\odot}\left(\frac{1-e}{q}\right)^3} \cdot (t - t_0) \quad ,$$

and, introducing a new variable

$$U = \sqrt{\frac{3e \cdot c_3(E)}{1-e}} \cdot E$$

we obtain the modified form

$$U + \frac{1}{3}U^3 = \sqrt{6ec_3(E)} \cdot \sqrt{\frac{GM_\odot}{2q^3}} \cdot (t - t_0) \quad .$$

This equation has the same general form as Barker's equation and may be solved for U, the right side being known. With $c_1(E) = \sin(E)/E$ and $c_2(E) = (1 - \cos(E))/E^2$ we further obtain

$$\begin{aligned}
x &= r\cos\nu = \frac{q}{1-e}(\cos E - e) &&= q \cdot \left(1 - \left[\frac{2c_2}{6ec_3}\right] U^2\right) \\
y &= r\sin\nu = \frac{q}{1-e}\sqrt{1-e^2}\sin E &&= 2q \cdot \sqrt{\frac{1+e}{2e} \cdot \left[\frac{1}{6c_3}\right]} \cdot c_1 \cdot U \qquad (4.16) \\
r & &&= q \cdot \left(1 + \left[\frac{2c_2}{6c_3}\right] U^2\right) \quad .
\end{aligned}$$

Later the equations for the velocity of a comet will be required, but their derivation is involved and will, therefore, be omitted here:

$$\begin{aligned}
\dot{x} &= -\sqrt{\frac{GM_\odot}{q(1+e)}}\left(\frac{y}{r}\right) \qquad (4.17) \\
\dot{y} &= +\sqrt{\frac{GM_\odot}{q(1+e)}}\left(\frac{x}{r} + e\right) \quad .
\end{aligned}$$

In addition, the relationship between U and the true anomaly ν is

$$\tan\left(\frac{\nu}{2}\right) = \left[\sqrt{\frac{1+e}{3ec_3}} \cdot \frac{c_2}{c_1}\right] \cdot U \quad ,$$

which is mentioned here for the sake of completeness.

All the terms that are enclosed in square brackets in the preceding equations have the property of tending to a value of 1 for $E \rightarrow 0$ and $e \rightarrow 1$. In this limiting case, the equations reduce to the equations for a parabolic orbit, which have already been discussed.

These rearranged forms do not, of course, alter the fact that Kepler's equation must still be solved iteratively. What is important, however, is that the iteration no longer presents mathematical difficulties. We begin with the assumption that $E \approx 0$ ($c_3(E) \approx 1/6$) and determine the appropriate value for U by solving Kepler's/Barker's equation

$$U = B - 1/B$$

where

$$B = \sqrt[3]{A + \sqrt{A^2 + 1}} \quad \text{and} \quad A = \frac{3}{2}\sqrt{6ec_3(E)}\sqrt{\frac{GM_\odot}{2q^3}} \cdot (t - t_0) \quad .$$

From this we obtain improved values

$$\hat{E} = U \cdot \sqrt{\frac{1-e}{3e \cdot c_3(E)}} \quad \text{and} \quad c_3(\hat{E}) = \frac{\hat{E} - \sin\hat{E}}{\hat{E}^2} \quad ,$$

with which we again obtain a more accurate value $\hat{u}$. These steps are repeated until U no longer varies beyond the desired degree of accuracy. Nevertheless it is important that the functions $c_1(E)$, $c_2(E)$ and $c_3(E)$ are correctly evaluated for small values of E. By virtue of the fact that individual terms cancel out, it is not possible to do this by directly evaluating the trigonometrical functions. Instead, we use the Taylor expansions of c_1, c_2 and c_3, which have the following form:

$$\begin{aligned}
c_1(E) &= \frac{\sin E}{E} &&= 1 - \frac{E^2}{3!} + \frac{E^4}{5!} - \ldots &&= \sum_{n=1}^{\infty} \alpha_n \\
c_2(E) &= \frac{1-\cos E}{E^2} &&= \frac{1}{2!} - \frac{E^2}{4!} + \frac{E^4}{6!} - \ldots &&= \sum_{n=1}^{\infty} \beta_n \\
c_3(E) &= \frac{E - \sin E}{E^3} &&= \frac{1}{3!} - \frac{E^2}{5!} + \frac{E^4}{7!} - \ldots &&= \sum_{n=1}^{\infty} \gamma_n \quad .
\end{aligned} \tag{4.18}$$

All three series contain even powers of E exclusively, and therefore converge very rapidly. The summands may conveniently be obtained by the following recursive procedure:

$$\begin{aligned}
&\alpha_1 = 1 \\
&\beta_n = \alpha_n \cdot \frac{1}{2n}, \quad \gamma_n = \beta_n \cdot \frac{1}{2n+1}, \quad \alpha_{n+1} = -E^2 \cdot \gamma_n \quad (n = 1, \ldots) \quad .
\end{aligned}$$

This process is carried out in the `Stumpff` function until a summand becomes smaller than the required accuracy eps.

```
//------------------------------------------------------------------------
// Stumpff: computes values for the Stumpff functions C1, C2 and C3
//   E2         Square of eccentric anomaly (E2=E*E) in [rad^2]
//   c1         Value of C1 = sin(E)/E
//   c2         Value of C2 = (1-cos(E))/(E*E)
//   c3         Value of C3 = (E-sin(E))/(E^3)
//------------------------------------------------------------------------
void Stumpff (double E2, double& c1, double& c2, double& c3)
{
  const double eps = 100.0*eps_mach;
  double       n, add;

  c1 = c2 = c3 = 0.0;
  add = n = 1.0;
  do {
    c1 += add;  add /= (2.0*n);
    c2 += add;  add /= (2.0*n+1.0);
    c3 += add;  add *= -E2;
```

```
    n += 1.0;
  }
  while (fabs(add) >= eps);
}
```

The remaining equations for parabolic and near-parabolic cometary orbits are coded in function Parab, which is called in the same way as Hyperb.

```
//------------------------------------------------------------------------------
// Parab: computes position and velocity vectors for parabolic and near
//        parabolic orbits
//   GM       Product of gravitational constant and centre mass [AU^3*d^-2]
//   t0       Time of perihelion passage
//   t        Time for calculation
//   q        Perihelion distance in [AU]
//   e        Eccentricity of the orbit (~1)
//   r        Position w.r.t. orbital plane in [AU]
//   v        Velocity w.r.t. orbital plane in [AU/d]
// Note: t0 and t in Julian centuries since J2000
//------------------------------------------------------------------------------
void Parab ( double GM, double t0, double t, double q, double e,
             Vec3D& r, Vec3D& v )
{
  const int maxit = 15;
  const double eps = 100.0*eps_mach;
  int     i=0;
  double  E2=0.0;
  double  E20, fac, c1, c2, c3, k, tau, A, B, u, u2, R;

  fac = 0.5*e;
  k   = sqrt(GM / (q*(1.0+e)));
  tau = sqrt(GM)*(t-t0);

  do {
    ++i;
    E20 = E2;
    A = 1.5*sqrt(fac/(q*q*q))*tau;
    B = pow ( sqrt(A*A+1.0)+A, 1.0/3.0 );
    u  = B - 1.0/B;
    u2 = u*u;
    E2 = u2*(1.0-e)/fac;
    Stumpff (E2, c1,c2,c3);
    fac = 3.0*e*c3;
    if (i == maxit) {
      cerr << " Convergence problems in Parab" << endl;
      break;
    }
  }
  while (fabs(E2-E20) >= eps);
  R  = q * ( 1.0 + u2*c2*e/fac );
  r = Vec3D (q*(1.0-u2*c2/fac), q*sqrt((1.0+e)/fac)*u*c1, 0.0);
  v = Vec3D (-k*r[y]/R, k*(r[x]/R+e), 0.0);
}
```

4.5 Gaussian Vectors

The various functions described so far enable us to calculate the position of a comet in the plane of its orbit. To be able to determine the ecliptic coordinates of the comet, we must first establish the orientation of the orbit relative to the ecliptic. For this we use three additional orbital elements (Fig. 4.6):

- i The *inclination* gives the angle of intersection between the orbital plane and the ecliptic. An inclination of more than 90° means that the comet is retrograde, its direction of revolution around the Sun being opposite to that of the planets.
- Ω The *longitude of the ascending node* indicates the angle between the vernal equinox and the point on the orbit at which the comet crosses the ecliptic from south to north.
- ω The *argument of perihelion* is the angle between the direction of the ascending node and the direction of the closest point of the orbit to the Sun.

Frequently, the last element ω, is replaced by the *longitude of perihelion* ($\varpi = \Omega + \omega$), defined as being the sum of the longitude of the ascending node and the argument of perihelion.

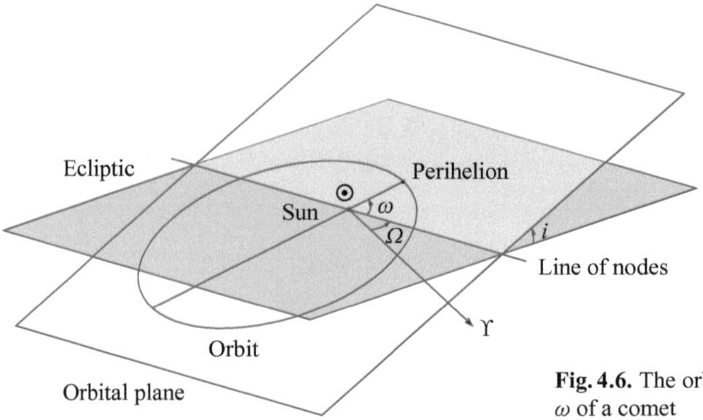

Fig. 4.6. The orbital elements i, Ω and ω of a comet

Because the position of the vernal equinox and the ecliptic are slowly altering as a result of precession, the equinox is given as well as the orbital elements. 'Equinox 1950', for example, means that the elements are referred to the 1950 vernal equinox and ecliptic.

To derive the transformation equations we first need to consider the perifocal coordinate system, which is aligned parallel to the orbital plane and the line of apsides. In this system, a point on the orbit with a true anomaly ν and distance from the Sun r has the coordinates

$$\tilde{\boldsymbol{r}} = \begin{pmatrix} \tilde{x} \\ \tilde{y} \\ \tilde{z} \end{pmatrix} = \begin{pmatrix} r\cos\nu \\ r\sin\nu \\ 0 \end{pmatrix} ,$$

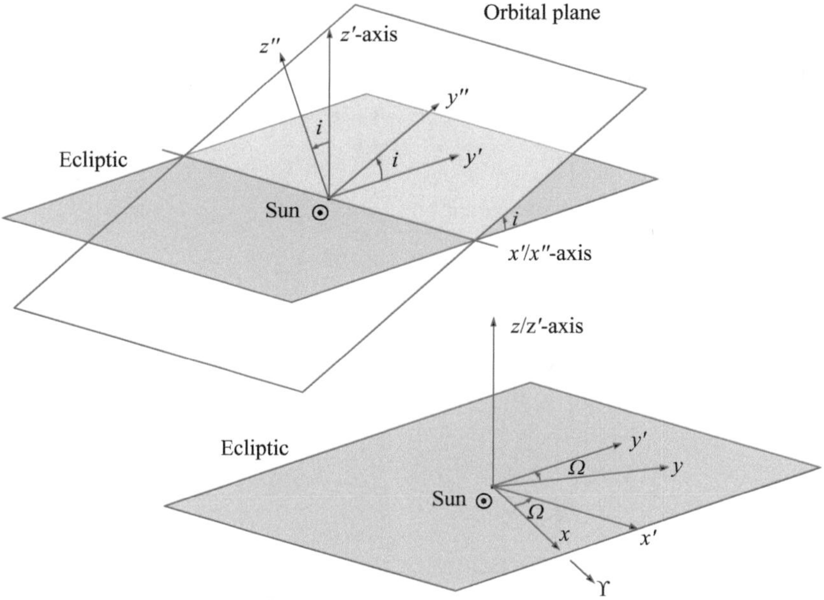

Fig. 4.7. Conversion to ecliptic coordinates

which are differentiated from the ecliptic coordinates (x, y, z) by the use of a tilde. The perifocal coordinate system is produced by a positive rotation through angle ω around the $\tilde{z}/z''$-axis in an (x'', y'', z'') coordinate system, where the x''-axis points in the direction of the ascending node, and where the x''-y''-plane coincides with the orbital plane (Fig.4.7). For the coordinates of a point on the orbit in this system we therefore have

$$\boldsymbol{r}'' = \boldsymbol{R}_z(-\omega)\,\tilde{\boldsymbol{r}} = \begin{pmatrix} +\cos\omega & -\sin\omega & 0 \\ +\sin\omega & +\cos\omega & 0 \\ 0 & 0 & 1 \end{pmatrix} \begin{pmatrix} r\cos\nu \\ r\sin\nu \\ 0 \end{pmatrix} = \begin{pmatrix} r\cos u \\ r\sin u \\ 0 \end{pmatrix} \quad ,$$

where the *argument of latitude* u is the sum of the argument of perihelion and the true anomaly:

$$u = \omega + \nu \quad . \tag{4.19}$$

We now consider the (x', y', z') system, where the x'-axis also indicates the ascending node, but where the x'-y'-plane lies in the ecliptic. Both coordinate systems are simply inclined to one another at the orbital inclination i:

$$\boldsymbol{r}' = \boldsymbol{R}_x(-i)\,\boldsymbol{r}'' = \begin{pmatrix} 1 & 0 & 0 \\ 0 & +\cos i & -\sin i \\ 0 & +\sin i & +\cos i \end{pmatrix} \begin{pmatrix} r\cos u \\ r\sin u \\ 0 \end{pmatrix} = \begin{pmatrix} r\cos u \\ r\sin u\cos i \\ r\sin u\sin i \end{pmatrix} \quad .$$

The coordinates thus obtained are referred to the ecliptic, but are still reckoned from the ascending node. These must therefore be altered in the following step into yet another coordinate system, where the (x, y, z)-axes are referred in the conventional manner to the vernal equinox and the ecliptic:

$$\boldsymbol{r} = \boldsymbol{R}_z(-\Omega)\,\boldsymbol{r}' = \begin{pmatrix} +\cos\Omega & -\sin\Omega & 0 \\ +\sin\Omega & +\cos\Omega & 0 \\ 0 & 0 & 1 \end{pmatrix} \begin{pmatrix} r\cos u \\ r\sin u\cos i \\ r\sin u\sin i \end{pmatrix} \quad .$$

We therefore obtain the combined equations

$$\begin{pmatrix} x \\ y \\ z \end{pmatrix} = r \begin{pmatrix} \cos u\cos\Omega - \sin u\cos i\sin\Omega \\ \cos u\sin\Omega + \sin u\cos i\cos\Omega \\ \sin u\sin i \end{pmatrix} \quad , \qquad (4.20)$$

with which we may derive the ecliptic coordinates (x, y, z) of a given point on the orbit at a distance r and true anomaly ν.

Alternatively, the rotations described above may also be combined into a single matrix

$$\boldsymbol{U} = \boldsymbol{R}_z(-\Omega)\boldsymbol{R}_x(-i)\boldsymbol{R}_z(-\omega) \quad , \qquad (4.21)$$

the use of which allows the entire transformation

$$\boldsymbol{r} = \boldsymbol{U}\tilde{\boldsymbol{r}} \qquad \dot{\boldsymbol{r}} = \boldsymbol{U}\dot{\tilde{\boldsymbol{r}}} \qquad (4.22)$$

between perifocal and ecliptic coordinates to be described in a particularly compact manner. Use of this relationship is recommended, for example, if we want to consider velocity in addition to position, or to calculate a whole series of heliocentric cometary positions at a time. The matrix elements need be determined only once at the beginning, because they depend on the orbital elements only, and not on time.

The $(\boldsymbol{P}, \boldsymbol{Q}, \boldsymbol{R})$ columns in the matrix $\boldsymbol{U}$ describe three unit vectors, which are perpendicular to one another and known as Gaussian vectors. Then

$$\boldsymbol{P} = \begin{pmatrix} +\cos\omega\cos\Omega - \sin\omega\cos i\sin\Omega \\ +\cos\omega\sin\Omega + \sin\omega\cos i\cos\Omega \\ +\sin\omega\sin i \end{pmatrix} \qquad (4.23)$$

and

$$\boldsymbol{Q} = \begin{pmatrix} -\sin\omega\cos\Omega - \cos\omega\cos i\sin\Omega \\ -\sin\omega\sin\Omega + \cos\omega\cos i\cos\Omega \\ +\cos\omega\sin i \end{pmatrix} \qquad (4.24)$$

lie in the orbital plane and point, respectively, in the direction of perihelion, and in the direction of a point with a true anomaly $\nu = 90°$. Finally, the third vector,

$$\boldsymbol{R} = \begin{pmatrix} +\sin i\sin\Omega \\ -\sin i\cos\Omega \\ +\cos i \end{pmatrix} \qquad (4.25)$$

is perpendicular to the orbital plane and thus solely depends on i and Ω, but not on the argument of perihelion ω.

The `GaussVec` function calculates the Gaussian vectors from the orbital elements in the form of a 3×3-matrix of class `Mat3D`. In a manner analagous to (4.21) the entire transformation matrix is constructed from three individual rotations. In comparison to the component-wise description of Gaussian vectors, this permits a particularly short and clear method of programming:

```
//------------------------------------------------------------------------------
// GaussVec: computes the transformation matrix from the orbital plane
//           coordinate system to the ecliptic
//   Omega     Longitude of the ascending node of the orbit in [rad]
//   i         Inclination of the orbit to the ecliptic in [rad]
//   omega     Argument of perihelion in [rad]
//   <return>  Transformation matrix containing the Gaussian vectors P,Q,R
//------------------------------------------------------------------------------
Mat3D GaussVec (double Omega, double i, double omega) {
  return R_z(-Omega) * R_x(-i) * R_z(-omega);
}
```

The various routines that have been developed may be combined into an overall function that is suitable for all types of orbit, and which yields heliocentric ecliptic coordinates directly.

```
//------------------------------------------------------------------------------
// Kepler: computes position and velocity vectors for Keplerian orbits
//         w.r.t. the ecliptic
//   GM       Product of gravitational constant and centre mass [AU^3*d^-2]
//   t0       Time of perihelion passage
//   t        Time for calculation
//   q        Perihelion distance in [AU]
//   e        Eccentricity of the orbit
//   PQR      Transformation orbital plane -> ecliptic (Gaussian vectors)
//   r        Heliocentric ecliptical position in [AU]
//   v        Heliocentric ecliptical velocity vector in [AU/d]
// Note: t0 and t in Julian centuries since J2000
//------------------------------------------------------------------------------
void Kepler ( double GM, double t0, double t,
              double q, double e, const Mat3D& PQR,
              Vec3D& r, Vec3D& v )
{
  const double M0  = 0.1; // [rad]
  const double eps = 0.1;
  double M, delta, tau, invax;
  Vec3D  r_orb,v_orb;
  delta = fabs(1.0-e);
  invax = delta / q;
  tau   = sqrt(GM)*(t-t0);
  M     = tau * sqrt(invax*invax*invax);
  if ( (M < M0) && (delta < eps) )
    Parab (GM, t0,t, q,e, r_orb, v_orb);
  else if ( e < 1.0 )
    Ellip (GM, M, 1.0/invax, e, r_orb, v_orb);
  else
```

```
    Hyperb (GM, t0,t, 1.0/invax, e, r_orb, v_orb);
  r = PQR*r_orb;
  v = PQR*v_orb;
}
```

4.6 Light-Time

The light observed from a comet requires a certain amount of time to reach us. Depending on the comet's distance, this time may range from a few minutes to a few hours. As the comet continues to move during this light-time, the observed position of the comet is slightly behind the actual geometrical position. The velocity of light is significantly greater than the typical velocity of a comet, which means that taking light-time into account generally involves a small correction of the order of only a few arc-seconds to about one arc-minute. We can therefore use a few approximations, which allow us to deal with light-time relatively simply. To calculate the position $\boldsymbol{r}(t)$ at which the comet will be seen at time t, we first calculate the comet's heliocentric coordinates $\boldsymbol{r}_\mathrm{k}(t)$ and the geocentric coordinates of the Sun $\boldsymbol{r}_\odot(t)$. The geometrical distance of the comet from the Earth at time t is then given by

$$\Delta_0 \;=\; |\boldsymbol{r}_0| \;=\; |\boldsymbol{r}_\odot(t)+\boldsymbol{r}_\mathrm{k}(t)| \;=\; \sqrt{(x_\odot+x_\mathrm{k})^2+(y_\odot+y_\mathrm{k})^2+(z_\odot+z_\mathrm{k})^2}\;.$$

It differs only slightly from the distance actually traversed by a ray of light, which is

$$\Delta \;=\; |\boldsymbol{r}_\odot(t)+\boldsymbol{r}_\mathrm{k}(t')| \text{ where } t' = t-\frac{\Delta}{c} \quad ,$$

but which may be calculated only iteratively, because the point in time t' at which the light is emitted by the comet cannot be known in advance. Light-time is therefore $\tau \approx \Delta_0/c$, and the heliocentric position of the comet at time t' may be obtained to a good approximation from

$$\boldsymbol{r}_\mathrm{k}(t') \approx \boldsymbol{r}_\mathrm{k}(t) - \boldsymbol{v}_\mathrm{k}(t)\cdot\frac{\Delta_0}{c} \quad .$$

Using these *retarded* heliocentric coordinates of the comet, the observed, geocentric position vector is

$$\boldsymbol{r} \approx \boldsymbol{r}_\odot(t) + \left\{\boldsymbol{r}_\mathrm{k}(t) - \boldsymbol{v}_\mathrm{k}(t)\cdot\frac{\Delta_0}{c}\right\} \quad . \tag{4.26}$$

With the units employed here (AU and d), the factor $1/c$ has a value of

$$1/c = 0\overset{\mathrm{d}}{.}00578/\mathrm{AU} \quad .$$

Cometary coordinates corrected in the manner described are known as *astrometric* coordinates. They are suitable for directly comparing the position of a comet with stellar coordinates taken from a star atlas. It should be noted that it is customary to give the geocentric distance as the value of the geometrical distance Δ_0 (and not of the light-path Δ).

4.7 The COMET Program

The Comet program calculates ephemerides of comets and minor planets on the basis of the two-body problem. Perturbations of the body under consideration by other planets are not taken into account. This simplification generally suffices for most practical applications. In comparing the results given by this program and those from other sources, however, we must remember that the data given in the latter may not have been calculated with the same assumptions. The user does not have to bother with the type of orbit (ellipse, parabola, or hyperbola). Comet recognizes this from the given eccentricity and chooses the appropriate method of calculation. Elongated, high eccentricity ellipses are handled by the mathematically sound method devised by Stumpff.

Comet calculates heliocentric ecliptic and geocentric equatorial coordinates. In both cases, precession to the required equinox is taken into account. In addition, the geocentric coordinates are corrected for light-time effects. Geocentric coordinates in this form (including precession and light-time effects) are designated astrometric coordinates. They are directly comparable with positions in a star catalogue, or may be used to plot the position of the object on a star chart with the same equinox!

In comparing the results with other sources, it is important to check whether the latter do really give astrometric coordinates with the same equinox, or even if perhaps they are apparent coordinates (in which nutation and aberration are also taken into account).

```
//------------------------------------------------------------------------
// File:    Comet.cpp
// Purpose: Calculation of unperturbed ephemeris with arbitrary
//          eccentricity for comets and minor planets
// (c) 1999 Oliver Montenbruck, Thomas Pfleger
//------------------------------------------------------------------------

#include <cmath>
#include <fstream>
#include <iomanip>
#include <iostream>
#include "APC_Const.h"
#include "APC_IO.h"
#include "APC_Kepler.h"
#include "APC_Math.h"
#include "APC_PrecNut.h"
#include "APC_Spheric.h"
#include "APC_Sun.h"
#include "APC_Time.h"
#include "APC_VecMat3D.h"
using namespace std;

//------------------------------------------------------------------------
// GetElm: Reads orbital elements from input file
//   Filename  Name of input file with orbital elements
//   Mjd0      Time of perihelion as Modified Julian Date
```

```
//   q          Perihelion distance in [AU]
//   e          Eccentricity
//   PQR        Transformation matrix orbital plane -> ecliptic
//   T_eqx0     Equinox for the orbital elements
//------------------------------------------------------------------------------
void GetElm ( char* Filename, double& Mjd0,
              double& q, double& e, Mat3D& PQR, double& T_eqx0 )
{
  int       year, month;
  double    Omega, i, omega, Year,day;
  ifstream  inp;
  inp.open(Filename);   // Open orbital elements file
  // Read elements and dump to screen
  cout << endl
       << " Orbital elements from file " << Filename << endl << endl;
  inp >> year >> month >> day; inp.ignore(81,'\n');
  Mjd0  = Mjd(year,month,int(day))+(day-int(day));
  cout << "  perihelion time (y m d) "
       << "  " << setprecision(2) << DateTime(Mjd0,DDd) << endl;
  inp >> q; inp.ignore(81,'\n');
  cout << "  perihelion distance (q) "
       << setprecision(7) << setw(14) << q << " AU" << endl;
  inp >> e; inp.ignore(81,'\n');
  cout << "  eccentricity (e)        "
       << setprecision(7) << setw(14) << e << endl;
  inp >> i; inp.ignore(81,'\n');
  cout << "  inclination (i)         "
       << setprecision(5) << setw(12) << i << " deg" << endl;
  inp >> Omega; inp.ignore(81,'\n');
  cout << "  long. of ascending node "
       << setprecision(5) << setw(12) << Omega << " deg" << endl;
  inp >> omega; inp.ignore(81,'\n');
  cout << "  argument of perihelion  "
       << setprecision(5) << setw(12) << omega << " deg" << endl;
  inp >> Year; inp.ignore(81,'\n');
  cout << "  equinox                 "
       << setprecision(2) << setw(9)  << Year << endl;
  inp.close();
  T_eqx0 = (Year-2000.0)/100.0;
  PQR = GaussVec(Rad*Omega, Rad*i, Rad*omega);
}
//------------------------------------------------------------------------------
// GetEph: Prompts user for date of start and end of ephemeris, step size
//         and desired equinox
//   MjdStart  Start time of ephemeris as Modified Julian Date
//   Step      Step size of ephemeris in days
//   MjdEnd    End time of ephemeris as Modified Julian Date
//   T_eqx     Desired equinox for ephemeris in Jul. cent. since J2000
//------------------------------------------------------------------------------
void GetEph (double& MjdStart, double& Step, double& MjdEnd, double& T_eqx)
{
  int    year, month, day;
  double hour, Year;
```

```
  cout << endl
       << " Begin and end of the ephemeris: " << endl << endl;
  // Query user for start and end date, step and desired equinox
  cout << "  first date (yyyy mm dd hh.hhh)            ... ";
  cin >> year >> month >> day >> hour; cin.ignore(81,'\n');
  MjdStart = Mjd(year, month, day) + hour/24.0 ;
  cout << "  final date (yyyy mm dd hh.hhh)            ... ";
  cin >> year >> month >> day >> hour; cin.ignore(81,'\n');
  MjdEnd = Mjd(year, month, day) + hour/24.0 ;
  cout << "  step size (dd hh.hh)                      ... ";
  cin >> day >> hour; cin.ignore(81,'\n');
  Step  = day + hour/24.0;
  cout << endl << " Desired equinox of the ephemeris (yyyy.y)  ... ";
  cin >> Year; cin.ignore(81,'\n');
  T_eqx = (Year-2000.0)/100.0;
}
//------------------------------------------------------------------------------
//
// Main program
//
//------------------------------------------------------------------------------
void main(int argc, char* argv[]) {

  // Variables
  int       n_line = 0;
  double    MjdStart, Step, MjdEnd, T_eqx;
  double    Mjd0, q, e, T_eqx0;
  Mat3D     PQR;
  double    Date, T;
  double    dist, fac;
  Vec3D     R_Sun, r_helioc, v_helioc, r_geoc, r_equ;
  char      InputFile[APC_MaxFilename]  = "";
  char      OutputFile[APC_MaxFilename] = "";
  bool      FoundInputfile  = false;
  bool      FoundOutputfile = false;
  ofstream  OutFile;

  // Title
  cout << endl
    << " COMET: ephemeris calculation for comets and minor planets" << endl
    << "        (c) 1999 Oliver Montenbruck, Thomas Pfleger        " << endl
    << endl;

  // Find input file
  GetFilenames( argc, argv, "Comet.dat", InputFile, FoundInputfile,
                OutputFile, FoundOutputfile );
    // Terminate program if input file could not be found
  if ( !FoundInputfile ) {
    cerr << " Terminating program." << endl;  exit(-1);
  }

  // Input orbital elements and prompt user
  GetElm ( InputFile, Mjd0, q, e, PQR, T_eqx0 );
```

```
  GetEph ( MjdStart, Step, MjdEnd, T_eqx );
  PQR = PrecMatrix_Ecl(T_eqx0, T_eqx) * PQR;

  // Redirect output if output file shall be created
  if (FoundOutputfile) {
    OutFile.open(OutputFile);
    if (OutFile.is_open()) cout = OutFile;
  }
  // Header
  cout << endl << endl
       << "    Date      ET   Sun      l       b      r"
       << "       RA          Dec       Distance " << endl
       << setw(43) << " " << "   h  m  s      o  '  \"     (AU) " << endl;

  // Compute ephemeris
  Date = MjdStart;
  // Time loop
  while ( Date < MjdEnd + Step/2 ) {

    // Geocentric ecliptic coordinates of the Sun, equinox T_eqx
    T = ( Date - 51544.5 ) / 36525.0;
    R_Sun = PrecMatrix_Ecl(T, T_eqx) * SunPos(T);
    // Heliocentric ecliptic coordinates of the comet, equinox T_eqx
    Kepler (GM_Sun, Mjd0, Date, q, e, PQR, r_helioc, v_helioc);
    // Geometric geocentric coordinates of the comet
    r_geoc = r_helioc + R_Sun;
    // First-order light-time correction
    dist = Norm(r_geoc);
    fac  = 0.00578*dist;
    r_geoc = r_geoc - fac*v_helioc;
    // Equatorial coordinates
    r_equ = Ecl2EquMatrix(T_eqx) * r_geoc;
    // Output
    cout<< DateTime(Date,HHh)
        << fixed << setprecision(1)
        << setw(7) << Deg*R_Sun[phi]
        << setw(7) << Deg*r_helioc[phi]
        << setw(6) << Deg*r_helioc[theta]
        << setprecision(3) << setw(7)  << r_helioc[r]
        << setprecision(1) << setw(12) << Angle(Deg*r_equ[phi]/15.0,DMMSSs)
        << "  " << showpos << setw(9)  << Angle(Deg*r_equ[theta],DMMSS)
        << noshowpos << setprecision(6) << setw(11) << dist
        << endl;
    ++n_line;
    if ( (n_line % 5) == 0 ) cout << endl; // add line feed every 5 lines
    Date += Step; // Next time step
  };  // End of time loop

  if (OutFile.is_open()) OutFile.close();
}
```

As an example for the use of Comet we will calculate an ephemeris for Comet 1P/Halley for its period of visibility during its last return (1P/1982 U1). First, the orbital elements should be entered into a data file with the name `Comet.dat`. The comments (the exclamation marks and succeeding words on each line) are there for information only, and may be omitted if not required. For Comet 1P/Halley at its 1985/1986 return the details are:

```
1986 2 9.43867  ! Time of perihelion (year month day.fraction)
   0.5870992    ! Perihelion distance q in AU
   0.9672725    ! Eccentricity e
 162.23932      ! Inclination i in [deg]
  58.14397      ! Longitude of the ascending node in [deg]
 111.84658      ! Argument of perihelion in [deg]
1950.0          ! Equinox for the orbital elements
```

When run, `Comet` displays the data read from the input file as a check, before asking the user to enter the time interval, the step width and the required equinox:

```
COMET: ephemeris calculation for comets and minor planets
       (c) 1999 Oliver Montenbruck, Thomas Pfleger

Using default input file Comet.dat

Orbital elements from file Comet.dat

 perihelion time (y m d)     1986/02/09.44
 perihelion distance (q)        0.5870992 AU
 eccentricity (e)               0.9672725
 inclination (i)              162.23932 deg
 long. of ascending node       58.14397 deg
 argument of perihelion       111.84658 deg
 equinox                     1950.00
```

We want an ephemeris for the period between 1985 November 15 and 1986 April 1, in ten-day steps. We choose 2000 for the equinox. The extreme dates for the ephemeris are to be given in the form: Year, Month, Day and Hour with decimal fraction. User input is shown in italic.

```
Begin and end of the ephemeris:

 first date (yyyy mm dd hh.hhh)              ... 1985 11 15  00.0
 final date (yyyy mm dd hh.hhh)              ... 1986 04 01  00.0
 step size (dd hh.hh)                        ... 10 00.0

Desired equinox of the ephemeris (yyyy.y)  ... 2000.0
```

`Comet` now calculates the ephemeris for the data entered:

```
    Date      ET    Sun      l       b      r       RA          Dec       Distance
                                                   h  m  s      o  '  "      (AU)
1985/11/15 00.0  232.8    56.9    0.6  1.720    4 00 40.3  +22 04 27  0.736822
1985/11/25 00.0  242.9    53.2    1.8  1.572    2 15 19.1  +18 24 11  0.623053
1985/12/05 00.0  253.0    48.7    3.2  1.422    0 28 47.5  +10 39 00  0.665665
1985/12/15 00.0  263.2    43.1    5.0  1.269   23 18 18.6  + 3 53 22  0.821652
1985/12/25 00.0  273.3    35.9    7.1  1.114   22 36 56.1  - 0 20 05  1.019722
```

```
1986/01/04 00.0  283.5   26.2   9.8  0.961  22 10 36.0  - 3 00 26  1.216950
1986/01/14 00.0  293.7   12.8  13.0  0.814  21 50 58.9  - 4 58 44  1.388260
1986/01/24 00.0  303.9  353.4  16.2  0.687  21 33 26.2  - 6 47 36  1.511970
1986/02/03 00.0  314.1  326.3  17.7  0.604  21 15 33.4  - 8 49 26  1.563479
1986/02/13 00.0  324.2  294.8  14.9  0.592  20 57 11.7  -11 15 58  1.520207

1986/02/23 00.0  334.3  267.2   8.7  0.658  20 39 33.7  -14 09 18  1.383204
1986/03/05 00.0  344.3  247.1   2.6  0.775  20 22 12.2  -17 39 45  1.179039
1986/03/15 00.0  354.3  232.8  -1.9  0.918  20 00 46.1  -22 28 16  0.936979
1986/03/25 00.0    4.2  222.5  -5.2  1.070  19 22 34.1  -30 13 41  0.685469
1986/04/04 00.0   14.1  214.7  -7.5  1.225  17 41 16.2  -42 56 51  0.475274
```

The first two columns contain date and time. The column headed `Sun` then gives the geocentric ecliptic longitude of the Sun. One can use this information to estimate the elongation of the comet (or minor planet) from the Sun. This gives a clue as to whether the object is accessible for observation. The three following columns (headed `l`, `b`, and `r`) give the heliocentric ecliptic coordinates of the body. The last three columns give the geocentric coordinates right ascension and declination, and the geocentric, geometric distance of the comet in astronomical units (not corrected for light-time.)

5. Special Perturbations

The calculation of cometary and planetary orbits, based on a solution of the unperturbed two-body problem, offers a simple and easily understood method of predicting the position of such a body at any desired time. In addition, the analytical description of the motion in terms of conic sections also provides a particularly clear way of describing the relationship between the various orbital elements and the position in the orbit.

There is, however, the disadvantage that the perturbations caused by the major planets cannot be incorporated. Although the gravitational attraction exerted by the planets is generally an order of magnitude smaller than that of the Sun, over a period of many years it may lead to considerable deviations from an unperturbed orbit. The perturbations are particularly significant when a comet or a minor planet repeatedly passes close to one of the major planets. In this case, substantial changes are possible – as may be observed with various periodic comets. Examples are Comet Pons-Winnecke (7P/1819 L1), whose orbital inclination increased from 10° to 22° between 1819 and 1976 because its aphelion lay close to Jupiter, and Comet Wolf (14P/1884 S1), whose orbital period increased from 6.8 years to 8.3 years, following an encounter with Jupiter in 1922.

Although the perturbations caused by the major planets to most minor-planet orbits are far smaller than those mentioned in these examples, they still generally prevent satisfactory predictions of the orbits over a period of many years on the basis of a simple Keplerian orbit. This means that to find and identify faint minor planets it is essential to take these perturbations into account in calculating the ephemerides.

Basically, there are two different procedures that may be used to do this. In the *special perturbations* method, we begin with the position and velocity of the minor planet at an initial epoch, and calculate the orbit step by step, using numerical integration, taking the accelerations caused by the Sun and all the planets into account, until we reach the desired end-point. The second possibility, the *general perturbations* method, consists of expressing the deviations from an unperturbed orbit as a periodic series, which then enables us to calculate the perturbed position at any given time. The advantage of an analytical method is, however, offset by having to devote considerable effort to developing the series expansions, which prevents the method from being applied to the several thousand minor planets. Analytical perturbation theories are therefore generally available for just the major planets (see Chap. 6).

5.1 Equation of Motion

The basis for describing a minor planet's orbit using numerical integration is what is known as the equation of motion, which expresses the acceleration as a function of position and time. According to Newton's law of gravity, a body of mass m in the gravitational field of just the Sun is subject to a force

$$F = ma = GM_{\odot}m\frac{1}{r^2} \quad , \tag{5.1}$$

which decreases as the square of the distance r from the Sun. Here, once again (see Chap. 4),

$$GM_{\odot} = k^2\mathrm{AU}^3\mathrm{d}^{-2} \qquad \text{where} \qquad k = 0.01720209895$$

is the product of the gravitational constant and the mass of the Sun. Taking into account the fact that the force is always directed towards the Sun, we find that the resultant acceleration, expressed in terms of the corresponding unit vector $-\boldsymbol{r}/r$ is

$$\boldsymbol{a} = -GM_{\odot}\frac{\boldsymbol{r}}{r^3} \quad . \tag{5.2}$$

As we can see, the acceleration is not dependent on the mass m of the body, because the gravitational force itself increases as m.

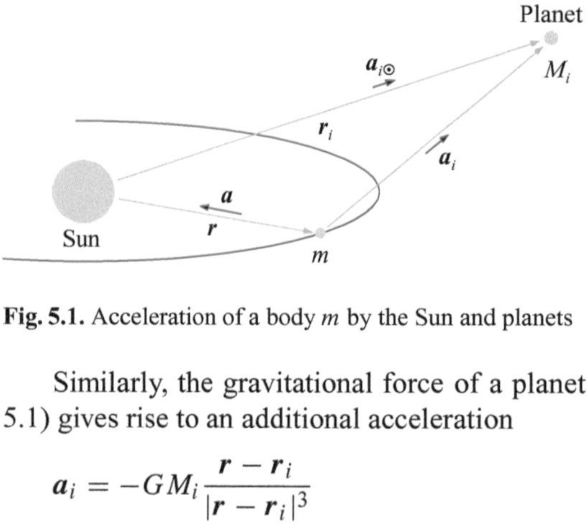

Fig. 5.1. Acceleration of a body m by the Sun and planets

Similarly, the gravitational force of a planet of mass M_i at point $\boldsymbol{r}_i$ (cf. Fig. 5.1) gives rise to an additional acceleration

$$\boldsymbol{a}_i = -GM_i\frac{\boldsymbol{r} - \boldsymbol{r}_i}{|\boldsymbol{r} - \boldsymbol{r}_i|^3} \tag{5.3}$$

of the body under consideration. However, because the planet also simultaneously imposes an acceleration of magnitude

$$\boldsymbol{a}_{i\odot} = -GM_i\frac{-\boldsymbol{r}_i}{|-\boldsymbol{r}_i|^3} \tag{5.4}$$

on the Sun, the overall acceleration acting on the mass m is

$$\Delta\boldsymbol{a}_i = \boldsymbol{a}_i - \boldsymbol{a}_{i\odot} = -GM_i\left(\frac{\boldsymbol{r} - \boldsymbol{r}_i}{|\boldsymbol{r} - \boldsymbol{r}_i|^3} + \frac{\boldsymbol{r}_i}{|\boldsymbol{r}_i|^3}\right) \quad , \tag{5.5}$$

relative to the Sun. Here, the first term is also known as the direct perturbation acceleration, and the second as the indirect acceleration.

If the contributions of all nine planets are taken together, then the following equation gives the overall acceleration of a comet or minor planet:

$$\boldsymbol{a} = -GM_{\odot}\frac{\boldsymbol{r}}{r^3} - \sum_{i=1}^{9} GM_i \left(\frac{\boldsymbol{r}-\boldsymbol{r}_i}{|\boldsymbol{r}-\boldsymbol{r}_i|^3} + \frac{\boldsymbol{r}_i}{|\boldsymbol{r}_i|^3} \right) \quad . \tag{5.6}$$

Table 5.1 gives the ratios of the masses of the Sun and the planets and from this we can see that, in general, even the largest planets, Jupiter and Saturn, will produce only minor perturbations. If, however, the distance of these planets is less than 1/10 AU, and certainly if less than 1/100 AU, then the corresponding acceleration may be of the same order of magnitude as that caused by the Sun's own gravitational attraction.

Table 5.1. Ratio of planetary masses to that of the Sun ($M_{\odot}$) (IAU 1976)

Planet	mass ratio
Mercury	$M_{\odot}/M_1$ =6023600.0
Venus	$M_{\odot}/M_2$ = 408523.5
Earth/Moon	$M_{\odot}/M_3$ = 328900.5
Mars	$M_{\odot}/M_4$ =3098710.0
Jupiter	$M_{\odot}/M_5$ = 1047.355
Saturn	$M_{\odot}/M_6$ = 3498.5
Uranus	$M_{\odot}/M_7$ = 22869.0
Neptune	$M_{\odot}/M_8$ = 19314.0
Pluto	$M_{\odot}/M_9$ =3000000.0

The acceleration $\boldsymbol{a} = (a_x, a_y, a_z)$ acting on a body at heliocentric coordinates $\boldsymbol{r} = (x, y, z)$ at a given time

$$T = \frac{\text{JD} - 2451545}{36525} = \frac{\text{MJD} - 51544.5}{36525}$$

may be calculated, using the `Accel` function. This assumes that there is an appropriate function

```
Vec3D Position (PlanetType Planet, double T);
```

with index

```
enum PlanetType  Sun, Mercury, Venus, Earth, Mars,
                  Jupiter, Saturn, Uranus, Neptune, Pluto ;
```

available to provide the corresponding, heliocentric planetary coordinates. In principle, there is freedom to choose any coordinate system, but in what follows, we assume that the coordinates are based on the current ecliptic and vernal equinox. `Accel` subsequently converts these into the system of the J2000 ecliptic and equinox, in which the equation of motion is formulated.

```
//------------------------------------------------------------------------------
// Accel: Calculates the acceleration vector for a minor body in
//        the solar system
//   Mjd      Time as Modified Julian Date
//   r        J2000 ecliptic coordinates of minor body
//   <return> J2000 ecliptic acceleration in [AU/d**2]
//------------------------------------------------------------------------------
Vec3D Accel (double Mjd, const Vec3D& r)
{
  // Grav. constant * solar and planetary masses in AU**3/d**2
  static const double  GM[10] = { GM_Sun,                     // Sun
                                  GM_Sun / 6023600.0,         // Mercury
                                  GM_Sun /  408523.5,         // Venus
                                  GM_Sun /  328900.5,         // Earth
                                  GM_Sun / 3098710.0,         // Mars
                                  GM_Sun /    1047.355,       // Jupiter
                                  GM_Sun /    3498.5,         // Saturn
                                  GM_Sun /   22869.0,         // Uranus
                                  GM_Sun /   19314.0,         // Neptune
                                  GM_Sun / 3000000.0 };       // Pluto
  // Variables
  int         iPlanet;
  PlanetType  Planet;
  Vec3D       a, r_p, d;
  Mat3D       P;
  double      T, D;
  // Solar attraction
  D = Norm(r);  a = - GM_Sun * r / (D*D*D);
  // Planetary perturbation
  T = ( Mjd - MJD_J2000 ) / 36525.0;
  P = PrecMatrix_Ecl ( T, T_J2000 );      // Precession
  for (iPlanet=Mercury; iPlanet<=Neptune; iPlanet++) {
    Planet = (PlanetType) iPlanet;
    // Planetary position (ecliptic and equinox of J2000)
//  r_p = P * KepPosition(Planet,T);      // Fast but low precision
    r_p = P * PertPosition(Planet,T);     // Accurate but slow
    d   = r - r_p;
    // Direct acceleration
    D = Norm(d);    a += - GM[iPlanet] * d / (D*D*D);
    // Indirect acceleration
    D = Norm(r_p); a += - GM[iPlanet] * r_p / (D*D*D);
  }
  return a;
}
```

5.2 Planetary Coordinates

To be able to calculate the acceleration at any given time using (5.6), we require the heliocentric coordinates $\boldsymbol{r}$ of the body concerned, and the corresponding coordinates $\boldsymbol{r}_i$ for the nine planets from Mercury to Pluto. For highly accurate calculations we may obtain the latter from previously calculated planetary ephemerides, such as those prepared by NASA's Jet Propulsion Laboratory and made available for scientific use. Although such ephemerides are highly accurate, they do, however, have the disadvantage of being extremely comprehensive (taking as many as 10 000 coefficients into account for each year). Luckily, in many cases, considerably simpler, approximate ephemerides are adequate for calculating the perturbations of a cometary or minor-planetary orbit to arc-second accuracy. Apart from close encounters, the perturbational acceleration is small in comparison with the acceleration produced by the Sun. As a result, small errors in the planetary coordinates normally produce negligible, second-order errors in the calculated coordinates of the perturbed body.

Table 5.2. Planetary orbital elements at epoch 2000 January 1.5 (JD 2451545.0) referred to the J2000 ecliptic and equinox

Planet	a (AU)	e	M	n	Ω	i	ω
Mercury	0.387099	0.205634	174.°794	149472.°6738/cy	48.°331	7.°004	77.°455
Venus	0.723332	0.006773	50.°407	58517.°8149/cy	76.°680	3.°394	131.°571
Earth	1.000000	0.016709	357.°525	35999.°3720/cy	174.°876	0.°000	102.°940
Mars	1.523692	0.093405	19.°387	19140.°3023/cy	49.°557	1.°849	336.°059
Jupiter	5.204267	0.048775	18.°819	3033.°6272/cy	100.°491	1.°305	15.°558
Saturn	9.582018	0.055723	320.°348	1213.°8664/cy	113.°643	2.°485	89.°657
Uranus	19.229412	0.044406	142.°956	426.°9282/cy	73.°989	0.°773	170.°531
Neptune	30.103658	0.011214	267.°765	217.°9599/cy	131.°794	1.°768	37.°444
Pluto	39.264230	0.244672	15.°023	146.°3183/cy	110.°287	17.°151	224.°050

The simplest possibility is obviously that of making an approximation of the planetary coordinates oneself, using Keplerian orbits. The accuracy obtained in this way is, however, limited, especially for the outer planets. For predictions over a period of ten years, for example, the errors may amount to a few minutes of arc. An appropriate set of orbital elements for the period 1995 to 2005 is given in Table 5.2. It is applied within the `State` and `KepPosition` functions to compute approximate positions (and velocities) for all planets.

```
//------------------------------------------------------------------------------
// State: calculates position and velocity based on Keplerian orbits
//        (used by KepPosition() and KepVelocity())
//   Planet  Identifies the planet
//   T       Time in Julian centuries since J2000
//   r       Heliocentric position [AU], true-of-date ecliptic and equinox
//   v       Heliocentric velocity [AU/d], true-of-date ecliptic and
/            equinox
//------------------------------------------------------------------------------

void State (PlanetType Planet, double T, Vec3D& r, Vec3D& v)
{
  const double p = 1.3970;          // Centennial precession [deg]
  double a, e, M0, O, i, w, n, T0;
  Mat3D  PQR;
```

```
  // Orbital elements; ecliptic and equinox of J2000
  switch ( Planet )
  {
    case Sun:
      r = v = Vec3D();  return;
    case Mercury:
      a =  0.387099; e = 0.205634; M0 = 174.7947; n = 149472.6738;
      O =  48.331;   i = 7.0048;   w  =  77.4552; T0 = 0.0;  break;
    case Venus:
      a =  0.723332; e = 0.006773; M0 =  50.4071; n  = 58517.8149;
      O =  76.680;   i = 3.3946;   w  = 131.5718; T0 = 0.0;  break;
    case Earth:
      a =  1.000000; e = 0.016709; M0 = 357.5256; n  = 35999.3720;
      O = 174.876;   i = 0.0000;   w  = 102.9400; T0 = 0.0;  break;
    case Mars:
      a =  1.523692; e = 0.093405; M0 =  19.3879; n  = 19140.3023;
      O =  49.557;   i = 1.8496;   w  = 336.0590; T0 = 0.0;  break;
    case Jupiter:
      a =  5.204267; e = 0.048775; M0 =  18.8185; n  =  3033.6272;
      O = 100.4908;  i = 1.3046;   w  =  15.5576; T0 = 0.0;  break;
    case Saturn:
      a =  9.582018; e = 0.055723; M0 = 320.3477; n  =  1213.8664;
      O = 113.6427;  i = 2.4852;   w  =  89.6567; T0 = 0.0;  break;
    case Uranus:
      a = 19.229412; e = 0.044406; M0 = 142.9559; n  =   426.9282;
      O =  73.9893;  i = 0.7726;   w  = 170.5310; T0 = 0.0;  break;
    case Neptune:
      a = 30.103658; e = 0.011214; M0 = 267.7649; n  =   217.9599;
      O = 131.7942;  i = 1.7680;   w  =  37.4435; T0 = 0.0;  break;
    case Pluto:
      a = 39.264230; e = 0.244672; M0 =  15.0233; n  =   146.3183;
      O = 110.2867;  i = 17.1514;  w =  224.0499; T0 = 0.0;
  }

  // State vector w.r.t. true-of-date ecliptic and equinox
  Ellip ( GM_Sun, Rad*(M0+n*(T-T0)), a, e, r, v );  // w.r.t. Orbital plane
  PQR = GaussVec ( Rad*(O+p*T), Rad*i, Rad*(w-O) ); // Transf. to ecliptic
  r = PQR*r;                                        // Ecliptic
  v = PQR*v;                                        // coordinates
}

//------------------------------------------------------------------------
// KepPosition: Planetary position from Keplerian elements
//   Planet    Identifies the planet
//   T         Time in Julian centuries since J2000
//   <return>  Heliocentric position [AU], true-of-date ecliptic & equinox
//------------------------------------------------------------------------
Vec3D KepPosition (PlanetType Planet, double T)
{
  Vec3D r, v;
  State ( Planet, T, r, v );
  return r;
}
```

As an alternative to the function `KepPosition`, we may – as implied earlier – also substitute the function `PertPosition` to calculate the positions of the perturbing planets in `Accel`. `PertPosition` is based on the series expansion described in Chap. 6, and provides, especially for the outer planets, significantly more accurate positions than `KepPosition`. This advantage is, however, at the expense of a considerably greater computation time.

5.3 Numerical Integration

To simplify further discussion, the position $\boldsymbol{r}$ and velocity $\boldsymbol{v}$ of the body concerned may be combined as a single, six-dimensional vector

$$\boldsymbol{y}(t) = \begin{pmatrix} \boldsymbol{r}(t) \\ \boldsymbol{v}(t) \end{pmatrix} = \begin{pmatrix} x(t) \\ y(t) \\ z(t) \\ \dot{x}(t) \\ \dot{y}(t) \\ \dot{z}(t) \end{pmatrix} , \tag{5.7}$$

known as the state vector. The change $\dot{\boldsymbol{y}}$ of position and velocity with time may therefore be expressed as a differential equation of the form

$$\dot{\boldsymbol{y}} = \boldsymbol{f}(t, \boldsymbol{y}) \quad , \tag{5.8}$$

where the

$$\boldsymbol{f}(t, \boldsymbol{y}) = \begin{pmatrix} \boldsymbol{v}(t) \\ \boldsymbol{a}(t, \boldsymbol{r}) \end{pmatrix} \tag{5.9}$$

function combines the velocity and acceleration, which are themselves dependent on time and position.

The basic, numerical integration procedure may be simply expressed using this convention. If we know the state vector $\boldsymbol{y}_0$ at a time t_0, then we can calculate the approximate state vector at a later time $t_1 = t_0 + h$ from the approximation

$$\boldsymbol{y}(t_1) \approx \boldsymbol{y}_1 = \boldsymbol{y}(t_0) + h \cdot \dot{\boldsymbol{y}}(t_0) = \boldsymbol{y}_0 + h \cdot \boldsymbol{f}(t_0, \boldsymbol{y}_0) \quad . \tag{5.10}$$

If this step is carried out, we can then obtain an approximation for the state vector at time $t_2 = t_0 + 2h$ from t_1 and $\boldsymbol{y}_1$, and, in a similar way, corresponding approximate values for the position and velocity at any other times $t_i = t_0 + ih$. This simple procedure is not, however, particularly suitable for practical use, because it gives accurate results only when the step values h are very small, which means that an extremely large number of steps must be taken to calculate a complete orbit.

To obtain usable results with larger step values, the procedure adopted is the so-called multistep method, with improved approximations, which are based on several earlier values of $\boldsymbol{f}$. To illustrate the principle behind this procedure, let us assume that we have approximations $\boldsymbol{y}_j$ for the state vector $\boldsymbol{y}(t_j)$ at times $t_j = t_0 + jh$, where $j = 0, 1, \ldots, i$. If we then integrate both sides of

$$\dot{\boldsymbol{y}} = \boldsymbol{f}(t, \boldsymbol{y}) \tag{5.11}$$

with respect to time t from t_i to t_{i+1}, we then obtain the equivalent equation

$$\boldsymbol{y}(t_{i+1}) = \boldsymbol{y}(t_i) + \int_{t_i}^{t_i+h} \boldsymbol{f}(t, \boldsymbol{y}(t))dt \quad . \tag{5.12}$$

The integral cannot be calculated directly in this form, however, because it depends on the unknown state vector $\boldsymbol{y}(t)$. To avoid this problem, the integrand is replaced by a polynomial $\boldsymbol{p}(t)$, that interpolates some of the values of the function

$$\boldsymbol{f}_j = \boldsymbol{f}(t_j, \boldsymbol{y}_j) \tag{5.13}$$

at earlier times t_j, which, in accordance with the assumption that we made above, are already known.

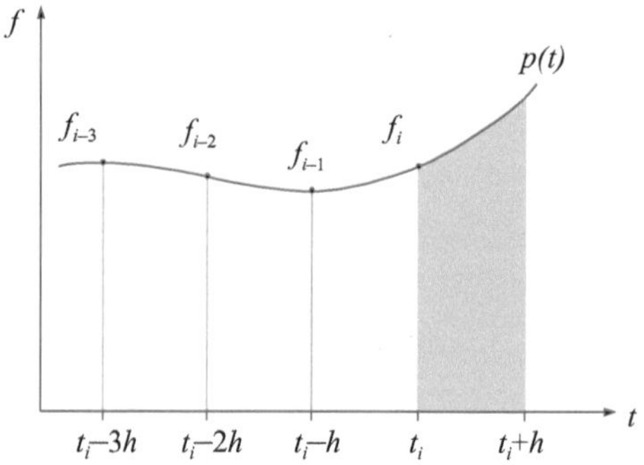

Fig. 5.2. Interpolating earlier values of a function and integrating for the next time interval

As an example, Fig. 5.2 shows a 3rd-order polynomial, which is determined by the four values $\boldsymbol{f}_{i-3}$, $\boldsymbol{f}_{i-2}$, $\boldsymbol{f}_{i-1}$ and $\boldsymbol{f}_i$ at times t_{i-3}, t_{i-2}, t_{i-1} and t_i. If this polynomial is written in the form

$$\boldsymbol{p}(t) = \boldsymbol{a}_0 + \boldsymbol{a}_1\sigma + \boldsymbol{a}_2\sigma^2 + \boldsymbol{a}_3\sigma^3 \qquad \text{with} \qquad \sigma = \frac{1}{h}(t - t_i) \tag{5.14}$$

and the appropriate coefficients

$$\begin{aligned}
\boldsymbol{a}_0 &= (&& && && 6\boldsymbol{f}_i)/6 \\
\boldsymbol{a}_1 &= (-2\boldsymbol{f}_{i-3} && +9\boldsymbol{f}_{i-2} && -18\boldsymbol{f}_{i-1} && +11\boldsymbol{f}_i)/6 \\
\boldsymbol{a}_2 &= (-3\boldsymbol{f}_{i-3} && +12\boldsymbol{f}_{i-2} && -15\boldsymbol{f}_{i-1} && +6\boldsymbol{f}_i)/6 \\
\boldsymbol{a}_3 &= (-1\boldsymbol{f}_{i-3} && +3\boldsymbol{f}_{i-2} && -3\boldsymbol{f}_{i-1} && +1\boldsymbol{f}_i)/6
\end{aligned} \tag{5.15}$$

are entered, we obtain the Adams-Bashforth 4th-order equation

$$\boldsymbol{y}_{i+1} = \boldsymbol{y}_i + \int_{t_i}^{t_i+h} \boldsymbol{p}(t)dt = \boldsymbol{y}_i + \frac{h}{24}(-9\boldsymbol{f}_{i-3} + 37\boldsymbol{f}_{i-2} - 59\boldsymbol{f}_{i-1} + 55\boldsymbol{f}_i) \ , \tag{5.16}$$

which gives a good approximation of the state vector for larger step sizes. Repeating the procedure enables us to obtain corresponding values for subsequent times $t_i + jh$. In general, the more terms are used in the interpolation polynomial (and thus the higher the degree), the higher the procedure's accuracy. With large step values, however, instabilities may occur, so the order of the polynomial and the step value must be appropriately chosen depending of the desired accuracy.

The DE method, which will be used in what follows to calculate perturbed minor-planetary orbits, is also, in principle, an Adams multistep procedure. In contrast to the simplified form that has just been outlined – i.e., a multistep procedure with fixed order and step-width – DE has some special features that are of particular value for practical applications:

- The number of earlier points, and thus the order of the procedure is not fixed, but is independently selected by DE itself. When extremely high accuracies are required, DE uses 12th-order polynomials.
- Similarly, the step-width employed may vary from one step to the next, and thus be chosen to agree with the changes in the state vector with time. This is particularly useful when dealing with highly eccentric orbits, which require significantly smaller steps near perihelion than at aphelion.
- In contrast to simple fixed-order, fixed-step-size, multistep procedures, no special initial calculation is required to obtain the first points $\boldsymbol{f}_0, \boldsymbol{f}_1, \ldots$. Instead, DE begins with a small step of order 1, and increases the order and step-width at each step until optimal values are reached. Consequently, DE is self-starting.
- The state vector $\boldsymbol{y}(t)$ may be obtained at arbitrary output times, and the values at the reference points used internally will be automatically interpolated. The choice of the step-size is not, therefore, affected by the choice of output points.

The DE method was developed by L.F. Shampine and M.K. Gordon, and implemented as a Fortran program. Because of space considerations, we can discuss neither the extremely comprehensive program code, nor details of the implementation. The reader is referred to the book *Computer Solution of Ordinary Differential Equations*, in which the basis of the code is thoroughly explained. For the current application, DE has been converted to C++, and adapted to the latter language's specific features and peculiarities. Great advantages were offered by the definition of a separate class SolverDE, which almost completely relieves the user from administrative tasks such as allocating the necessary working space:

```
enum DE_STATE {                       // Status codes of the SolverDE integrator
  DE_INIT                  = 1,  // Starting step
  DE_DONE                  = 2,  // Successful integration step
  DE_ACCURACY_NOT_ACHIEVED = 3,  // Too stringent accuracy requirements
  DE_TOO_MANY_STEPS        = 4,  // Too many steps required
  DE_STIFF                 = 5,  // Suspect of stiff differential equation
  DE_INVALID_PARAMS        = 6   // Invalid input
};
```

```
// SolverDE integrator class
class SolverDE {
  public:
    // Constructor, Destructor
    SolverDE (DEfunct F, int Neqn);    ~SolverDE ();
    // Integration
    void Integ (
      double y[],         // Solution vector; updated on output
      double& t,          // Value of independent variable; updated on output
      double tout,        // Value of independent variable to integrate to
      double& relerr,     // Desired relative accuracy of solution; updated
      double& abserr,     // Desired absolute accuracy of solution; updated
      DE_STATE& State,    // State code
      bool PermitTOUT = true
    );
  private:
    // Data members
    DEfunct  f;
    int      neqn,ns,k,kold;
    double   *yy,*wt,*p,*yp,*ypout, **phi;
    double   alpha[13],beta[13],v[13],w[13],psi[13],sig[14],g[14];
    double   x,h,hold,told,delsgn;
    bool     OldPermit, phase1,start,nornd;
    // Interpolation
    void Intrp ( double x, const double y[], double xout,
                 double yout[], double ypout[] );
    // Elementary integration step
    void Step (double& x, double y[], double& eps, bool& crash);
};
```

Altogether the integration class incorporates the three methods `Integ`, `Step` and `Intrp`, of which `Integ` alone is directly called by the user. The remaining two member functions are employed within `Integ` and serve to calculate a single integration step (`Step`) and to interpolate the state vector at the output times (`Intrp`).

One peculiarity of the C++ implementation should be noted, in that all arrays within the class `SolverDE` are addressed, as in the original Fortran code, beginning with 1 as the lowest index. Because C++ does not allow any freedom in the choice of array bounds, vectors with a (logical) dimension n are declared as arrays with the (physical) dimension `[n+1]`, where the first array element (index `[0]`) is not employed.

The constructor `SolverDE` expects as parameters the differential equation system to be integrated, and its dimension. The first argument is a function of the form

```
// DEfunct prototype: void f (double x, double y[], double yp[])
typedef void (*DEfunct)(
  double x,     // independent variable
  double y[],   // vector with components of function
  double yp[]   // vector with components of derivative w.r.t. x
);
```

which calculates the derivative yp $= dy/dx$, of a differential equation $dy/dx = f(x, y)$, with given arguments x $= x$ and y $= y$. The second parameter (Neqn) in the constructor's call indicates the number of differential equations and is identical with the upper array index for the individual vectors. A corresponding function for the equation of motion of a minor planet, using the expressions for the acceleration as a function of position and time, as described earlier, may be written as follows:

```
//------------------------------------------------------------------------
// F: Equation of motion in a form compatible with the SolverDE integrator
//   X         Independent variable (Time as Modified Julian Date)
//   Y[]       State vector (position and velocity)
//   dYdX[]    Derivative of state vector w.r.t. X (veloc. and acceleration)
//------------------------------------------------------------------------
void F (double X, double Y[], double dYdX[]) {
  Vec3D a, r;
  r = Vec3D(Y[1], Y[2], Y[3]);
  a = Accel(X, r);
  dYdX[0] = 0.0;
  dYdX[1] = Y[4]; dYdX[2] = Y[5];  dYdX[3] = Y[6];  // velocity
  dYdX[4] = a[x]; dYdX[5] = a[y];  dYdX[6] = a[z];  // acceleration
}
```

The state vector and its derivative are here values with a dimension $n_{\text{eqn}} = 6$. As explained earlier, however, this requires the declaration of Y and dYdX as arrays with dimension Neqn+1. The array components with index [0] have no further significance, but to avoid having non-initialized variables, they are set to zero. If, as in the example

```
SolverDE  Orbit(F,Neqn);
```

we have allocated an object of the class SolverDE, then the relevant differential equations may be subsequently integrated by calling the method Integ:

```
Orbit.Integ(y,t,tout,relerr,abserr,State,PermitTOUT);
```

The individual parameters denote the state vector (y), the time (t), the output time (tout), the relative and the absolute accuracies required (relerr, abserr), as well as the status variable (State). The parameter PermitTOUT is optional, and denotes whether, during the internal integration, the process may be carried out past the time t_{out}. This is always the case if the differential equation to be integrated has no discontinuities or poles. By default, therefore, the parameter is set to true.

The first time Integ is called, the starting time and the corresponding state vector are given by t and y. The desired output time tout and the required relative and absolute accuracies relerr and abserr also have to be entered. The status variable State is set to DE_INIT to indicate to the integrator that the integration of a new problem is to be started.

Under normal circumstances, y eventually contains the desired state vector at time tout. Simultaneously, t is assigned the value of tout, and the status variable State is set to DE_DONE (= successful step). In all subsequent steps, only tout is assigned a new value by the user, while all the other variables (in particular, `State`) remain unaltered. Only at the beginning of a new problem or if the direction of integration is reversed does `State` have to be reset to `DE_INIT`.

If, following a call to `DE`, `State` returns any value other than `DE_DONE`, the significance is as follows:

`DE_ACCURACY_NOT_ACHIEVED`	`tout` was not reached, because either the required tolerance `relerr` or `abserr` was too small. Both values should be appropriately increased for subsequent calculations.
`DE_TOO_MANY_STEPS`	`tout` was not reached, because more than 500 (`maxnum`) steps were required internally.
`DE_STIFF`	`tout` was not reached, because the differential equation appears to be too difficult to solve. With problems of celestial mechanics, this will probably not arise.
`DE_INVALID_PARAMS`	Inadmissible input parameter (e.g., `t=tout`).

In all cases, except when `State=DE_INVALID_PARAMS`, the integration may be directly resumed by calling `Integ` again without altering any parameters. The interruption mainly serves to alert the user to any problems and to avoid (for example) non-terminating integrations.

5.4 Osculating Elements

For an unperturbed Keplerian orbit, the motion of a body around the Sun is specified by six orbital elements: semi-major axis a, eccentricity e, inclination i, longitude of the ascending node Ω, longitude of perihelion ω, and mean anomaly M. The orbital elements are therefore constants, from which the position and orbital velocity may be calculated at any required time (cf. Chap. 4).

Because of their geometrical significance, these orbital elements are particularly easy to understand, so it is advisable to use them to describe a perturbed minor-planet or cometary orbit. We are able to utilize the fact that the orbital elements and the state vector are linked in an unambiguous way. If, at a specific time t we have the position $\boldsymbol{r} = (x, y, z)$ and velocity $\dot{\boldsymbol{r}} = (\dot{x}, \dot{y}, \dot{z})$ of a celestial body, then there is a corresponding set of specific orbital elements $(a, e, i, \Omega, \omega, M)$, from which we can obtain the state vector $(\boldsymbol{r}, \boldsymbol{v})$, using the unperturbed Kepler equations. For perturbed motion, orbital elements may be defined in a similar way. These are no longer constant, however, but exhibit various fluctuations with time depending on the form and degree of perturbation. These orbital elements derived from the state vector are normally described as *osculating* elements, because the corresponding Kepler orbit 'kisses' the true orbit at the appropriate epoch, and is very close to it for a certain period of time (Fig. 5.3).

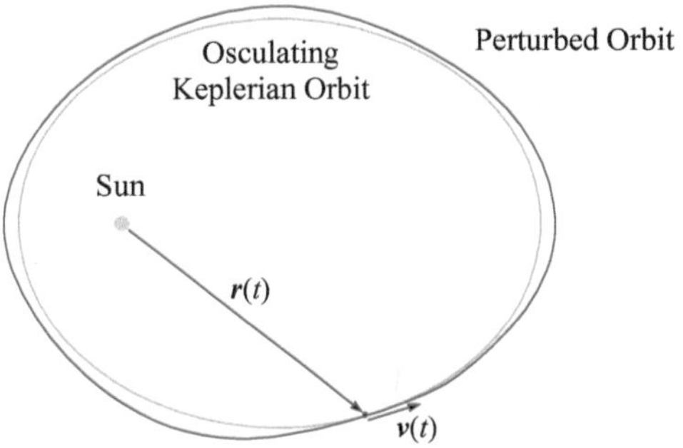

Fig. 5.3. Using osculating elements to define a perturbed orbit

As for unperturbed motion, the osculating orbital plane is defined at any point in time by the position and velocity vector (see also Fig. 4.1). The cross product of $\boldsymbol{r}(t)$ and $\dot{\boldsymbol{r}}(t)$:

$$\boldsymbol{h} = \boldsymbol{r} \times \dot{\boldsymbol{r}} = \begin{pmatrix} y\dot{z} - z\dot{y} \\ z\dot{x} - x\dot{z} \\ x\dot{y} - y\dot{x} \end{pmatrix} \tag{5.17}$$

therefore defines a vector $\boldsymbol{h}(t)$, that is perpendicular to the corresponding orbital plane, and is therefore parallel to the Gaussian vector normal to the orbit:

$$\boldsymbol{R} = \begin{pmatrix} R_x \\ R_y \\ R_z \end{pmatrix} = \begin{pmatrix} +\sin i \sin \Omega \\ -\sin i \cos \Omega \\ +\cos i \end{pmatrix} = \begin{pmatrix} h_x/h \\ h_y/h \\ h_z/h \end{pmatrix} . \tag{5.18}$$

From this we may obtain the osculating orbital inclination i and the longitude of the ascending node Ω as

$$i = \tan^{-1}\left(\frac{\sqrt{h_x^2 + h_y^2}}{h_z}\right) \qquad \Omega = \tan^{-1}\left(\frac{h_x}{-h_y}\right) . \tag{5.19}$$

By solving (4.20) for u we also obtain the additional expression

$$u = \nu + \omega = \tan^{-1}\left(\frac{zh}{-xh_y + yh_x}\right) \tag{5.20}$$

as the argument for the latitude u, which we shall require later to calculate the argument of perihelion.

As may be seen by the relevant equations (4.5) and (4.10), the semi-major axis of the orbit depends solely on the distance r from the Sun, and the magnitude $v = |\dot{\boldsymbol{r}}|$ of the velocity:

$$a = \left(\frac{2}{r} - \frac{v^2}{GM_\odot}\right)^{-1} . \tag{5.21}$$

Accordingly, we always obtain an ellipse with a positive semi-major axis as an osculating Keplerian orbit if the velocity of the body relative to the Sun is smaller than the corresponding escape velocity $v_\infty = \sqrt{2GM_\odot/r}$.

In this case we may rewrite the equations (4.5) and (4.10) for position and velocity as functions of the eccentric anomaly, to give us

$$\begin{aligned} e\cos(E) &= 1 - r/a \\ e\sin(E) &= (x\dot{x} + y\dot{y} + z\dot{z})/\sqrt{GM_\odot a} \end{aligned} \tag{5.22}$$

and enable us to determine the eccentricity e and the eccentric anomaly E. From Kepler's equation (4.7) we then obtain the required mean anomaly

$$M = E - \frac{180°}{\pi} e\sin(E) \tag{5.23}$$

at time t. Finally, from the true anomaly

$$\nu = \tan^{-1} \frac{\sqrt{1-e^2}\sin(E)}{\cos(E) - e} \tag{5.24}$$

corresponding to E, we obtain the sixth orbital element, the argument of perihelion

$$\omega = \nu - u \quad . \tag{5.25}$$

There are corresponding expressions for parabolic and hyperbolic orbits, which are, however, not required for calculation of the orbits of minor planets or periodic comets. The following `Elements` function is therefore restricted to elliptical orbits alone.

```
//------------------------------------------------------------------------------
// Elements:  computes the elements of an elliptical orbit from position
//            and velocity vectors
//   GM       Product of gravitational constant and centre mass [AU^3*d^-2]
//   r        Heliocentric ecliptic position in [AU]
//   v        Heliocentric ecliptic velocity vector in [AU/d]
//   a        Semi-major axis of the orbit in [AU]
//   e        Eccentricity of the orbit
//   i        Inclination of the orbit to the ecliptic in [rad]
//   Omega    Longitude of the ascending node of the orbit in [rad]
//   omega    Argument of perihelion in [rad]
//   M        Mean anomaly in [rad]
//------------------------------------------------------------------------------
void Elements ( double GM, const Vec3D& r, const Vec3D& v,
                double& a, double& e, double& i,
                double& Omega, double& omega, double& M )
{
  Vec3D  h;
  double H, u, R, v2;
  double eCosE, eSinE, e2, E, nu;

  h = Cross(r,v);                                       // Areal velocity
  H = Norm(h);
  Omega = atan2(h[x], -h[y]);                           // Long. ascend. node
  i     = atan2(sqrt(h[x]*h[x]+h[y]*h[y]), h[z]);       // Inclination
```

```
  u     = atan2(r[z]*H, -r[x]*h[y]+r[y]*h[x]);        // Arg. of latitude

  R  = Norm(r);                                        // Distance
  v2 = Dot(v, v);                                      // Velocity squared
  a = 1.0 / (2.0/R-v2/GM);                             // Semi-major axis
  eCosE = 1.0-R/a;                                     // e*cos(E)
  eSinE = Dot(r, v)/sqrt(GM*a);                        // e*sin(E)
  e2 = eCosE*eCosE +eSinE*eSinE;
  e  = sqrt(e2);                                       // Eccentricity

  E  = atan2(eSinE,eCosE);                             // Eccentric anomaly
  M  = E - eSinE;                                      // Mean anomaly
  nu = atan2(sqrt(1.0-e2)*eSinE, eCosE-e2);            // True anomaly
  omega = u - nu;                                      // Arg. of perihelion
  if (Omega<0.0) Omega += 2.0*pi;
  if (omega<0.0) omega += 2.0*pi;
  if (M    <0.0) M     += 2.0*pi;
};
```

5.5 The NUMINT Program

The `Numint` program may be used for accurate calculation of the orbits of minor planets or periodic comets. Its structure and use closely resemble the `Comet` program discussed in Chap. 4. Through the use of a numerical integration procedure both the Sun's gravitational attraction and perturbations caused by the major planets may be taken into account in orbital predictions. In particular, significantly more accurate results may be obtained for predictions covering long periods of time. Against this higher accuracy there is the disadvantage that the time required for the integration increases in proportion to the span of the ephemerides required, and is generally significantly greater than for the `Comet` program.

When the program is called, the orbital elements and the corresponding equinox are read from the data file `Numint.dat` and displayed on the screen for checking purposes, before the user enters the time-span, the step value, and the equinox required for the ephemerides. Using these values, the position and velocity of the minor planet are determined relative to the fixed equinox of J2000, and then predicted for the beginning of the ephemerides by numerical integration from the epoch of the orbital elements. If one is working with orbital elements whose epoch is some years earlier than the starting date, then the integration may take some tens of seconds, depending on the computer's processing power.

Using the state vector thus obtained, `Numint` then calculates the corresponding osculating elements and displays these on the screen. From the changes in the individual elements relative to those for the original epoch one can gain a good idea of the effect of the perturbations caused by the major planets. In addition, the orbital elements thus obtained may be used, when repeating calculations with old orbital elements, to avoid having to span major intervals of time prior to the start of the ephemerides.

Finally, as in the Comet program, an ephemeris of the minor planet is calculated, using the required step-width and duration, and output. Light-time is taken into account, so the resulting, astrometric coordinates may be directly compared with star charts or catalogues. The position of the Earth relative to the Sun is calculated – again as in Comet – using the SunPos function, to avoid errors in the conversion of heliocentric to geocentric coordinates for the minor planet.

```
//------------------------------------------------------------------------------
// File:    Numint.cpp
// Purpose: Numerical integration of perturbed minor planet orbits
// (c) 1999 Oliver Montenbruck, Thomas Pfleger
//------------------------------------------------------------------------------

#include <cmath>
#include <fstream>
#include <iomanip>
#include <iostream>

#include "APC_Const.h"
#include "APC_DE.h"
#include "APC_IO.h"
#include "APC_Kepler.h"
#include "APC_Math.h"
#include "APC_Planets.h"
#include "APC_PrecNut.h"
#include "APC_Spheric.h"
#include "APC_Sun.h"
#include "APC_Time.h"
#include "APC_VecMat3D.h"

using namespace std;

//------------------------------------------------------------------------------
// WriteElm: Screen output of orbital elements
//   MjdEpoch Epoch as Modified Julian Date
//   a        Semi-major axis of the orbit in [AU]
//   e        Eccentricity of the orbit
//   i        Inclination of the orbit to the ecliptic in [rad]
//   Omega    Longitude of the ascending node of the orbit in [rad]
//   omega    Argument of perihelion in [rad]
//   M        Mean anomaly in [rad]
//   T_eqx0   Equinox in Julian centuries since J2000
//------------------------------------------------------------------------------
void WriteElm ( double MjdEpoch, double a, double e, double i,
                double Omega, double omega, double M, double T_eqx0 )
{
  cout << "  Epoch (y m d)              "
       << "  " << setprecision(2) << DateTime(MjdEpoch,DDd) << endl;
  cout << "  Semi-major axis (a)        "
       << setprecision(7) << setw(14) << a << " AU" << endl;
  cout << "  Eccentricity (e)           "
       << setprecision(7) << setw(14) << e << endl;
```

```
  cout << "  Inclination (i)         "
       << setprecision(5) << setw(12) << i*Deg << " deg" << endl;
  cout << "  Long. of ascending node "
       << setprecision(5) << setw(12) << Omega*Deg << " deg" << endl;
  cout << "  Argument of perihelion  "
       << setprecision(5) << setw(12) << omega*Deg << " deg" << endl;
  cout << "  Mean anomaly (M)        "
       << setprecision(5) << setw(12) << M*Deg     << " deg" << endl;
  cout << "  Equinox                 "
       << setprecision(2) << setw(9)  << 2000.0+100*T_eqx0 << endl << endl;
};

//------------------------------------------------------------------------------
// GetElm: Reads orbital elements from input file
//   Filename  Name of input file with orbital elements
//   MjdEpoch Epoch as Modified Julian Date
//   a         Semi-major axis of the orbit [AU]
//   e         Eccentricity of the orbit
//   M         Mean anomaly in [rad]
//   PQR       Transformation matrix orbital plane -> ecliptic
//   T_eqx0    Equinox for the orbital elements
//------------------------------------------------------------------------------
void GetElm ( char* Filename,
              double& MjdEpoch,
              double& a, double& e, double& M, Mat3D& PQR, double& T_eqx0 )
{
  int       year, month;
  double    Omega, i, omega, Year, day;
  ifstream  inp;
  // Read elements
  inp.open(Filename);
  inp >> year >> month >> day; inp.ignore(81,'\n');
  inp >> a; inp.ignore(81,'\n');
  inp >> e; inp.ignore(81,'\n');
  inp >> i; inp.ignore(81,'\n');
  inp >> Omega; inp.ignore(81,'\n');
  inp >> omega; inp.ignore(81,'\n');
  inp >> M; inp.ignore(81,'\n');
  inp >> Year; inp.ignore(81,'\n');
  inp.close();
  // Compute derived data
  i*=Rad;  Omega*=Rad;  omega*=Rad;  M*=Rad;
  MjdEpoch = Mjd(year,month,int(day))+(day-int(day));
  T_eqx0   = (Year-2000.0)/100.0;
  PQR      = GaussVec(Omega,i,omega);
  // Print elements
  cout << " Orbital elements from file " << Filename << ":" << endl<< endl;
  WriteElm ( MjdEpoch, a,e,i,Omega,omega,M, T_eqx0 );
}

//------------------------------------------------------------------------------
// GetEph: Prompts user for date of start and end of ephemeris, step size
//         and desired equinox
```

```
//   MjdStart  Start time of ephemeris as Modified Julian Date
//   Step      Step size of ephemeris in days
//   MjdEnd    End time of ephemeris as Modified Julian Date
//   T_eqx     Desired equinox for ephemeris in Jul. cent. since J2000
//
//------------------------------------------------------------------------
void GetEph (double& MjdStart, double& Step, double& MjdEnd, double& T_eqx)
{
  int    year, month, day;
  double hour, Year;
  cout << " Begin and end of the ephemeris: " << endl;
  cout << "  First date (yyyy mm dd hh.hhh)            ... ";
  cin >> year >> month >> day >> hour; cin.ignore(81,'\n');
  MjdStart = Mjd(year,month,day) + hour/24.0 ;
  cout << "  Final date (yyyy mm dd hh.hhh)            ... ";
  cin >> year >> month >> day >> hour; cin.ignore(81,'\n');
  MjdEnd = Mjd(year,month,day) + hour/24.0 ;
  cout << "  Step size (dd hh.hh)                      ... ";
  cin >> day >> hour; cin.ignore(81,'\n');
  Step  = day + hour/24.0;
  cout << endl << " Desired equinox of the ephemeris (yyyy.y) ... ";
  cin >> Year; cin.ignore(81,'\n');
  T_eqx = (Year-2000.0) /100.0;
}

//------------------------------------------------------------------------
//
// Main program
//
//------------------------------------------------------------------------
void main(int argc, char* argv[])
{
  // Constants
  const pol_index R     = pol_index(2);   // Index for vector length
  const int       Neqn = 6;               // Number of differential eqns.
  const double    eps  = 1.0e-10;         // Relative accuracy

  // Variables
  int        n_line;
  double     MjdStart, Step, MjdEnd, T_eqx;
  double     MjdEpoch, a, e, M, T_eqx0;
  double     Omega, i, omega;
  Mat3D      PQR, P;
  double     Mjd, MjdStep, T;
  SolverDE   Orbit(F, Neqn);
  double     Y[Neqn+1];
  double     relerr,abserr;
  DE_STATE   State;
  Vec3D      r, v;
  Vec3D      R_Sun, r_helioc, v_helioc, r_geoc, r_equ;
  double     dist, fac;
  char       InputFile[APC_MaxFilename] = "";
  char       OutputFile[APC_MaxFilename] = "";
```

```
bool      FoundInputfile = false;
bool      FoundOutputfile = false;
ofstream  OutFile;

// Title
cout << endl
   << "  NUMINT: numerical integration of perturbed minor planet orbits"
   << endl
   << "          (c) 1999 Oliver Montenbruck, Thomas Pfleger          "
   << endl << endl;

// Find input and optional output files
GetFilenames( argc, argv, "Numint.dat", InputFile, FoundInputfile,
              OutputFile, FoundOutputfile );
// Terminate program if input file could not be found
if (!FoundInputfile) {
  cerr << " Terminating program." << endl;
  exit(-1);
}
// Read orbital elements from file and prompt user for ephemeris data
GetElm ( InputFile, MjdEpoch, a, e, M, PQR, T_eqx0 );
GetEph ( MjdStart, Step, MjdEnd, T_eqx );

// Redirect output if output file shall be created
if (FoundOutputfile) {
  OutFile.open(OutputFile);
  if (OutFile.is_open()) cout = OutFile;
}

P = PrecMatrix_Ecl(T_J2000,T_eqx);

// Initial state vector (ecliptic and equinox of J2000)
PQR = PrecMatrix_Ecl(T_eqx0,T_J2000) * PQR;
Ellip ( GM_Sun, M,a,e, r,v );
r = PQR*r;
v = PQR*v;
Y[0]=0.0;  // Unused
Y[1]=r[x]; Y[2]=r[y]; Y[3]=r[z];
Y[4]=v[x]; Y[5]=v[y]; Y[6]=v[z];
Mjd = MjdEpoch;

// Start integration: propagate state vector from epoch
// to start of ephemeris
State = DE_INIT;
relerr = eps;  abserr = 0.0;
do {
  Orbit.Integ(Y, Mjd, MjdStart, relerr, abserr, State);
  if ( State==DE_INVALID_PARAMS ) {
    cerr << "Exit (invalid parameters)" << endl;
    exit(1);
  }
}
while ( State > DE_DONE );
```

```
// Orbital elements at start of ephemeris (equinox T_eqx)
r = P*Vec3D(Y[1],Y[2],Y[3]);
v = P*Vec3D(Y[4],Y[5],Y[6]);
Elements (GM_Sun, r, v, a, e, i, Omega, omega, M);

// Print elements
cout << endl << endl
     << " Orbital elements at start epoch:" << endl
     << endl;
WriteElm ( Mjd, a,e,i,Omega,omega,M, T_eqx );

// Ephemeris header
cout << endl << endl
     << "    Date     ET    Sun      l       b     r"
     << "       RA            Dec       Distance " << endl
     << setw(43) << " " << "   h  m  s      o  '  \"    (AU) " << endl;

// Compute ephemeris
n_line = 0;
MjdStep = MjdStart;

// Time loop
while ( MjdStep < MjdEnd + Step/2 ) {

  // Integrate orbit to MjdStep
  do {
    Orbit.Integ(Y, Mjd, MjdStep, relerr, abserr, State);
    if (State==DE_INVALID_PARAMS) {
      cerr << "Exit (invalid parameters)" << endl;
      exit(1);
    }
  }
  while ( State > DE_DONE );
  // Heliocentric ecliptic coordinates, equinox T_eqx
  r_helioc = P*Vec3D(Y[1],Y[2],Y[3]);
  v_helioc = P*Vec3D(Y[4],Y[5],Y[6]);
  // Geocentric ecliptic coordinates of the Sun, equinox T_eqx
  T = ( Mjd - MJD_J2000 ) / 36525.0;
  R_Sun = PrecMatrix_Ecl(T,T_eqx) * SunPos(T);
  // Geometric geocentric coordinates
  r_geoc = r_helioc + R_Sun;
  // First-order light-time correction
  dist = Norm(r_geoc);
  fac  = 0.00578*dist;
  r_geoc = r_geoc - fac*v_helioc;
  // Equatorial coordinates
  r_equ = Ecl2EquMatrix(T_eqx) * r_geoc;
  // Output
  cout << DateTime(Mjd,HHh)
       << fixed << setprecision(1)
       << setw(7) << Deg*R_Sun[phi]
       << setw(7) << Deg*r_helioc[phi]
       << setw(6) << Deg*r_helioc[theta]
```

```
        << setprecision(3) << setw(7) << r_helioc[R]
        << setprecision(1) << setw(12)<< Angle(Deg*r_equ[phi]/15.0,DMMSSs)
        << "  " << showpos << setw(9) << Angle(Deg*r_equ[theta],DMMSS)
        << noshowpos << setprecision(6) << setw(11) << dist
        << endl;
    ++n_line;
    if ( (n_line % 5) ==0 ) cout << endl; // insert line feed every 5 lines
    MjdStep += Step;                      // Next time step

  } // End time loop

  if (OutFile.is_open()) OutFile.close();

}
```

As an example of how this program may be applied we will calculate an ephemeris for the minor planet 1, Ceres around the time of opposition in 1992, from osculating elements for the year 1983 (cf. Table 5.3). The first line of the file `Numint.dat` contains the epoch of the elements, followed by the six orbital elements a, e, i, Ω, ω, M in sequence, and finally the equinox of the elements.

```
1983 09 23.0     ! Epoch (year month day.fraction) Minor planet 1 Ceres
   2.7657991     ! Semi-major axis a in AU
   0.0785650     ! Eccentricity e
  10.60646       ! Inclination i (deg)
  80.05225       ! Longitude of the ascending node (deg)
  73.07274       ! Argument of perihelion (deg)
 174.19016       ! Mean anomaly (deg)
1950.0           ! Equinox of the orbital elements
```

Corresponding data for other numbered minor planets are published yearly in the *Ephemerides of Minor Planets* (EMP) by the Institute for Theoretical Astronomy in St. Petersburg (formerly Leningrad), and in the Smithsonian Astrophysical Observatory's *Minor Planet Circulars*, as well as in various other yearbooks.

Table 5.3. Osculating orbital elements for minor planet Ceres

Element	1983 Sep. 23.0 ET	1992 Jun. 27.0 ET
a	2.7657991 AU	2.7674385 AE
e	0.0785650	0.0765551
i (1950)	10°.60646	10°.59950
(2000)	10°.60695	10°.59999
Ω (1950)	80°.05225	80°.01288
(2000)	80°.71588	80°.67649
ω (1950)	73°.07274	71°.07929
(2000)	73°.10812	71°.11469
M	174°.19016	141°.46349

The method of entering the required time-span and the step-size for the ephemeris and the equinox is similar to that in the `Comet` program. As usual, the user's input is indicated in italics.

```
 NUMINT: numerical integration of perturbed minor planet orbits
         (c) 1999 Oliver Montenbruck, Thomas Pfleger

Using default input file Numint.dat
Orbital elements from file Numint.dat:

 Epoch (y m d)               1983/09/23.00
 Semi-major axis (a)            2.7657991 AU
 Eccentricity (e)               0.0785650
 Inclination (i)               10.60646 deg
 Long. of ascending node       80.05225 deg
 Argument of perihelion        73.07274 deg
 Mean anomaly (M)             174.19016 deg
 Equinox                     1950.00
Begin and end of the ephemeris:
 First date (yyyy mm dd hh.hhh)            ... 1992 06 27 00.0
 Final date (yyyy mm dd hh.hhh)            ... 1992 07 25 00.0
 Step size (dd hh.hh)                      ... 2 00.0

Desired equinox of the ephemeris (yyyy.y)  ... 2000.0

Orbital elements at start epoch:

 Epoch (y m d)               1992/06/27.00
 Semi-major axis (a)            2.7674372 AU
 Eccentricity (e)               0.0765610
 Inclination (i)               10.59999 deg
 Long. of ascending node       80.67653 deg
 Argument of perihelion        71.11641 deg
 Mean anomaly (M)             141.46191 deg
 Equinox                     2000.00

  Date      ET     Sun      l      b      r        RA           Dec      Distance
                                                 h  m  s       o  '  "     (AU)
1992/06/27 00.0   95.7  297.9  -6.5  2.939   20 55 54.5   -27 00 46   2.042458
1992/06/29 00.0   97.6  298.2  -6.5  2.940   20 54 53.4   -27 13 47   2.029192
1992/07/01 00.0   99.5  298.6  -6.6  2.941   20 53 46.7   -27 26 56   2.016828
1992/07/03 00.0  101.4  299.0  -6.6  2.942   20 52 34.7   -27 40 13   2.005397
1992/07/05 00.0  103.3  299.4  -6.7  2.942   20 51 17.5   -27 53 32   1.994924

1992/07/07 00.0  105.2  299.7  -6.7  2.943   20 49 55.6   -28 06 53   1.985432
1992/07/09 00.0  107.1  300.1  -6.8  2.944   20 48 29.0   -28 20 11   1.976941
1992/07/11 00.0  109.0  300.5  -6.8  2.945   20 46 58.3   -28 33 24   1.969468
1992/07/13 00.0  111.0  300.9  -6.9  2.946   20 45 23.6   -28 46 28   1.963030
1992/07/15 00.0  112.9  301.2  -6.9  2.946   20 43 45.4   -28 59 22   1.957641

1992/07/17 00.0  114.8  301.6  -7.0  2.947   20 42 03.9   -29 12 03   1.953316
1992/07/19 00.0  116.7  302.0  -7.0  2.948   20 40 19.7   -29 24 26   1.950066
1992/07/21 00.0  118.6  302.4  -7.1  2.949   20 38 33.1   -29 36 31   1.947903
1992/07/23 00.0  120.5  302.7  -7.1  2.950   20 36 44.6   -29 48 14   1.946836
1992/07/25 00.0  122.4  303.1  -7.2  2.950   20 34 54.5   -29 59 32   1.946871
```

The above example assumes that accurate planetary positions as provided by e.g. the `PertPosition` function are used in the computation of planetary perturbations within the `Accel` function. Following the data input, a certain time is required for the orbit to be integrated over nine years, before the osculating elements for 1992 Jun.27 are displayed. Subsequent calculation of the ephemeris over a period of one month takes merely a fraction of a second.

Comparison with a yearbook shows that the achieved accuracy of the ephemeris is about 1″, and is therefore of the same degree of accuracy as the solar coordinates used in converting between heliocentric and geocentric coordinates.

If the calculation is carried out without the perturbations caused by the planets, such as would be possible, for example, by commenting out the appropriate lines in the `Accel` function, then the following ephemeris is obtained:

```
    Date      ET     Sun      l       b      r        RA           Dec       Distance
                                                    h  m  s      o  '  "      (AU)
1992/06/27 00.0    95.7  298.6  -6.6  2.938  21 00 35.9  -26 49 55  2.049522
1992/06/29 00.0    97.6  299.0  -6.6  2.939  20 59 37.9  -27 02 56  2.035891
...
1992/07/23 00.0   120.5  303.5  -7.2  2.950  20 41 55.4  -29 39 16  1.948277
1992/07/25 00.0   122.4  303.9  -7.3  2.951  20 40 06.3  -29 50 51  1.947831
```

It will be seen that the perturbations mainly affect the heliocentric longitude of the minor planet, and amount to 0°.7–0°.8, while the latitude shows smaller, but still significant differences amounting to about 0°.1. Because of the smaller distance to the Earth, the right ascension of the unperturbed ephemeris differs from the actual orbit by about five minutes of time or 1°.3 angular degrees.

5.6 The Asteroid Orbital Elements Database

To date, far more than 100 000 minor planets have been discovered, whose orbits are regularly improved from new observations. The responsibility for designating and cataloguing them rests with the Minor Planet Center at the Smithsonian Astrophysical Observatory in Massachusetts. In addition, the Lowell Observatory maintains

Table 5.4. Structure of the Asteroid Orbital Elements Database (Summary)

Columns	Description
1– 5	Number of the minor planet
7– 24	Name or designation
42– 46	Brightness [mag]
106–113	Epoch of the orbital elements (yyyymmdd)
115–124	Mean anomaly [°]
126–135	Argument of latitude (J2000) [°]
137–146	Longitude of the ascending node (J2000) [°]
148–156	Orbital inclination (J2000) [°]
158–167	Eccentricity [°]
169–180	Semi-major axis [AU]

a freely accessible catalogue of orbital elements and other information, which is currently prepared by Ted Bowell, and continuously updated. With the full permission of the originators, this catalogue is included on the accompanying CD-ROM, together with a full description[1]. The current version of the file `astorb.dat` (as of Oct. 2003) contains around 230 000 records, each with a line length of 266 characters, and may be read and modified directly with an appropriate text editor. The size of the data file is approximately 60 MB. The meaning of the most important entries is summarized in Table 5.4.

To facilitate use of the database, individual asteroids may be selected, using the `AOEcat` program, according to their number or designation:

```
AOEcat: Selection of Data from the Asteroid Orbital Elements Database
          (c) 1999 Oliver Montenbruck, Thomas Pfleger

Name/number of the minor planet: Lutetia

2002 01 26.0   ! Epoch        Minor planet(21) Lutetia
  2.43621894   ! a [AU]
  0.16290770   ! e
    3.064760   ! i [deg]
   80.937852   ! Omega [deg]
  250.338617   ! omega [deg]
  149.947257   ! M [deg]
      2000.0   ! Equinox
```

The output format is so chosen that the orbital elements may be directly further processed by the `Numint` program. For this reason, on calling `AOEcat`, the name of an output file may also be specified (see Sect. A.1.3). In entering a minor planet's number, it should be noted that there should be no spaces before or after the number. For names (e.g. `Tycho Brahe`) and other catalogue designations (e.g. `5073 T-3`), on the other hand, spaces are obligatory. Letters are accepted in either upper or lower case. The full source code of `AOEcat` is, together with all the other programs, included on the book's CD-ROM.

[1]Current versions of the AOE Database may be obtained, as required, over the Internet, from the address `ftp://ftp.lowell.edu/pub/elgb/astorb.dat` or `.../astorb.gz+`.

6. Planetary Orbits

Excluding Mercury and Pluto, the major planets are very similar in the low eccentricities and low inclinations of their orbits relative to the ecliptic. Cometary orbits, by contrast, are distributed at random in space and have very varied shapes. The near-circular orbits of the planets are so widely separated that they never come close to one another.

These properties of planetary orbits have two important consequences for the calculation of ephemerides. The first concerns the way of handling the non-perturbed Kepler orbit, which was described in Chap. 4 for the various general forms of orbit. Because of the low eccentricity, the true anomaly shows only slight periodic departures from the mean anomaly. It may therefore be determined quicker by using a series expansion than by solving Kepler's equation. The same applies to the distance from the Sun, which in general differs only by a range of about one percent from the value of the semi-major axis. Furthermore, the orbit's departure from the ecliptic may be described by a brief approximation, thus doing away with the complication of calculating Gaussian vectors. These approximations will be described in greater detail in the next section.

The second point concerns the alterations to planetary orbits as a result of their mutual gravitational forces. Because close approaches of the planets do not occur, these forces are always several orders of magnitude smaller than the Sun's gravitational attraction. Major changes in orbits, like those that occur when comets approach Jupiter or Saturn, are not found among planetary motions. Their orbits may therefore be described by average Kepler ellipses, on which are superimposed small periodic perturbations.

In general, planetary ephemerides are nowadays calculated by using numerical integration methods, the results being stored in an appropriate form on magnetic media. Thanks to the introduction of modern computers, this method is remarkable for its simplicity and accuracy. There is a corresponding disadvantage, however, in that an analytical description of the planetary motions is no longer available, as was the case with the purely two-body problem. One of the greatest achievements of celestial mechanics was, and remains, the derivation of series expansions from which the deviations of planetary orbits from unperturbed motion may be calculated for any given time. As may be seen from the length of these series, their derivation is an exceptionally laborious and long-winded process, which is far beyond the scope of this book. For our purposes – the accurate calculation of planetary positions – however, it suffices to have an understanding of the basic structure of the series expansions and how they may be calculated numerically. Anyone taking a greater interest in the development of perturbation series should refer to the various textbooks on celestial mechanics.

Among the best-known analytical descriptions of planetary motions are the tables of the motions of the inner planets from Mercury to Mars, prepared by Simon Newcomb at the U.S. Naval Observatory. Although they were produced around the end of the 19th century, until 1984 they still formed the basis of the corresponding ephemerides in the *Astronomical Almanac*. There are similar tables by G.W. Hill for Jupiter and Saturn, and by Newcomb for Uranus and Neptune. These were soon replaced by the first numerically integrated ephemerides, however, because they did not attain the same accuracy over long periods of time as the tables for the inner planets.

The works mentioned, together with the series expansion for the orbit of Pluto by E. Goffin, J. Meeus and C. Steyart, form the basis for this chapter. The various formulae for the perturbation terms sometimes differ so greatly for individual planets, however, that it is not possible to convert them directly into a program in a simple and uniform manner. Because of this, all the series were first processed by a formula-manipulation program. This enabled the Keplerian motion and the perturbation terms to be expressed independently of their original description as periodic motion in longitude, latitude, and radius. In addition, a large number of terms were excluded, which, because of their low values (generally $\ll 1''$), have no influence on the observed accuracy of the series. This reorganisation allows comparatively short and compact programming, without reducing the accuracy of the expansions to any significant degree.

6.1 Series Expansion of the Kepler Problem

The orbit of a planet in an unperturbed Kepler ellipse is a strictly periodic function of the mean anomaly M. If the mean anomaly alters by 360°, the planet has completed one orbit and reached exactly the same point. Any function of the orbital position may therefore be expressed as a Fourier series, i.e., as the sum of the sine and cosine functions for multiple arguments of M. Examples of this are the series expansions

$$\nu - M = a_1 \sin(M) + a_2 \sin(2M) + a_3 \sin(3M) + \ldots$$

for the difference between true and mean anomaly and that for the distance r in units of the semi-major axis a:

$$r/a = b_0 + b_1 \cos(M) + b_2 \cos(2M) + \ldots \quad .$$

The coefficients a_i and b_i in these series are functions of the orbital eccentricity e. The most important terms for $\nu - M$ (known as the *equation of the centre*) and r/a up to powers of e^2 are:

$$\nu - M = 2e \sin(M) + \frac{5}{4}e^2 \sin(2M) + \ldots \quad \text{(radians)} \tag{6.1}$$

and

$$r/a = \left(1 + \frac{1}{2}e^2\right) - e\cos(M) - \frac{1}{2}e^2\cos(2M) + \ldots \quad . \tag{6.2}$$

For small eccentricities both series converge very quickly, because from term to term the individual components become smaller by a factor that is of the order of e. For the Earth's orbit, where $e = 0.0167$, we have

$$\begin{aligned}\nu &= M + 0.0334\sin(M) + 0.00035\sin(2M) + \ldots \quad \text{(radians)}\\ &= M + 1^\circ\!.916\sin(M) + 0^\circ\!.020\sin(2M) + \ldots\\ r &= a\cdot(1.00014 - 0.0167\cos(M) - 0.00014\cos(2M) + \ldots) \quad .\end{aligned}$$

The error in these shortened formulae amounts to about 1 arc-second only or 10^{-5} AU, and may be reduced even further if one includes additional terms.

Because the full description of these formulae requires a rather deeper understanding of the mathematics involved, we will derive the first term only of the equation of the centre. We begin by taking Kepler's equation (4.7):

$$E - e\sin E = M \quad .$$

As E and M differ by a term of order e only, we may substitute $e\sin M$ for $e\sin E$ to a close approximation, and obtain

$$E - M \approx e\sin M$$

for the difference between the eccentric and the true anomaly. In addition, from the equation of a conic section $r = a(1 - e^2)/(1 + e\cos\nu)$, where $r\cos\nu = a(\cos E - e)$, it follows that

$$\cos\nu \approx \cos E - e(1 - \cos\nu\cos E) \quad .$$

In the second term on the right-hand side we can now again substitute $\cos E$ for $\cos\nu$, obtaining

$$\cos\nu \approx \cos E - \{e\sin E\}\sin E \quad .$$

If we compare this with the Taylor expansion of the cosine function,

$$\cos(x + \Delta x) \approx \cos x - \Delta x\sin x \quad ,$$

then we can see that for small eccentricities, $\nu - E$ is approximately equal to $e\sin E$. Therefore, as indicated, the first component of the equation of the centre is

$$\nu - M = (\nu - E) + (E - M) \approx e\sin E + e\sin M \approx 2e\sin M \quad .$$

For small orbital inclinations i, the orbits are always close to the ecliptic. A planet's ecliptic longitude l therefore differs from the sum $\varpi + \nu$ of the perihelion longitude and the true anomaly only by a small correction R (known as the *reduction to the ecliptic*). R disappears for $i = 0$, and for small inclinations has, to a first approximation, a value of

$$R = l - (\nu + \varpi) \approx -\tan^2\left(\frac{i}{2}\right) \cdot \sin(2u) \quad \text{(radians)} \tag{6.3}$$

where $u = \nu + \omega = \nu + \varpi - \Omega$. As $\nu \approx M$ ($u \approx M + \omega$), R may also be expressed in the form

$$R \approx -\tan^2\left(\frac{i}{2}\right) \cdot \{\sin(2\omega)\cos(2M) + \cos(2\omega)\sin(2M)\} \quad .$$

The value of R in degrees may be obtained by multiplying by $180°/\pi$.

Finally, the ecliptic latitude b of the planets may also be handled in a similar manner. Taking equation (4.20)

$$z = r \cdot (\sin(\nu + \omega)\sin(i)) \tag{6.4}$$

for the z-coordinate as a function of the orbital elements and

$$\sin(b) = z/r \tag{6.5}$$

for small orbital inclinations i and ecliptic latitudes b, we have the approximation

$$b \approx i\sin(\nu + \omega) = \{i\sin\omega\}\cos(\nu) + \{i\cos\omega\}\sin(\nu) \quad .$$

If we replace ν by $\nu \approx M + 2e\sin M$, then the Taylor expansion gives

$$\begin{aligned} \cos\nu &\approx \cos M - (2e\sin M)\sin M = \cos(M) - e(1 - \cos(2M)) \\ \sin\nu &\approx \sin M + (2e\sin M)\cos M = \sin(M) + e\sin(2M) \end{aligned}$$

and thus

$$\begin{aligned} b \approx\ & \{i\sin\omega\} \cdot \{\cos(M) - e(1 - \cos(2M))\} + \\ & \{i\cos\omega\} \cdot \{\sin(M) + e\sin(2M)\} \quad . \end{aligned}$$

The most important components in a description of unperturbed planetary orbits using trigonometric series therefore are:

$$\begin{aligned} l =\ & \varpi + M \\ & + \frac{180°}{\pi} \cdot \{2e\} \cdot \sin(M) \\ & + \frac{180°}{\pi} \cdot \left\{\frac{5}{4}e^2 - \tan^2\left(\frac{i}{2}\right)\cos(2\omega)\right\} \cdot \sin(2M) \\ & + \frac{180°}{\pi} \cdot \left\{-\tan^2\left(\frac{i}{2}\right)\sin(2\omega)\right\} \cdot \cos(2M) \end{aligned} \tag{6.6}$$

$$\begin{aligned} b =\ & -\{ie\sin\omega\} \\ & + \{i\sin\omega\}\cos(M) + \{i\cos\omega\}\sin(M) \\ & + \{ie\sin\omega\}\cos(2M) + \{ie\cos\omega\}\sin(2M) \end{aligned} \tag{6.7}$$

$$r = \{a(1 + e^2/2)\} - \{ae\} \cdot \cos(M) - \{ae^2/2\} \cdot \cos(2M) \quad . \tag{6.8}$$

If specific values for the orbital elements are entered in these equations, then we obtain very compact formulae for ecliptic coordinates (l, b, r) as functions of the mean anomaly M. Taking Jupiter as an example, where

$$\begin{array}{lllllllll} a & = & 5.2\text{AU} & \Omega & = & 100^\circ\!.0 & \omega & = & 274^\circ\!.0 \\ e & = & 0.048 & i & = & 1^\circ\!.31 & \varpi & = & 14^\circ\!.0 \end{array}$$

we obtain the following series:

$$\begin{aligned} l &= M + 14^\circ\!.0 + 19800'' \sin(M) + 620'' \sin(2M) + 4'' \cos(2M) \\ b &= +226'' - 4700'' \cos(M) + 329'' \sin(M) - 226'' \cos(2M) + 16'' \sin(2M) \\ r &= (5.206 - 0.250 \cos(M) - 0.006 \cos(2M)) \text{ AU} \quad . \end{aligned}$$

6.2 Perturbation Terms

The mutual perturbations of the planets, which are responsible for the departures from a keplerian orbit, result in additional terms in the series expansions, which contain both the anomaly of the perturbed planet and that of the perturbing planet. In the following, M_n indicates the mean anomaly of the n-th planet, M_5 being that of Jupiter, for example, and M_6 that of Saturn:

$$\begin{array}{lcll} M_1 & = & 0^r\!.4855407 + 415^r\!.2014314 \cdot T & \text{(Mercury)} \\ M_2 & = & 0^r\!.1400197 + 162^r\!.5494552 \cdot T & \text{(Venus)} \\ M_3 & = & 0^r\!.9931266 + 99^r\!.9973604 \cdot T & \text{(Sun, Earth)} \\ M_4 & = & 0^r\!.0538553 + 53^r\!.1662736 \cdot T & \text{(Mars)} \\ M_5 & = & 0^r\!.0565314 + 8^r\!.4302963 \cdot T & \text{(Jupiter)} \\ M_6 & = & 0^r\!.8829867 + 3^r\!.3947688 \cdot T & \text{(Saturn)} \\ M_7 & = & 0^r\!.3967117 + 1^r\!.1902849 \cdot T & \text{(Uranus)} \\ M_8 & = & 0^r\!.7214906 + 0^r\!.6068526 \cdot T & \text{(Neptune)} \\ M_9 & = & 0^r\!.0385795 + 0^r\!.4026667 \cdot T & \text{(Pluto)} \quad . \end{array} \qquad (6.9)$$

The values are given here in multiples of an orbit (1^r)[1]. If they are multiplied by 360° or 2π, then we obtain the mean anomalies in the customary degrees or radians. Time is reckoned, as usual, in Julian centuries since epoch J2000:

$$T = (\text{JD} - 2451545)/36525 \quad .$$

The various perturbation terms may best be represented by a table, such as Table 6.1, which applies to Jupiter. This shows the complete periodic variations in ecliptic longitude (dl) and latitude (db), as well as distance from the Sun (dr). Terms with equal arguments in the trigonometric series are brought together in a single line,

[1] In some cases the values used in the programs differ slightly from the values given here. This is not an error, but a consequence of the different theories that form the basis of the series expansions.

Table 6.1. Periodic Terms in the Orbit of Jupiter

			$dl['']$		$dr[10^{-5}$ AU]		$db['']$		
i_5		i_T	cos	sin	cos	sin	cos	sin	Remarks
1		0	-113.1	19998.6	-25208.2	-142.2	-4670.7	288.9	Keplerian terms
1		1	-76.1	66.9	-84.2	-95.8	21.6	29.4	
1		2	-0.5	-0.3	0.4	-0.7	0.1	-0.1	
2		0	-3.4	632.0	-610.6	-6.5	-226.8	12.7	
2		1	-4.2	3.8	-4.1	-4.5	0.2	0.6	
3		0	-0.1	28.0	-22.1	-0.2	-12.5	0.7	
4		0	0.0	1.4	-1.0	0.0	-0.6	0.0	
i_5	i_6	i_T							
-1	-1	0	-0.2	1.4	2.0	0.6	0.1	-0.2	Saturn
0	-1	0	9.4	8.9	3.9	-8.3	-0.4	-1.4	
0	-2	0	5.6	-3.0	-5.4	-5.7	-2.0	0.0	
0	-3	0	-4.0	-0.1	0.0	5.5	0.0	0.0	
0	-5	0	3.3	-1.6	-1.6	-3.1	-0.5	-1.2	
1	-1	0	78.8	-14.5	11.5	64.4	-0.2	0.2	
1	-2	0	-2.0	-132.4	28.8	4.3	-1.7	0.4	
1	-2	1	-1.1	-0.7	0.2	-0.3	0.0	0.0	
1	-3	0	-7.5	-6.8	-0.4	-1.1	0.6	-0.9	
1	-4	0	0.7	0.7	0.6	-1.1	0.0	-0.2	
1	-5	0	51.5	-26.0	-32.5	-64.4	-4.9	-12.4	
1	-5	1	-1.2	-2.2	-2.7	1.5	-0.4	0.3	
2	-1	0	5.3	-0.7	0.7	6.1	0.2	1.1	
2	-2	0	-76.4	-185.1	260.2	-108.0	1.6	0.0	
2	-3	0	66.7	47.8	-51.4	69.8	0.9	0.3	
2	-3	1	0.6	-1.0	1.0	0.6	0.0	0.0	
2	-4	0	17.0	1.4	-1.8	9.6	0.0	-0.1	
2	-5	0	1066.2	-518.3	-1.3	-23.9	1.8	-0.3	
2	-5	1	-25.4	-40.3	-0.9	0.3	0.0	0.0	
2	-5	2	-0.7	0.5	0.0	0.0	0.0	0.0	
3	-2	0	-5.0	-11.5	11.7	-5.4	2.1	-1.0	
3	-3	0	16.9	-6.4	13.4	26.9	-0.5	0.8	
3	-4	0	7.2	-13.3	20.9	10.5	0.1	-0.1	
3	-5	0	68.5	134.3	-166.9	86.5	7.1	15.2	*
3	-5	1	3.5	-2.7	3.4	4.3	0.5	-0.4	
3	-6	0	0.6	1.0	-0.9	0.5	0.0	0.0	
3	-7	0	-1.1	1.7	-0.4	-0.2	0.0	0.0	
4	-2	0	-0.3	-0.7	0.4	-0.2	0.2	-0.1	
4	-3	0	1.1	-0.6	0.9	1.2	0.1	0.2	
4	-4	0	3.2	1.7	-4.1	5.8	0.2	0.1	
4	-5	0	6.7	8.7	-9.3	8.7	-1.1	1.6	
4	-6	0	1.5	-0.3	0.6	2.4	0.0	0.0	
4	-7	0	-1.9	2.3	-3.2	-2.7	0.0	-0.1	
4	-8	0	0.4	-1.8	1.9	0.5	0.0	0.0	
4	-9	0	-0.2	-0.5	0.3	-0.1	0.0	0.0	
4	-10	0	-8.6	-6.8	-0.4	0.1	0.0	0.0	
4	-10	1	-0.5	0.6	0.0	0.0	0.0	0.0	
5	-5	0	-0.1	1.5	-2.5	-0.8	-0.1	0.1	
5	-6	0	0.1	0.8	-1.6	0.1	0.0	0.0	
5	-9	0	-0.5	-0.1	0.1	-0.8	0.0	0.0	
5	-10	0	2.5	-2.2	2.8	3.1	0.1	-0.2	
i_5	i_7	i_T							
1	-1	0	0.4	0.9	0.0	0.0	0.0	0.0	Uranus
1	-2	0	0.4	0.4	-0.4	0.3	0.0	0.0	
i_5	i_6	i_7							
2	-6	3	-0.8	8.5	-0.1	0.0	0.0	0.0	Saturn & Uranus
3	-6	3	0.4	0.5	-0.7	0.5	-0.1	0.0	

which is of assistance in later calculations. Expressed in full, and taking the line marked with an asterisk (∗) as an example, the values

$$\begin{aligned} dl &= 68''\!.5\cdot\cos(3M_5 - 5M_6) + 134''\!.3\cdot\sin(3M_5 - 5M_6) \\ dr &= (-166.9\cdot\cos(3M_5 - 5M_6) + 86.5\cdot\sin(3M_5 - 5M_6))\cdot 10^{-5}\mathrm{AU} \\ db &= 7''\!.1\cdot\cos(3M_5 - 5M_6) + 15''\!.2\cdot\sin(3M_5 - 5M_6) \end{aligned}$$

are to be added to the longitude, latitude and radius. Terms that are additionally to be multiplied by T or T^2, are indicated by 1 or 2 in the column headed i_T.

The influence of other planets is not confined to periodic perturbations but is also detectable in long-term (*secular*) variations in the mean orbital elements. This means that now components dependent on time T appear even in terms that actually describe purely Keplerian motion. In the `JupPosition` procedure the following equations are used to calculate Jupiter's coordinates:

$$\begin{aligned} l &= M_5 + 14^\circ\!.00076 + (5025''\!.2 + 0''\!.8T)T + dl \\ b &= 227''\!.3 - 0''\!.3T + db \\ r &= (5.208873 + 0.000041T)\mathrm{AU} + dr \quad . \end{aligned}$$

where dl, db and dr are the complete periodic contributions to the orbit, and therefore include the Kepler terms and perturbations by Saturn and Uranus. The values of ecliptic longitude and latitude determined in this manner are referred to the *mean equinox of date*. The formula for l therefore contains the conspicuously large secular term $+5025''\!.2\cdot T$, which reflects the motion of the vernal equinox.

Reasons of space forbid us from giving all the perturbation terms for every planet. Because the individual series expansions have a uniform structure, however, the details of Jupiter's orbit should suffice for readers to recognize the formulae and terms employed in the programs.

There are, nevertheless, differences in the way in which the Earth's orbit is handled from the treatment of the orbits of the other planets. The Earth orbits the Sun in company with the Moon, and one can imagine the two bodies being combined at their common centre of gravity (the barycentre). The motion of this centre of gravity (which moreover defines the mean ecliptic), may be envisaged as a perturbed Kepler orbit, like those of the other planets, and expressed as an appropriate series expansion. Because of the relative masses of the Earth and the Moon, the distance between the centre of the Earth and the barycentre is 1/81 of the distance of the Moon from the Earth, and therefore amounts to about 3/4 Earth radii (see Fig. 6.1). Seen from the Sun, the ecliptic longitude of the Earth therefore varies

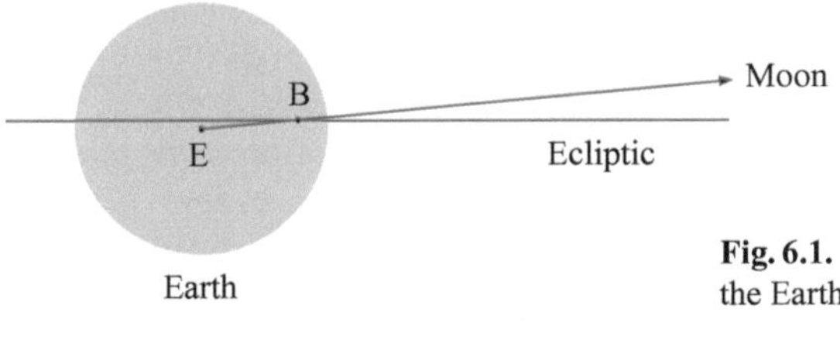

Fig. 6.1. The position of the barycentre (B) of the Earth (E) and the Moon

over the course of a month by as much as 7 arc-seconds from its mean position. At the same time the Earth's ecliptic latitude varies by 0.5 arc-seconds.

6.3 Numerical Treatment of the Series Expansions

To obtain accuracies that are of the order of seconds of arc, several hundred perturbation terms are required for each planet. If these were all specifically programmed, then not only would the code for the program be very long and unintelligible, but we would also have to accept unnecessarily long calculation times. This is because the evaluation of trigonometrical functions is relatively complex and slow when compared with simple computational processes such as addition and multiplication. Because of the large number of angular functions in the series expansions, this overhead becomes quite obvious even with fast computers.

We can, however, avoid a major part of this work, if we use the sine and cosine addition theorems. They enable angular functions to be calculated as the sum or difference of angles, the functions of the individual angles being known:

$$\begin{aligned} \cos(\alpha_1+\alpha_2) &= \cos\alpha_1\cos\alpha_2 - \sin\alpha_1\sin\alpha_2 \\ \sin(\alpha_1+\alpha_2) &= \sin\alpha_1\cos\alpha_2 + \cos\alpha_1\sin\alpha_2 \quad . \end{aligned} \tag{6.10}$$

Using this method, we can very conveniently compute and store $\cos(iM)$ and $\sin(iM)$ for multiples $i = 2, 3, \ldots$ of the mean anomaly M, from $\cos M$ and $\sin M$. To this end we first implement the function `AddThe`:

```
//------------------------------------------------------------------------
// AddThe: Calculates cos(alpha+beta) and sin(alpha+beta) using addition
//         theorems
//   c1,s1      cos(alpha), sin(alpha)
//   c2,s2      cos(beta), sin(beta)
//   c,s        cos(alpha+beta), sin(alpha+beta)
//------------------------------------------------------------------------
void AddThe ( double c1, double s1, double c2, double s2,
              double& c, double& s ) {
  c = c1 * c2 - s1 * s2;
  s = s1 * c2 + c1 * s2;
}
```

All further operations as well as the necessary auxiliary arrays for the storage of the previously calculated values have been included in a separate class `Pert`. Most importantly, it provides the two basic functions `Init` and `Term`. If we define M as the mean anomaly of the perturbed planet, and m the mean anomaly of the perturbing planet, then `Init` calculates the values $\sin(IM+\phi)$ and $\cos(IM+\phi)$ for multiples $I_{\min} \le I \le I_{\max}$, as well as $\sin(im)$ and $\cos(im)$ for multiples $i_{\min} \le i \le i_{\max}$. In this, the angle ϕ is an auxiliary value, which enables perturbation terms that are dependent on the mean anomaly of a third planet to be handled in a uniform manner. In evaluating an individual perturbation term, the `Term` method takes the previously calculated trigonometric functions and sums the contributions to longitude, radius, and latitude in corresponding class members. The accumulated perturbation terms may then be retrieved using the methods dl, dr and db.

```
// Constants
const int o   = 16;          // Index offset
const int dim = 2*o+1;       // Work array dimension

// Definition of Pert class for summing up trigonometric perturbation
// series
class Pert {
  public:
    // Set time, mean anomalies and index range
    void Init ( double T,
                double M, int I_min, int I_max,
                double m, int i_min, int i_max,
                double phi=0.0 );
    // Sum-up perturbations in longitude, radius and latitude
    void Term ( int I, int i, int iT,
                double dlc, double dls,
                double drc, double drs,
                double dbc, double dbs );
    // Retrieve perturbations in longitude, radius and latitude
    double dl();
    double dr();
    double db();
  private:
    double m_T;
    double m_cosM, m_sinM;
    double m_C[dim], m_S[dim], m_c[dim], m_s[dim];
    double m_dl, m_db, m_dr;
    double m_u, m_v;
};

// Set time, mean anomalies and index range
void Pert::Init ( double T,
                  double M, int I_min, int I_max,
                  double m, int i_min, int i_max,
                  double phi )
{
  int i;
  m_dl=0.0; m_dr=0.0; m_db=0.0; // reset perturbations
  m_T=T;                        // set time
  // cosine and sine of multiples of M
  m_cosM=cos(M); m_sinM=sin(M);
  m_C[o]=cos(phi);
  m_S[o]=sin(phi);
    for (i=0; i<I_max; i++)
    AddThe ( m_C[o+i],m_S[o+i], +m_cosM,+m_sinM, m_C[o+i+1],m_S[o+i+1] );
  for (i=0; i>I_min; i--)
    AddThe ( m_C[o+i],m_S[o+i], +m_cosM,-m_sinM, m_C[o+i-1],m_S[o+i-1] );
  // cosine and sine of multiples of m
  m_c[o]=1.0; m_c[o+1]=cos(m); m_c[o-1]=+m_c[o+1];
  m_s[o]=0.0; m_s[o+1]=sin(m); m_s[o-1]=-m_s[o+1];
```

```
    for (i=1; i<i_max; i++)
      AddThe( m_c[o+i],m_s[o+i], m_c[o+1],m_s[o+1], m_c[o+i+1],m_s[o+i+1]);
    for (i=-1; i>i_min; i--)
      AddThe( m_c[o+i],m_s[o+i], m_c[o-1],m_s[o-1], m_c[o+i-1],m_s[o+i-1]);
  }

  // Sum-up perturbations in longitude, radius and latitude
  void Pert::Term ( int I, int i, int iT,
                    double dlc, double dls,
                    double drc, double drs,
                    double dbc, double dbs )
  {
    if (iT == 0)
      AddThe ( m_C[o+I],m_S[o+I], m_c[o+i],m_s[o+i], m_u,m_v );
    else
      { m_u *= m_T; m_v *= m_T; };
    m_dl += ( dlc*m_u + dls*m_v );
    m_dr += ( drc*m_u + drs*m_v );
    m_db += ( dbc*m_u + dbs*m_v );
  }

  // Retrieve perturbations in longitude, radius and latitude
  double Pert::dl() { return m_dl; }
  double Pert::dr() { return m_dr; }
  double Pert::db() { return m_db; }
```

The following example of the JupiterPos function shows how all the perturbation terms may be evaluated and summed in this way. The Term method calculates the values of $u = \cos(i_5 M_5 + i_j M_j)$ and $v = \sin(i_5 M_5 + i_j M_j)$ with the indices i_5 and i_j, multiplies these with the corresponding coefficients and finally adds them to dl, dr and db. If we make the assumption that all terms with equal i_5 and i_j, but different $i_T = 0, 1, \ldots$, are evaluated one after the other, it suffices – as is done in Term – to calculate u and v just for $i_T = 0$, and simply multiply by T each time to obtain $i_T = 1$ or $i_T = 2$. Although at first sight this may suggest a greater amount of labour, it soon pays off when – as in this case – one is working with a large number of terms.

```
//------------------------------------------------------------------------------
// JupiterPos: Position of Jupiter from analytical series
//   T         Time in Julian centuries since J2000
//   <return>  Heliocentric position [AU], true-of-date ecliptic & equinox
//------------------------------------------------------------------------------
Vec3D JupiterPos (double T)
{
  // Variables
  double  M5,M6,M7;                // Mean anomalies
  Pert    Sat,Ura;                 // Perturbations
  double  phi,c,s;
  double  dl, dr, db;              // Corrections in longitude ["],
                                   // radius [AU] and latitude ["]
  double  l,b,r;                   // Ecliptic coordinates
```

```
// Mean anomalies of planets in [rad]
M5 = pi2 * Frac ( 0.0565314 + 8.4302963*T );
M6 = pi2 * Frac ( 0.8829867 + 3.3947688*T );
M7 = pi2 * Frac ( 0.3969537 + 1.1902586*T );

// Keplerian motion and perturbations by Saturn
Sat.Init ( T, M5,-1,5, M6,-10,0 );
Sat.Term (-1, -1,0,  -0.2,     1.4,      2.0,   0.6,    0.1, -0.2);
Sat.Term ( 0, -1,0,   9.4,     8.9,      3.9,  -8.3,   -0.4, -1.4);
Sat.Term ( 0, -2,0,   5.6,    -3.0,     -5.4,  -5.7,   -2.0,  0.0);
Sat.Term ( 0, -3,0,  -4.0,    -0.1,      0.0,   5.5,    0.0,  0.0);
Sat.Term ( 0, -5,0,   3.3,    -1.6,     -1.6,  -3.1,   -0.5, -1.2);
Sat.Term ( 1,  0,0,-113.1,19998.6,-25208.2,-142.2,-4670.7,288.9);
Sat.Term ( 1,  0,1, -76.1,    66.9,    -84.2, -95.8,   21.6, 29.4);
Sat.Term ( 1,  0,2,  -0.5,    -0.3,      0.4,  -0.7,    0.1, -0.1);
Sat.Term ( 1, -1,0,  78.8,   -14.5,     11.5,  64.4,   -0.2,  0.2);
Sat.Term ( 1, -2,0,  -2.0,  -132.4,     28.8,   4.3,   -1.7,  0.4);
Sat.Term ( 1, -2,1,  -1.1,    -0.7,      0.2,  -0.3,    0.0,  0.0);
Sat.Term ( 1, -3,0,  -7.5,    -6.8,     -0.4,  -1.1,    0.6, -0.9);
Sat.Term ( 1, -4,0,   0.7,     0.7,      0.6,  -1.1,    0.0, -0.2);
Sat.Term ( 1, -5,0,  51.5,   -26.0,    -32.5, -64.4,   -4.9,-12.4);
Sat.Term ( 1, -5,1,  -1.2,    -2.2,     -2.7,   1.5,   -0.4,  0.3);
Sat.Term ( 2,  0,0,  -3.4,   632.0,   -610.6,  -6.5, -226.8, 12.7);
Sat.Term ( 2,  0,1,  -4.2,     3.8,     -4.1,  -4.5,    0.2,  0.6);
Sat.Term ( 2, -1,0,   5.3,    -0.7,      0.7,   6.1,    0.2,  1.1);
Sat.Term ( 2, -2,0, -76.4,  -185.1,    260.2,-108.0,    1.6,  0.0);
Sat.Term ( 2, -3,0,  66.7,    47.8,    -51.4,  69.8,    0.9,  0.3);
Sat.Term ( 2, -3,1,   0.6,    -1.0,      1.0,   0.6,    0.0,  0.0);
Sat.Term ( 2, -4,0,  17.0,     1.4,     -1.8,   9.6,    0.0, -0.1);
Sat.Term ( 2, -5,0,1066.2,  -518.3,     -1.3, -23.9,    1.8, -0.3);
Sat.Term ( 2, -5,1, -25.4,   -40.3,     -0.9,   0.3,    0.0,  0.0);
Sat.Term ( 2, -5,2,  -0.7,     0.5,      0.0,   0.0,    0.0,  0.0);
Sat.Term ( 3,  0,0,  -0.1,    28.0,    -22.1,  -0.2,  -12.5,  0.7);
Sat.Term ( 3, -2,0,  -5.0,   -11.5,     11.7,  -5.4,    2.1, -1.0);
Sat.Term ( 3, -3,0,  16.9,    -6.4,     13.4,  26.9,   -0.5,  0.8);
Sat.Term ( 3, -4,0,   7.2,   -13.3,     20.9,  10.5,    0.1, -0.1);
Sat.Term ( 3, -5,0,  68.5,   134.3,   -166.9,  86.5,    7.1, 15.2);
Sat.Term ( 3, -5,1,   3.5,    -2.7,      3.4,   4.3,    0.5, -0.4);
Sat.Term ( 3, -6,0,   0.6,     1.0,     -0.9,   0.5,    0.0,  0.0);
Sat.Term ( 3, -7,0,  -1.1,     1.7,     -0.4,  -0.2,    0.0,  0.0);
Sat.Term ( 4,  0,0,   0.0,     1.4,     -1.0,   0.0,   -0.6,  0.0);
Sat.Term ( 4, -2,0,  -0.3,    -0.7,      0.4,  -0.2,    0.2, -0.1);
Sat.Term ( 4, -3,0,   1.1,    -0.6,      0.9,   1.2,    0.1,  0.2);
Sat.Term ( 4, -4,0,   3.2,     1.7,     -4.1,   5.8,    0.2,  0.1);
Sat.Term ( 4, -5,0,   6.7,     8.7,     -9.3,   8.7,   -1.1,  1.6);
Sat.Term ( 4, -6,0,   1.5,    -0.3,      0.6,   2.4,    0.0,  0.0);
Sat.Term ( 4, -7,0,  -1.9,     2.3,     -3.2,  -2.7,    0.0, -0.1);
Sat.Term ( 4, -8,0,   0.4,    -1.8,      1.9,   0.5,    0.0,  0.0);
Sat.Term ( 4, -9,0,  -0.2,    -0.5,      0.3,  -0.1,    0.0,  0.0);
Sat.Term ( 4,-10,0,  -8.6,    -6.8,     -0.4,   0.1,    0.0,  0.0);
Sat.Term ( 4,-10,1,  -0.5,     0.6,      0.0,   0.0,    0.0,  0.0);
Sat.Term ( 5, -5,0,  -0.1,     1.5,     -2.5,  -0.8,   -0.1,  0.1);
Sat.Term ( 5, -6,0,   0.1,     0.8,     -1.6,   0.1,    0.0,  0.0);
```

```
    Sat.Term ( 5, -9,0,  -0.5,    -0.1,      0.1,  -0.8,     0.0,  0.0);
    Sat.Term ( 5,-10,0,   2.5,    -2.2,      2.8,   3.1,     0.1, -0.2);
    dl = Sat.dl();  dr = Sat.dr();  db = Sat.db();

    // Perturbations by Uranus
    Ura.Init ( T, M5,1,1, M7,-2,-1 );
    Ura.Term ( 1, -1,0,   0.4,     0.9,      0.0,   0.0,     0.0,  0.0);
    Ura.Term ( 1, -2,0,   0.4,     0.4,     -0.4,   0.3,     0.0,  0.0);
    dl += Ura.dl();  dr += Ura.dr();  db += Ura.db();

    // Perturbations by Saturn and Uranus
    phi = (2*M5-6*M6+3*M7); c=cos(phi); s=sin(phi);
    dl += -0.8*c+8.5*s;  dr += -0.1*c;
    phi = (3*M5-6*M6+3*M7); c=cos(phi); s=sin(phi);
    dl += +0.4*c+0.5*s;  dr += -0.7*c+0.5*s;  db += -0.1*c;

    // Ecliptic coordinates ([rad],[AU])
    l = pi2 * Frac (0.0388910 + M5/pi2 + ((5025.2+0.8*T)*T+dl)/1296.0e3 );
    r = 5.208873 + 0.000041*T  +  dr * 1.0E-5;
    b = ( 227.3 - 0.3*T + db ) / Arcs;

    return Vec3D ( Polar(l,b,r) );  // Position vector

}
```

For a given time T, `JupiterPos` and the analogous `MercuryPos`, `VenusPos`, ..., `PlutoPos` functions return the heliocentric longitude l, latitude b and radius r of the planet concerned, relative to the ecliptic and vernal equinox of date. The time is measured in Julian centuries since epoch J2000:

$$T = (\text{JD} - 2451545)/36525 \quad .$$

In calculating the Julian Date JD, care must be taken to enter Ephemeris Time ET (or Dynamical Time TDT/TDB) rather than Universal Time UT (see Sect. 3.4). An appropriate routine for calculating geocentric solar coordinates has already been given in Chap. 2.

The structure of the `PlutoPos` routine differs from the other procedures. The coordinates are first calculated relative to the fixed ecliptic of 1950, and then transformed to the equinox of date. This method is necessary because of the high inclination of Pluto's orbit. Complete expansion of the coordinates would lead to numerous secular terms. It should also be noted that `PlutoPos` may be used only for calculations relating to the years between 1890 and 2100. The reason for this is that the series expansion used was not derived from perturbation theory, but from a Fourier analysis of a numerically integrated ephemeris covering that period of time. Even a few years before 1890 or after 2100, the errors in the calculated coordinates grow very sharply, reaching values of more than $0\overset{\circ}{.}5$.

For the ease of use, the individual routines are combined into the function `PertPosition`. In analogy with the `KepPosition` function introduced earlier, the heliocentric coordinates of all planets may thus be computed by a single function.

```
//------------------------------------------------------------------------------
// PertPosition: Planetary position from analytical perturbation series
//   Planet     Identifies the planet
//   T          Time in Julian centuries since J2000
//   <return>   Heliocentric position [AU], true-of-date ecliptic & equinox
//------------------------------------------------------------------------------
Vec3D PertPosition (PlanetType Planet, double T)
{
  Vec3D r; // Null vector;
  switch ( Planet )
  {
    case Sun:       break;
    case Mercury:   r = MercuryPos(T);  break;
    case Venus:     r = VenusPos(T);    break;
    case Earth:     r = (-SunPos(T));   break;
    case Mars:      r = MarsPos(T);     break;
    case Jupiter:   r = JupiterPos(T);  break;
    case Saturn:    r = SaturnPos(T);   break;
    case Uranus:    r = UranusPos(T);   break;
    case Neptune:   r = NeptunePos(T);  break;
    case Pluto:     r = PlutoPos(T);
  }
  return r;
}
```

The choice of a planet is performed by the `Planet` parameter of type

```
enum PlanetType { Sun, Mercury, Venus, Earth, Mars,
                  Jupiter, Saturn, Uranus, Neptune, Pluto };
```

which may also be mapped directly to a loop counter of values 0 to 9.

Table 6.2. Mean errors of the programs `MercuryPosPlutoPos` for the period 1750–2250 or 1890–2100 (Pluto)

Planet	Δl [″]	Δb [″]	Δr [AU]
Mercury	1.0 - 1.5	0.7 - 1.0	1.0 - 1.5$\cdot 10^{-6}$
Venus	0.5 - 1.0	0.2 - 2.5	0.5 - 1.5$\cdot 10^{-6}$
Sun Earth	0.5 - 2.0	0.0 - 0.1	0.4 - 1.5$\cdot 10^{-6}$
Mars	0.5 - 2.5	0.1 - 1.0	3 - 10$\cdot 10^{-6}$
Jupiter	2 - 8	0.7 - 1.0	20 - 30$\cdot 10^{-6}$
Saturn	2 - 11	0.8 - 1.5	40 - 100$\cdot 10^{-6}$
Uranus	3 - 8	0.7 - 1.0	50 - 200$\cdot 10^{-6}$
Neptune	3 - 40	1.0 - 2.0	500 - 1000$\cdot 10^{-6}$
Pluto	1 - 9	0.2 - 2.2	200 - 1000$\cdot 10^{-6}$

For the 19th and 20th centuries the accuracy of the programs is within a few seconds of arc (see Table 6.2). For more remote dates, one must expect less accurate results. First, the basic theories were developed around the end of the 19th century and therefore could only be devised to fit the observations available at that time. Second, perturbation terms that increase as T^3 have been neglected in the programs. Currently these values are very small ($< 0\overset{\prime\prime}{.}1$), but over historical periods of time they may easily increase to one thousand times this value.

6.4 Apparent and Astrometric Coordinates

If one consults a planetary ephemeris in a good yearbook, one usually finds details of whether the values included are *geometric*, *apparent*, or *astrometric* coordinates. The difference between these three variants is negligible, as long as one is concerned only with accuracies of 1 arc-minute or worse. To really use the accuracy of the programs given in the last section, however, we need to recognize and understand the most important corrections to the planetary coordinates.

While geometric coordinates indicate the point in space at which a planet is *located* at a specific time t, the apparent and astrometric coordinates indicate the direction in which the planet is *observed*.

Geometric coordinates are normally chosen only to express heliocentric planetary positions. The coordinate system being used is unambiguously determined by giving the equinox. The set of functions `MercuryPos...PlutoPos` yields *geometric heliocentric ecliptic coordinates referred to the equinox of date (i.e., of the calculated epoch)*. Naturally this does not apply to `SunPos`, which calculates the corresponding *geocentric* position of the Sun.

In contrast, geocentric positions (except distances) are generally given in the form of astrometric or apparent positions. The reason for this is that because of its motion and light-time, a planet is not observed at the position it occupies at the calculated time. This effect has already been mentioned in connection with calculating cometary orbits. Because the length of the path actually travelled by the light is not directly observable, geocentric ephemerides are also normally restricted to giving the geometric distance.

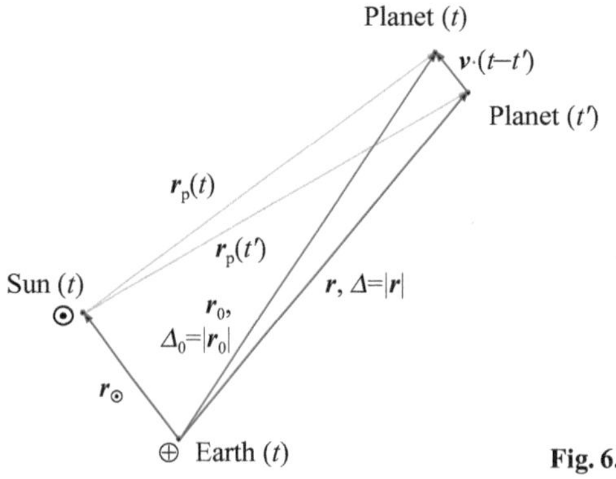

Fig. 6.2. The motion of a planet during light-time

6.4.1 Aberration and Light-Time

Astrometric coordinates indicate the direction from which an observed ray of light from a planet has reached the Earth. This ray of light links the position of the Earth at the time of observation t and the position of the planet at the time t' at which the light was emitted. The light-time $\tau = t - t'$ is given approximately by the geometric distance

$$\Delta_0 = |\boldsymbol{r}_0| \quad \text{where} \quad \boldsymbol{r}_0 = \boldsymbol{r}_\odot(t) + \boldsymbol{r}_\mathrm{P}(t) \tag{6.11}$$

whence $\tau \approx \Delta_0/c$ (see Fig. 6.2 for the definition of the individual quantities). Because the planet moves in approximately a straight line during this time, the astrometric coordinates of a planet at the time of observation t are given by the expression

$$\begin{aligned} \boldsymbol{r} &= \boldsymbol{r}_\odot(t) + \boldsymbol{r}_\mathrm{P}(t') \\ &\approx \boldsymbol{r}_\odot(t) + \left\{ \boldsymbol{r}_\mathrm{P}(t) - \boldsymbol{v}_\mathrm{P}(t) \cdot \frac{\Delta_0}{c} \right\} \quad . \end{aligned} \tag{6.12}$$

The difference between the *astrometric* coordinates $\boldsymbol{r}$ and geometric coordinates $\boldsymbol{r}_0$ is known as *light-time correction*. Astrometric coordinates may be directly marked on a celestial atlas or may be compared with known star positions on a photographic plate.

If, however, we point an equatorially mounted telescope at a planet, using astrometric coordinates for setting the circles, then we find a difference between the setting position and the observed position. This difference amounts to about 20 arc-seconds. This error is caused because astrometric coordinates indicate the direction of a ray of light falling on the Earth in a reference system that is *at rest* with respect to the Sun. However, the observer is subject to the daily rotation of the Earth and the yearly revolution of the Earth around the Sun. Because of these motions and the finite velocity of light the apparent direction in which the light arrives from the planet is altered. A frequently quoted example of the way in which appearances differ between a moving observer and one at rest may easily be observed by anybody: because of their velocity relative to observers in a vehicle, raindrops that are falling vertically towards the ground appear to be coming more or less from the direction in which the vehicle is moving.

In addition to the planets and comets the coordinate correction required in changing from a stationary to a moving reference system also affects the fixed stars. We therefore also speak of *stellar* aberration. Star positions in catalogues and celestial atlases are given as they would appear to an observer stationary in the Solar System. They therefore correspond to the astrometric coordinates of the planets, with which they may be directly compared. An observer on Earth, on the other hand, using setting circles, measures apparent stellar positions, which vary periodically over the course of the year from the catalogued positions.

A rigorous description of stellar aberration would take us rather too far, because it actually is a phenomenon governed by special relativity. Because the velocity of the Earth is relatively small when compared with the velocity of light, however, the semi-classical description that follows produces the same result within the degree of accuracy that concerns us here.

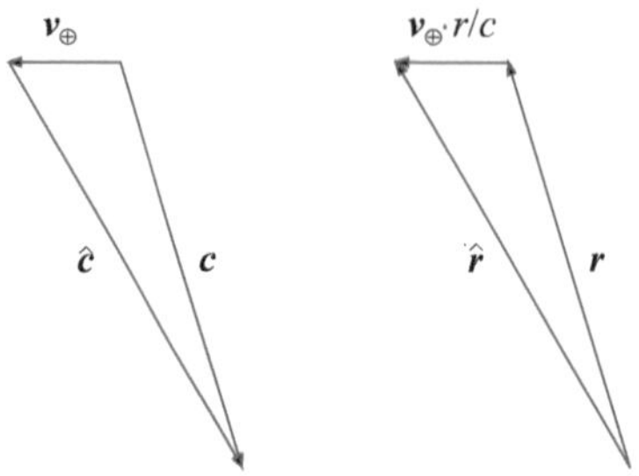

Fig. 6.3. The effect of stellar aberration

If $\boldsymbol{r}$ is the astrometric-coordinate vector of the planet, then the ray of light falling on the Earth may be described by the light-velocity vector

$$\boldsymbol{c} = -c \cdot \frac{\boldsymbol{r}}{|\boldsymbol{r}|}$$

(see Fig. 6.3). We obtain the corresponding vector $\hat{\boldsymbol{c}}$ in the observer's Earth-centred system if we subtract the velocity $\boldsymbol{v}_\oplus$ of the Earth relative to the Sun:

$$\hat{\boldsymbol{c}} = \boldsymbol{c} - \boldsymbol{v}_\oplus \quad .$$

An observer therefore sees the planet in the direction

$$\begin{aligned} \hat{\boldsymbol{r}} = -|\boldsymbol{r}| \cdot \frac{\hat{\boldsymbol{c}}}{c} &= \boldsymbol{r} + \boldsymbol{v}_\oplus \cdot \frac{|\boldsymbol{r}|}{c} \\ &\approx \boldsymbol{r} + \boldsymbol{v}_\oplus \cdot \frac{\Delta_0}{c} \quad . \end{aligned} \tag{6.13}$$

The coordinates thus determined are known as the *apparent* position of the planet, if the coordinate system chosen is additionally oriented in exactly the same direction as the current position of the Earth's axis. The choice of equinox required for this to be the case, and the allowance for nutation are the subject of the next section.

Before that, however, we still need to show how the heliocentric velocity vector for individual planets may be calculated, which is required to take account of the various aberration effects. Because the velocity v of an individual planet is small relative to the velocity of light c, the corrections caused by light-time and stellar aberration are of the order of one arc-minute only. As a result it suffices to know the velocities of the planets to about two decimal places, if we want to determine aberration to about one arc-second. Instead of complex series expansions, it thus suffices here to calculate the velocities on the basis of approximate Keplerian orbits. In a similar manner to the `KepPosition` function described earlier (see Sect. 5.2), `KepVelocity` yields the heliocentric velocity vector of a planet.

```
//------------------------------------------------------------------------------
// KepVelocity: Planetary velocity from Keplerian elements
//   Planet     Identifies the planet
//   T          Time in Julian centuries since J2000
// <return>:    Heliocentric velocity [AU/d], true-of-date ecliptic&equinox
//------------------------------------------------------------------------------
```

```
Vec3D KepVelocity (PlanetType Planet, double T)
{
  Vec3D r, v;
  State (Planet, T, r, v);  return v;
}
```

The orbital elements employed in `State` are precisely correct for the beginning of the year 2000, but will also provide sufficiently accurate results for a couple of years before and after that point in time.

6.4.2 Nutation

Apparent coordinates are coordinates that are designed to be used with a telescope fitted with setting circles. Because the telescope's mounting is always rigidly pointing in the same direction as the Earth's axis, the coordinate system employed must have the corresponding orientation. This chiefly demands that the apparent coordinates always relate to the equinox of date, i.e., to the current position of the equator, ecliptic, and vernal equinox. Apart from precession, there is another effect on the position of the Earth's axis to be taken into account, that has not yet been mentioned. While precession indicates the secular – i.e., the long-term – position of the Earth's axis, this additional, small, periodic variation is *nutation*. As a result of nutation, the Earth's *true* pole rotates once in every 18.6 years around the *mean* pole, the motion of which is described by precession. The nutation period is determined by the time required for the ascending node Ω of the Moon's orbit to complete one rotation. Nutation in longitude ($\Delta\psi$), i.e., the difference in longitude between the true and mean vernal equinoxes, has an amplitude of about 17 arc-seconds. The obliquity of the ecliptic simultaneously varies by $\Delta\varepsilon \approx 9''$ (see Fig. 6.4).

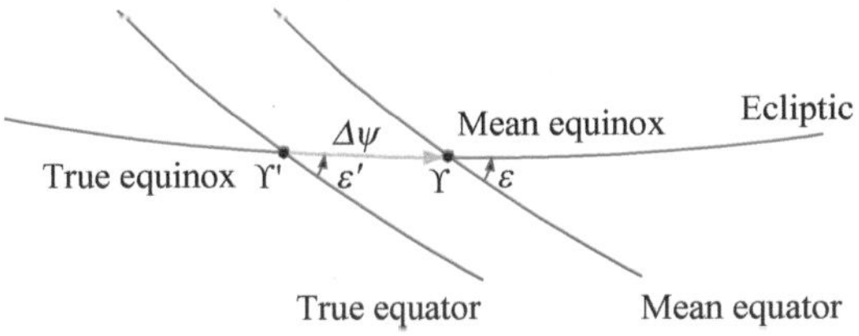

Fig. 6.4. The shift in the positions of the equator, the ecliptic and the vernal equinox, caused by nutation

Both of these quantities consist of around one hundred individual terms. For most purposes, however, only a few are required:

$$
\begin{aligned}
\Delta\psi &= -17''\!.200 \cdot \sin(\Omega) + 0''\!.206 \cdot \sin(2\Omega) + 0''\!.143 \cdot \sin(l') \\
&\quad -1''\!.319 \cdot \sin(2(F - D + \Omega)) - 0''\!.227 \cdot \sin(2(F + \Omega)) \\
\Delta\varepsilon &= +9''\!.203 \cdot \cos(\Omega) - 0''\!.090 \cdot \cos(2\Omega) \\
&\quad +0''\!.574 \cdot \cos(2(F - D + \Omega)) + 0''\!.098 \cdot \cos(2(F + \Omega)) \quad .
\end{aligned}
\tag{6.14}
$$

Apart from the longitude of the ascending node Ω, these terms depend on various additional arguments (F, D, l'), derived from the mean longitudes and anomalies of the Sun and Moon (see (8.2)…(8.5)).

The true equatorial coordinates $\boldsymbol{r}' = (x', y', z')$ are obtained from the mean coordinates $\boldsymbol{r} = (x, y, z)$ previously employed by multiplication by the nutation matrix

$$\boldsymbol{N} = \boldsymbol{R}_x(-\varepsilon-\Delta\varepsilon)\, \boldsymbol{R}_z(\Delta\psi)\, \boldsymbol{R}_x(\varepsilon) \quad , \tag{6.15}$$

with the components

$$\begin{aligned}
n_{11} &= +\cos(\Delta\psi) \\
n_{21} &= +\cos(\varepsilon')\cdot\sin(\Delta\psi) \\
n_{31} &= +\sin(\varepsilon')\cdot\sin(\Delta\psi) \\
\\
n_{12} &= -\cos(\varepsilon)\cdot\sin(\Delta\psi) \\
n_{22} &= +\cos(\varepsilon)\cdot\cos(\varepsilon')\cdot\cos(\Delta\psi)+\sin(\varepsilon)\cdot\sin(\varepsilon') \\
n_{32} &= +\cos(\varepsilon)\cdot\sin(\varepsilon')\cdot\cos(\Delta\psi)-\sin(\varepsilon)\cdot\cos(\varepsilon') \\
\\
n_{13} &= -\sin(\varepsilon)\cdot\sin(\Delta\psi) \\
n_{23} &= +\sin(\varepsilon)\cdot\cos(\varepsilon')\cdot\cos(\Delta\psi)-\cos(\varepsilon)\cdot\sin(\varepsilon') \\
n_{33} &= +\sin(\varepsilon)\cdot\sin(\varepsilon')\cdot\cos(\Delta\psi)+\cos(\varepsilon)\cdot\cos(\varepsilon')
\end{aligned} \tag{6.16}$$

which itself comprises three basic rotations (see Fig. 6.4). Here ε is the mean obliquity of the ecliptic from (2.9) and $\varepsilon' = \varepsilon + \Delta\varepsilon$ is the true obliquity.

The `NutMatrix` function expects the time T in Julian centuries since J2000 as input, and thence calculates the transformation from mean to true coordinates, using the simplified nutation angles.

```
//------------------------------------------------------------------------
//
// NutMatrix: Transformation from mean to true equator and equinox
//   T          Time in Julian centuries since epoch J2000
//   <return>   Nutation matrix
//
//------------------------------------------------------------------------
Mat3D NutMatrix (double T)
{
  // Variables
  double  ls, D, F, N;
  double  eps, dpsi, deps;
  // Mean arguments of lunar and solar motion
  ls = pi2*Frac(0.993133+  99.997306*T);   // mean anomaly Sun
  D  = pi2*Frac(0.827362+1236.853087*T);   // diff. longitude Moon-Sun
  F  = pi2*Frac(0.259089+1342.227826*T);   // mean argument of latitude
  N  = pi2*Frac(0.347346-   5.372447*T);   // longit. ascending node
  // Nutation angles
  dpsi = ( -17.200*sin(N)   - 1.319*sin(2*(F-D+N)) - 0.227*sin(2*(F+N))
          + 0.206*sin(2*N) + 0.143*sin(ls) ) / Arcs;
  deps = ( + 9.203*cos(N)   + 0.574*cos(2*(F-D+N)) + 0.098*cos(2*(F+N))
          - 0.090*cos(2*N)                 ) / Arcs;
```

```
  // Mean obliquity of the ecliptic
  eps  = 0.4090928-2.2696E-4*T;
  // Nutation matrix
  return
    R_x(-eps-deps)*R_z(-dpsi)*R_x(+eps);
}
```

6.5 The PLANPOS Program

The `Planpos` program calculates the position of the Sun and of the nine major planets for a given date. Because of the restricted validity of the `PlutoPos` function, Pluto's coordinates are given only between the years 1890 and 2100. As well as the date, the equinox to which the calculated positions are referred may be specified. There are three choices: apparent coordinates and astrometric coordinates for either of the two equinoxes currently employed, B1950 and J2000. Apparent coordinates are referred to the true equinox of date and contain corrections for precession, nutation, aberration and light-time. They may, for example, be either set on the circles of an equatorially mounted telescope, or read from such circles. Astrometric coordinates are required if we want to plot planetary positions on a star chart for the appropriate equinox. They take precession and light-time into account.

```
//------------------------------------------------------------------------
// File:    Planpos.cpp
// Purpose: Ephemerides of the major planets
// (c) 1999 Oliver Montenbruck, Thomas Pfleger
//------------------------------------------------------------------------

#include <cmath>
#include <fstream>
#include <iomanip>
#include <iostream>
#include "APC_Const.h"
#include "APC_Kepler.h"
#include "APC_Math.h"
#include "APC_Planets.h"
#include "APC_PrecNut.h"
#include "APC_Spheric.h"
#include "APC_Sun.h"
#include "APC_Time.h"
#include "APC_VecMat3D.h"
using namespace std;

//------------------------------------------------------------------------------
// Main program
//------------------------------------------------------------------------------
void main()
{
  // Body names
  const char* Name[] = { "Sun", "Mercury", "Venus", "Earth", "Mars",
                         "Jupiter", "Saturn", "Uranus", "Neptune", "Pluto" };
```

```
// Variables
PlanetType Planet;
char       Mode;
int        iPlanet, last, year, month, day;
double     Hour, MJD, T, dist, fac;
Vec3D      R_Sun, r_helioc, r_geoc, r_equ;
Mat3D      P;
bool       End = false;
// Title
cout << endl
<< "   PLANPOS: geocentric and heliocentric planetary positions  " << endl
<< "            (c) 1999 Oliver Montenbruck, Thomas Pfleger      " << endl
<< endl;

// Command loop
do {

  // Command input
  cout << endl
       << " (J) J2000 astrometric        (B) B1950 astrometric  " << endl
       << " (A) apparent coordinates     (E) exit               " << endl
       << endl
       << " Enter option ... ";
  cin >> Mode; cin.ignore(81,'\n'); Mode = tolower(Mode); cout << endl;
  // Terminate, if Mode=Exit
  if (Mode=='e') { End=true; break; }
  // Cycle, if Mode is invalid
  if (Mode!='j' && Mode!='b' && Mode!='a') continue;

  // Epoch date
  cout << " Date (yyyy mm dd hh.hhh) ... ";
  cin >> year >> month >> day >> Hour; cin.ignore(81,'\n');
  cout << endl << endl << endl;
  MJD = Mjd(year,month,day)+Hour/24.0;
  T   = ( MJD - MJD_J2000 ) / 36525.0;
  // Header
  cout << " Date:  " << DateTime(MJD,HHh) << " (ET)"
       << "     JD: " << (MJD+2400000.5)
       << setw(18) << "equinox ";
  switch (Mode) {
    case 'a': cout << "of date" << endl; break;
    case 'j': cout << "J2000"   << endl; break;
    case 'b': cout << "B1950"   << endl;
  }
  cout << endl << endl
       << "               l             b          r          "
       << "RA          Dec        delta"
       << endl
       << "            o  ' \"      o  ' \"       AU          "
       << "h  m  s      o  ' \"       AU" << endl;

  // Ecliptic coordinates of the Sun, equinox of date
  R_Sun = SunPos(T);
```

```
// Loop over planets
if ( (-1.1<T) && (T<1.0) ) last=Pluto; else last=Neptune;
for (iPlanet=Sun; iPlanet<=last; iPlanet++) {
  // Heliocentric ecliptic coordinates of the planet; equinox of date
  Planet = (PlanetType) iPlanet;
  r_helioc = PertPosition(Planet, T);
  // Geocentric ecliptic coordinates (equinox of date)
  r_geoc = r_helioc + R_Sun;
  // First-order light-time/aberration correction
  dist = Norm(r_geoc);
  fac = dist/c_light;
  if (Mode=='a')
    r_geoc -= fac*(KepVelocity(Planet,T)-KepVelocity(Earth,T));
  else
    r_geoc -= fac*KepVelocity(Planet,T);
  // Precession and equatorial coordinates;
  switch (Mode) {
    case 'a':  r_equ    = NutMatrix(T) * Ecl2EquMatrix(T) * r_geoc;
               break;
    case 'j':  P        = PrecMatrix_Ecl(T,T_J2000);
               r_helioc = P * r_helioc;
               r_geoc   = P * r_geoc;
               r_equ    = Ecl2EquMatrix(T_J2000) * r_geoc;
               break;
    case 'b':  P        = PrecMatrix_Ecl(T,T_B1950);
               r_helioc = P * r_helioc;
               r_geoc   = P * r_geoc;
               r_equ    = Ecl2EquMatrix(T_B1950) * r_geoc;
  }

  // Output
  cout << " " << setw(8) << left << Name[iPlanet] << right;
  cout << fixed << setprecision(1)
       << setw(11) << Angle(Deg*r_helioc[phi],DMMSSs) << " "
       << setw(11) << Angle(Deg*r_helioc[theta],DMMSSs);
  if (Planet<=Earth)
    cout << setprecision(6) << setw(11) << r_helioc[r];
  else
    cout << setprecision(5) << setw(10) << r_helioc[r] << " ";
  cout <<setprecision(2)<< setw(13)<< Angle(Deg*r_equ[phi]/15.0,DMMSSs)
       << " " << showpos
       << setprecision(1) << setw(11) << Angle(Deg*r_equ[theta],DMMSSs)
       << noshowpos;
  if (Planet<=Earth)
    cout << setprecision(6) << setw(11) << dist;
  else
    cout << setprecision(5) << setw(10) << dist << " ";
  cout << endl;
};

// Trailer
cout << endl
     << " l,b,r:   heliocentric ecliptic (geometric) " << endl
```

```
          << " RA,Dec:  geocentric equatorial ";
    if (Mode=='a')
      cout << "(apparent)" << endl;
    else
      cout << "(astrometric)" << endl;
    cout << " delta:   geocentric distance   (geometric)" << endl
          << endl;
  }
  while (!End);
}
```

To begin with, Planpos asks for the equinox to which the positions to be calculated should be referred. Possible entries are A (apparent coordinates, to the equinox of date), J (J2000, astrometric) and B (B1950, astrometric). This choice is always redisplayed after the coordinates have been computed, not just at the beginning of the program. It is always possible to exit from the program at this point by entering (E).

In the following example, we first choose apparent coordinates. After the appropriate response, we enter the exact time for which the coordinates are to be calculated. Note that the time must be given in Ephemeris Time (ET)! In the example, we select 1989 January 1 at 00:00 ET. (As usual, all data entries are shown in italic.) Planpos now returns the desired coordinates.

```
  PLANPOS: geocentric and heliocentric planetary positions
           (c) 1999 Oliver Montenbruck, Thomas Pfleger

(J) J2000 astrometric          (B) B1950 astrometric
(A) apparent coordinates       (E) exit

Enter option ... a

Date (yyyy mm dd hh.hhh) ... 1989 01 01 0.0

Date:  1989/01/01 00.0 (ET)     JD: 2447527.5          equinox of date

              l             b          r           RA           Dec        delta
          o  '   "      o  '   "      AU       h  m  s      o  '   "       AU
Sun        0 00 00.0    0 00 00.0  0.000000 18 45 53.66 -23 01 25.6  0.983309
Mercury 347 56 35.7 - 6 05 21.5  0.370100 19 59 16.60 -22 34 12.1  1.175632
Venus   226 05 15.0    1 43 26.9  0.723666 17 07 15.21 -22 03 57.5  1.522289
Earth   100 33 09.4    0 00 00.2  0.983309  0 00 00.00 + 0 00 00.0  0.000000
Mars     60 32 51.9    0 21 19.3  1.49745   1 13 47.44 + 8 24 05.2  0.97649
Jupiter  64 30 07.8 - 0 45 53.2  5.03195   3 38 35.14 +18 33 07.5  4.27637
Saturn  275 06 53.5    0 47 14.5 10.04354  18 24 14.50 -22 36 28.2 11.02274
Uranus  271 18 42.6 - 0 13 47.8 19.31483  18 07 39.22 -23 39 00.8 20.28599
Neptune 279 54 35.7    0 55 55.8 30.21917  18 42 54.09 -22 10 16.6 31.20230
Pluto   222 55 15.8   15 52 34.4 29.65861  15 06 35.06 - 1 16 18.1 30.17673

l,b,r:    heliocentric ecliptic (geometric)
RA,Dec:   geocentric equatorial (apparent)
delta:    geocentric distance   (geometric)
```

It should be noted that the heliocentric ecliptic coordinates of the Sun, and the geocentric equatorial coordinates of the Earth are always set to zero.

For comparison purposes, we now want to calculate the positions of the planets for the same time, but referred to equinox J2000. The small differences that result are caused by precession over a period of 11 years.

```
(J) J2000 astrometric          (B) B1950 astrometric
(A) apparent coordinates       (E) exit

Enter option ... j

Date (yyyy mm dd hh.hhh) ...  1989 01 01 0.0

Date:  1989/01/01 00.0 (ET)      JD: 2447527.5           equinox J2000

                   l            b            r          RA          Dec         delta
              o  '  "      o  '  "          AU      h  m  s      o  '  "         AU
Sun          0 00 00.0    0 00 00.0  0.000000 18 46 34.57 -23 00 32.4  0.983309
Mercury  348 05 49.4 - 6 05 22.1  0.370100 19 59 56.54 -22 32 12.4  1.175632
Venus    226 14 28.3    1 43 22.9  0.723666 17 07 55.78 -22 04 40.7  1.522289
Earth    100 42 22.5    0 00 05.2  0.983309  0 00 00.00 + 0 00 00.0  0.000000
Mars      60 42 05.0    0 21 24.1  1.49745   1 14 21.40 + 8 27 27.7  0.97649
Jupiter   64 39 21.0 - 0 45 48.3  5.03195   3 39 11.61 +18 35 03.6  4.27637
Saturn   275 16 06.7    0 47 09.5 10.04354  18 24 55.40 -22 35 56.1 11.02274
Uranus   271 27 55.8 - 0 13 52.9 19.31483  18 08 20.48 -23 38 44.9 20.28599
Neptune  280 03 48.9    0 55 50.8 30.21917  18 43 34.77 -22 09 26.4 31.20230
Pluto    223 04 29.9   15 52 30.6 29.65861  15 07 09.55 - 1 18 40.5 30.17673

l,b,r:    heliocentric ecliptic (geometric)
RA,Dec:   geocentric equatorial (astrometric)
delta:    geocentric distance   (geometric)
```

It will be seen that the differences are very small, but still to be taken into account. In comparing such results with yearbooks some care should be taken. Sometimes only the positions of the planets from Mercury to Neptune are given in apparent coordinates, while the ephemeris for Pluto uses astrometric coordinates. The reason for this is that in observing this faint planet it is advisable to use an atlas or a finder chart, to be sure of identifying it.

```
(J) J2000 astrometric          (B) B1950 astrometric
(A) apparent coordinates       (E) exit

enter option:  e
```

After the next response *E*, the program terminates.

7. Physical Ephemerides of the Planets

In the previous chapter we have provided the means of accurately calculating the positions of the major planets. Anyone who frequently observes the planets, the Sun, or the Moon, is also interested in data that describe the aspect of the various bodies. The visual magnitude and the apparent angular diameter are data of fundamental importance. Anyone who wants to observe and plot surface markings or atmospheric structure also needs to know the orientation of the planet's axis of rotation (and thus of the planetographic coordinate system), relative to the line of sight, together with the conditions of illumination. We intend to show how these data may be calculated. In doing so, we shall try to achieve an accuracy that should suffice, in most instances, for practical observations.

The data on the size, form and rotation of the planets that are required for calculation are generally taken from the *Report of the IAU/IAG/COSPAR Working Group on Cartographic Coordinates and Rotational Elements* of 1994, which is listed in the Bibliography. The physical ephemerides calculated using these data are consistent with the values currently given in the *Astronomical Almanac*.

Because the physical ephemerides depend on the relative position of the Sun, the Earth, and a planet, we require the heliocentric coordinates of the Earth and the planet. For this purpose we make use of the `PertPosition` function, which was already presented in Chap. 6 and yields accurate planetary coordinates even for extended time scales.

7.1 Rotation

Just as the rotation of the Earth is determined by the direction of the North Celestial Pole and sidereal time, the rotation of the other planets may be described by the position of the corresponding north poles and rotation through a specific angle. Following a recommendation of the International Astronomical Union (IAU) the north pole is defined as the pole of a planet, minor planet, or satellite, that lies on the northern side of what is known as the *invariable plane* of the Solar System. This plane is normal to the Solar System's combined angular momentum vector, but for simplicity it may be visualized as the mean orbital plane of all the planets. Because the orbits of the major planets are only slightly inclined to the Earth's orbital plane, the invariable plane is very nearly identical with the ecliptic.

According to the definition just given, not all the planets rotate like the Earth. The exceptions are Venus, Uranus, and Pluto, which rotate clockwise when looking

Table 7.1. Orientation of the rotational axes of the Sun and planets (IAU 1994)

Body	α_0 (J2000)	δ_0 (J2000)	Notes
Sun	$286^\circ\!.13$	$+63^\circ\!.87$	
Mercury	$281^\circ\!.01 \ -0^\circ\!.033T$	$+61^\circ\!.45 \ -0^\circ\!.005T$	
Venus	$272^\circ\!.76$	$+67^\circ\!.16$	
Earth	$0^\circ\!.00 \ -0^\circ\!.641T$	$+90^\circ\!.00 \ -0^\circ\!.557T$	
Mars	$317^\circ\!.681 - 0^\circ\!.108T$	$+52^\circ\!.886 - 0^\circ\!.061T$	
Jupiter	$268^\circ\!.05 \ -0^\circ\!.009T$	$+64^\circ\!.49 \ +0^\circ\!.003T$	
Saturn	$40^\circ\!.589 - 0^\circ\!.036T$	$+83^\circ\!.537 - 0^\circ\!.004T$	
Uranus	$257^\circ\!.311$	$-15^\circ\!.175$	
Neptune	$299^\circ\!.36 \ +0^\circ\!.70 \sin N$	$+43^\circ\!.46 \ -0^\circ\!.51 \cos N$	$N = 357^\circ\!.85 + 52^\circ\!.316T$
Pluto	$313^\circ\!.02$	$+9^\circ\!.09$	

down on the north pole. Unlike the *direct rotation* of most of the planets, their rotation is *retrograde.*

As well as calculating the rotational phase of the major planets, we also want to determine similar data for the Sun. In what follows, when we discuss planets, it should be borne in mind that the procedure described for determining the rotational phase may also be applied to the Sun. Factors specific to that particular body are discussed in a separate section.

Table 7.1 gives a summary of the orientations of the rotational axes of the Sun and planets. The columns show the right ascension α_0 and declination δ_0 of the north pole, referred to the J2000 equinox. The argument

$$T = (\text{JD} - 2451545)/36525 \tag{7.1}$$

indicates, as usual, the number of Julian centuries since the J2000 epoch.

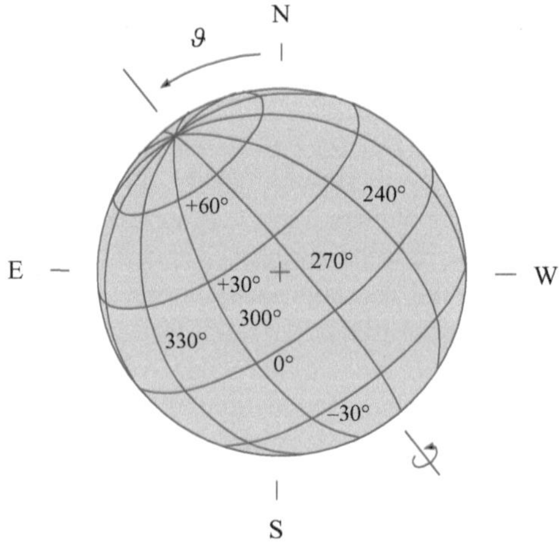

Fig. 7.1. Position angle of the rotational axis and planetographic coordinates (aspect of the planet Mars on 1993 Oct.01, 0^h ET; $\vartheta = 37^\circ\!.9$, $\lambda_\oplus = 276^\circ\!.1$, $\varphi_\oplus = 20^\circ\!.4$)

7.1.1 The Position Angle of the Axis

The position of the rotational axis of a planet, as seen by an observer, is usually described by giving the position angle of the axis and the latitude of the Earth relative to the planet's equator (see Fig. 7.1). The position angle ϑ of the planet's axis is taken to be the angle between the projection of the rotational axis onto the celestial sphere and the observer's North. The position angle is measured from North ($\vartheta = 0°$) through East ($\vartheta = 90°$).

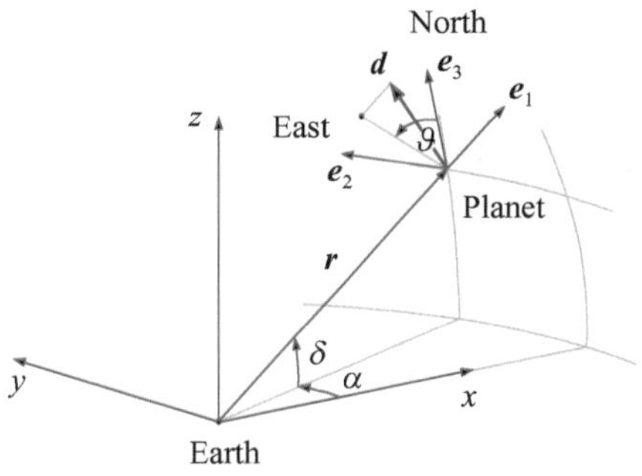

Fig. 7.2. Calculating the position angle

To calculate the position angle of the planet's axis we begin with the geocentric equatorial coordinates $\boldsymbol{r} = (x, y, z)$ of the planet. In addition, let

$$\boldsymbol{d} = \begin{pmatrix} d_x \\ d_y \\ d_z \end{pmatrix} = \begin{pmatrix} \cos\alpha_0 \cos\delta_0 \\ \sin\alpha_0 \cos\delta_0 \\ \sin\delta_0 \end{pmatrix} \tag{7.2}$$

be the vector, taken from the centre of the planet, describing the direction of the axis. With the aid of the vector $\boldsymbol{r}$, we now form the three, mutually perpendicular, unit vectors $(\boldsymbol{e}_1, \boldsymbol{e}_2, \boldsymbol{e}_3)$, so that $\boldsymbol{e}_1$ is directed towards the planet, $\boldsymbol{e}_2$ is directed towards increasing right ascension (i.e., towards East), and $\boldsymbol{e}_3 = \boldsymbol{e}_1 \times \boldsymbol{e}_2$ is directed towards increasing declination (towards North) as shown in Fig. 7.2. With this construction, we have

$$\boldsymbol{e}_1 = \frac{1}{r}\begin{pmatrix} x \\ y \\ z \end{pmatrix} \qquad \boldsymbol{e}_2 = \frac{1}{\varrho}\begin{pmatrix} -y \\ +x \\ 0 \end{pmatrix} \qquad \boldsymbol{e}_3 = \frac{1}{r\varrho}\begin{pmatrix} -xz \\ -yz \\ x^2 + y^2 \end{pmatrix} \quad , \tag{7.3}$$

whence

$$r = \sqrt{x^2 + y^2 + z^2} \qquad \text{and} \qquad \varrho = \sqrt{x^2 + y^2}$$

represent the value of the radius vector and its projection onto the plane of the equator. The position angle ϑ, at which we see the direction vector $\boldsymbol{d}$, is measured in the plane defined by $\boldsymbol{e}_2$ and $\boldsymbol{e}_3$, which is normal to the line of sight. By projecting the direction vector $\boldsymbol{d}$ onto $\boldsymbol{e}_2$ and $\boldsymbol{e}_3$ we obtain the following expressions for determining the position angle ϑ:

$$\begin{aligned} \cos\vartheta &= \boldsymbol{e}_3\cdot\boldsymbol{d} &= ((-xz)d_x+(-yz)d_y+(x^2+y^2)d_z)/(r\varrho) \\ \sin\vartheta &= \boldsymbol{e}_2\cdot\boldsymbol{d} &= ((-y)d_x+(x)d_y)/\varrho \quad . \end{aligned} \tag{7.4}$$

The PosAng function utilizes these equations to calculate the position angle:

```
//------------------------------------------------------------------------------
// PosAng: Computes the position angle of direction d w.r.t. the position r
//   r          Position of observed body
//   d          Direction for which the position angle is to be calculated
//   <return>   Position angle in [rad]; range [0, 2*pi]
// Notes: r and d must refer to the same coordinate system and equinox
//------------------------------------------------------------------------------
double PosAng (const Vec3D& r, const Vec3D& d)
{
  Vec3D e_1,e_2,e_3;  double  c,s,phi;
  // Unit vectors in radial, East, North direction
  e_1 = r / Norm(r);
  e_2 = Cross ( Vec3D(0, 0, 1), r );  e_2 = e_2 / Norm(e_2);
  e_3 = Cross (e_1, e_2);
  // Position angle
  c = Dot(d, e_3);  s = Dot(d, e_2);  phi = atan2(s, c);
  return Modulo(phi, 2.0*pi);
}
```

The PosAng function may also be used to calculate the position angle of the Sun, if we substitute the planetocentric coordinates of the Sun instead of the direction vector $\boldsymbol{d}$. This will be discussed in more detail in Sect. 7.2 in connection with the illumination geometry of the planets.

7.1.2 Planetographic Coordinates

Like the geographical coordinate system used on Earth, we may apply a system of spherical coordinates, fixed with reference to the surface, to each of the other planets. This may then be used for position measurements. The equator of the respective body serves as the reference plane for the system. Because most of the planets in the Solar System are flattened as a result of their rotation, a distinction is drawn between two definitions of latitude:

- The planeto*centric* latitude φ' indicates the angle between the direction vector of a point and the planet's equator. It therefore corresponds to declination in the geocentric equatorial coordinate system.
- The planeto*graphic* latitude φ differs from the planetocentric latitude φ' when the planet is flattened. If from a given point, we use a plumb-line to drop a normal to the surface of a planet, the planetographic latitude is the angle between this line and the equatorial plane. In the case of the Earth, it thus corresponds to geographical latitude (see Fig. 9.5).

The relationship between geocentric latitude φ' and geographic latitude φ is described in Sect. 9.3. We simply need to substitute the equatorial radius R_{Eq} of the planet for the Earth's radius $R_{\oplus}$, whence the flattening of the planetary ellipsoid is defined as

$$f = \frac{R_{\text{Eq}} - R_{\text{Pol}}}{R_{\text{Eq}}} \quad . \tag{7.5}$$

Conversion between planetocentric latitude φ' and planetographic latitude φ may be made by using the equation

$$\tan\varphi' = (1 - f)^2 \tan\varphi \quad . \tag{7.6}$$

To calculate physical ephemerides we set the appropriate values for the equatorial radius and flattening of the major planets using the Shape function, so that the data are independent of the calculation routines.

```
//------------------------------------------------------------------------------
// Shape: Gives equatorial radius and flattening of the planets
//   Planet     Identifies the planet
//   R_equ      Equatorial radius in [km]
//   f          Geometrical flattening
//------------------------------------------------------------------------------
void Shape (PlanetType Planet, double& R_equ, double& f) {
  switch (Planet) {
    case Sun:      R_equ =696000.0;  f = 0.0;         break;
    case Mercury:  R_equ =   2439.0;  f = 0.0;         break;
    case Venus:    R_equ =   6051.0;  f = 0.0;         break;
    case Earth:    R_equ =   6378.14; f = 0.00335281; break;
    case Mars:     R_equ =   3393.4;  f = 0.0051865;  break;
    case Jupiter:  R_equ = 71398.0;  f = 0.0648088;  break;
    case Saturn:   R_equ = 60000.0;  f = 0.1076209;  break;
    case Uranus:   R_equ = 25400.0;  f = 0.030;      break;
    case Neptune:  R_equ = 24300.0;  f = 0.0259;     break;
    case Pluto:    R_equ =   1500.0;  f = 0.0; }
}
```

Here the previously defined enumeration type

```
  enum PlanetType { Sun, Mercury, Venus, Earth, Mars,
                    Jupiter, Saturn, Uranus, Neptune, Pluto };
```

is employed to designate the individual planet.

As with the determination of the latitude of a point, we differentiate between planetocentric and planetographic longitude:

- Planeto*centric* longitude λ' is measured positively towards the east, irrespective of the direction of rotation.
- Planeto*graphic* longitude λ is reckoned in the opposite sense to the rotation. Because of this definition, the longitude of what is known as the *central meridian*, which runs through the centre of the disk as seen from Earth, always increases with increasing time.

For the planets Venus, Uranus, and Pluto, which have retrograde rotations, both longitude definitions are identical ($\lambda = \lambda'$). On the other hand, the planetocentric and planetographic longitudes differ in sign ($\lambda = -\lambda'$) for the planets Mercury, Mars, Jupiter, Saturn and Neptune, which, like the Earth, rotate from West to East. In the case of the Sun, however, it should be noted that what is known as heliographic longitude increases in the direction of rotation, and is thus identical with heliocentric longitude, despite the fact that the Sun has direct rotation.

Establishment of a reference point (the prime meridian) for longitude measurements either makes use of prominent features on the surface, or, failing these, is defined in an appropriate manner. With the giant planets Jupiter, Saturn, Uranus, and Neptune, which do not rotate like solid bodies with a rigid surface, rotation systems with constant angular velocities have been introduced. For Jupiter's atmosphere, which rotates with a period of 9^h30^m at the equator, and 9^h56^m at higher latitudes, System I applies to regions near the equator ($|\varphi| \leq 9°$), and System II for all higher latitudes. Radio-astronomical studies of the outer planets have shown that these planets' magnetic fields rotate with periods that do not agree with the optically determined periods for System I or II. The reason for this is that the basic causes of the radio emission lie in the planetary interiors, which essentially rotate as solid bodies, quite independently of the atmospheric layers. System III has therefore been introduced to describe the rotation relative to the radio emission. For Uranus and Neptune only System III is used nowadays. For Saturn, in addition to System III, we also give details for System I, so that comparisons with older drawings will still be possible.

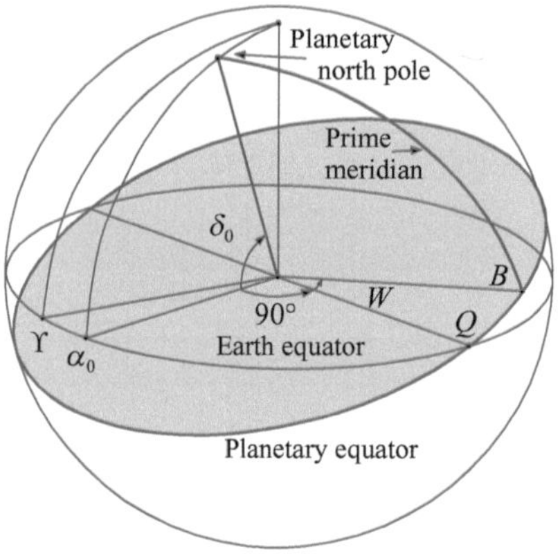

Fig. 7.3. Orientation of the prime meridian

We use the angle W to describe the position of a planet's prime meridian. It is measured

- from the intersection of the planes of the Earth's equator (J2000) and the planet's equator (point Q)
- along the planet's equator
- increasing in the direction of rotation
- to the intersection of the prime meridian and the planet's equator (point B).

Of the two points at which the planes of the Earth's and the planet's equators intersect, point Q is thus defined as the one at which a co-rotating point on the planet's equator crosses the plane of the Earth's equator from south to north (see Fig. 7.3). The angle W increases linearly with time and may be calculated from the data given in Table 7.2. The term

$$d = \mathrm{JD} - 2451545 \tag{7.7}$$

indicates the number of days that have elapsed since the standard epoch J2000.

Table 7.2. Orientation of the prime meridian of the Sun and planets (IAU 1994)

Body	System	W (J2000)	Notes
Sun		$84.^\circ182 + 14.^\circ1844000d$	
Mercury		$329.^\circ68 + 6.^\circ1385025d$	
Venus		$160.^\circ20 - 1.^\circ4813688d$	
Earth		$190.^\circ16 + 360.^\circ9856235d$	
Mars		$176.^\circ901 + 350.^\circ8919830d$	
Jupiter	System I	$67.^\circ1 + 877.^\circ900d$	
	System II	$43.^\circ3 + 870.^\circ270d$	
	System III	$284.^\circ695 + 870.^\circ536d$	
Saturn	System I	$227.^\circ2037 + 844.^\circ300d$	
	System III	$38.^\circ90 + 810.^\circ7939024d$	
Uranus	System III	$203.^\circ81 - 501.^\circ1600928d$	
Neptune	System III	$253.^\circ18 + 536.^\circ3128492d - 0.^\circ48\sin(N)$	$N = 357.^\circ85 + 52.^\circ316T$
Pluto		$236.^\circ77 - 56.^\circ3623195d$	

The numerical values of the parameters α_0, δ_0 and W, used to describe the position of the rotational axis and the prime meridian, are brought together in the `Orient` function. As a result, any future changes in the rotational parameters may be taken into account very easily. As well as the time and the name of the relevant planet, `Orient` requires the rotation system to be entered. For this we use the variable `System` which is of type

```
enum SystemType    { Sys_I, Sys_II, Sys_III };
```

Depending on the value of this parameter, and for the planets Jupiter and Saturn, `Orient` computes the position of the prime meridian for the equatorial regions, for higher latitudes, or for the radio emission. Furthermore, the parameter `Sense` of type

```
enum RotationType { Direct, Retrograde };      // Sense of rotation
```

specifies if the respective planet rotates in a direct or retrograde sense.

```
//------------------------------------------------------------------------------
// Orient: Calculates orientation of planetocentric reference frame and
//         sense of rotation
//   Planet    Identifies the planet (see APC_Planets.h)
//   System    Specifies rotation system (applies to Jupiter and Saturn)
//   T         Time in Julian centuries since epoch J2000
```

```
//   E          Transformation matrix from Earth mean equator and equinox
//              of J2000 to body fixed equator and prime meridian system
//   Sense      Sense of rotation (direct or retrograde)
//------------------------------------------------------------------------------
void Orient ( PlanetType Planet, SystemType System, double T,
              Mat3D& E, RotationType& Sense )
{
  double N, RA, Dec, W;
  double d = 36525.0*T;
  // Right ascension and declination of the axis of rotation
  // with respect to the equator and equinox of J2000;
  // Prime meridian orientation
  switch (Planet) {
    case Sun:      RA  = 286.13;
                   Dec =  63.87;
                   W   =  84.182 +  14.1844000*d; break;
    case Mercury:  RA  = 281.01  -   0.033*T;
                   Dec =  61.45  -   0.005*T;
                   W   = 329.68  +   6.1385025*d; break;
    case Venus:    RA  = 272.76;
                   Dec =  67.16;
                   W   = 160.20  -   1.4813688*d; break;
    case Earth:    RA  =   0.00  -   0.641*T;
                   Dec =  90.00  -   0.557*T;
                   W   = 190.16  + 360.9856235*d; break;
    case Mars:     RA  = 317.681 -   0.108*T;
                   Dec =  52.886 -   0.061*T;
                   W   = 176.901 + 350.8919830*d; break;
    case Jupiter:  RA  = 268.05  -   0.009*T;
                   Dec =  64.49  +   0.003*T;
                   switch (System) {
                     case Sys_I   : W =  67.10  + 877.900*d; break;
                     case Sys_II  : W =  43.30  + 870.270*d; break;
                     case Sys_III : W = 284.695 + 870.536*d; break;
                   } break;
    case Saturn:   RA  =  40.589 -   0.036*T;
                   Dec =  83.537 -   0.004*T;
                   switch (System) {
                     case Sys_I   :
                     case Sys_II  : W = 227.2037 + 844.3000000*d; break;
                     case Sys_III : W =  38.90   + 810.7939024*d; break;
                   } break;
    case Uranus:   RA  = 257.311;
                   Dec = -15.175;
                   W   = 203.81  - 501.1600928*d; break; // System III
    case Neptune:  N   = Rad*(357.85+52.316*T);
                   RA  = 299.36  + 0.70*sin(N);
                   Dec =  43.46  - 0.51*cos(N);
                   W   = 253.18  + 536.3128492*d - 0.48*sin(N); break;
    case Pluto:    RA  = 313.02;
                   Dec =   9.09;
                   W   = 236.77  -  56.3623195*d;
  }
```

```
  RA*=Rad; Dec*=Rad; W=Rad*Modulo(W,360.0);
  // Transformation from Earth mean equator and equinox of J2000 to
  // body fixed equator and prime meridian system
  E = R_z(W) * R_x(pi/2.0-Dec) * R_z(pi/2.0+RA);
  // Sense of rotation
  if ( Planet==Venus || Planet==Uranus || Planet==Pluto )
    Sense = Retrograde;
  else
    Sense = Direct;
}
```

If we know the position of a planet's rotational axis and prime meridian, we can determine the planetocentric and planetographic coordinates of the Sun and Earth, which are of particular importance for observation of a planet. These values describe, for example, how far the planet's north pole is inclined towards the Earth; which point of the planet's surface is at the centre of the apparent disk; and which point is vertically illuminated by the Sun.

To calculate the planetocentric longitude and latitude of the Earth or the Sun, we employ a fixed coordinate system that rotates with the planet and that is specified by the three, mutually perpendicular, unit vectors ($\boldsymbol{e}_1, \boldsymbol{e}_2, \boldsymbol{e}_3$). The vector $\boldsymbol{e}_3$ lies parallel to the rotational axis, while $\boldsymbol{e}_1$ is directed from the centre of the planet to the intersection of the equator and the prime meridian. Expressed in full, the components of these vectors, in equatorial coordinates, are

$$\begin{aligned}
\boldsymbol{e}_1 &= \begin{pmatrix} -\cos W \sin\alpha_0 - \sin W \sin\delta_0 \cos\alpha_0 \\ +\cos W \cos\alpha_0 - \sin W \sin\delta_0 \sin\alpha_0 \\ +\sin W \cos\delta_0 \end{pmatrix} \\
\boldsymbol{e}_2 &= \begin{pmatrix} +\sin W \sin\alpha_0 - \cos W \sin\delta_0 \cos\alpha_0 \\ -\sin W \cos\alpha_0 - \cos W \sin\delta_0 \sin\alpha_0 \\ +\cos W \cos\delta_0 \end{pmatrix} \qquad (7.8) \\
\boldsymbol{e}_3 &= \begin{pmatrix} +\cos\delta_0 \cos\alpha_0 \\ +\cos\delta_0 \sin\alpha_0 \\ +\sin\delta_0 \end{pmatrix} \quad .
\end{aligned}$$

These expressions are easily obtained from the known relationships between the Gaussian vectors $\boldsymbol{P}$, $\boldsymbol{Q}$ and $\boldsymbol{R}$ (see Sect. 4.5), if we replace the orbital elements i, Ω and ω, respectively, by: the angle $90° - \delta_0$ between the Earth's and the planet's equator; the right ascension $\alpha_0 + 90°$ of the line of intersection of the Earth's and the planet's equator; and the angle W between this line of intersection and the prime meridian. Similarly, the matrix

$$\boldsymbol{E} = \begin{pmatrix} \boldsymbol{e}_1^T \\ \boldsymbol{e}_2^T \\ \boldsymbol{e}_3^T \end{pmatrix} = \boldsymbol{R}_z(W)\boldsymbol{R}_x(\pi/2-\delta)\boldsymbol{R}_z(\pi/2+\alpha) \qquad (7.9)$$

for converting equatorial coordinates into the planet-fixed system may be represented by a sequence of three elementary rotation matrices.

If we represent the equatorial coordinates of the Earth relative to the centre of the planet by $\boldsymbol{r} = (x, y, z)$, then, by projecting $\boldsymbol{r}$ onto the direction vectors $\boldsymbol{e}_1$, $\boldsymbol{e}_2$ and $\boldsymbol{e}_3$, we obtain the Cartesian planetocentric coordinates

$$\begin{pmatrix} s_x \\ s_y \\ s_z \end{pmatrix} = \begin{pmatrix} -\boldsymbol{e}_1 \cdot \boldsymbol{r} \\ -\boldsymbol{e}_2 \cdot \boldsymbol{r} \\ -\boldsymbol{e}_3 \cdot \boldsymbol{r} \end{pmatrix} = \begin{pmatrix} r \cos\varphi' \cos\lambda' \\ r \cos\varphi' \sin\lambda' \\ r \sin\varphi' \end{pmatrix} \quad . \tag{7.10}$$

From these, by using

$$\begin{aligned} \tan\varphi' &= \frac{s_z}{\sqrt{s_x^2 + s_y^2}} \\ \tan\lambda' &= \frac{s_y}{s_x} \\ \tan\varphi &= \frac{\tan\varphi'}{(1-f)^2} \\ \lambda &= \begin{cases} +\lambda' & \text{(retrograde rotation)} \\ -\lambda' & \text{(direct rotation)} \end{cases} \end{aligned} \tag{7.11}$$

we may determine the corresponding values for the planetocentric and planetographic longitude and latitude of the Earth.

The planetocentric and planetographic coordinates of the Sun may be calculated in precisely the same way. We simply need to replace the planetocentric coordinates of the Earth by those of the Sun.

The steps required to calculate the planetographic coordinates that have been mentioned are all combined in the `Rotation` function. Apart from planetographic longitude and latitude, which are normally given in yearbooks, the routine also returns the planetocentric latitude, which will be required later when we come to calculate the illumination. If the planetocentric coordinates of the Sun are entered instead of the Earth's, we obtain the planetographic coordinates of the Sun.

```
//------------------------------------------------------------------------------
// Rotation: Calculates rotational parameters for the planets
//   r        Planetocentric coordinates of the Earth referred to the Earth
//            equator and equinox J2000
//   E        Transformation matrix from Earth mean equator and equinox of
//            J2000 to body fixed equator and prime meridian system
//   Sense    Sense of rotation (direct or retrograde)
//   f        Geometrical flattening
//   lon      Planetographic longitude of the Earth in [rad]
//   lat_g    Planetographic latitude of the Earth in [rad]
//   lat_c    Planetocentric latitude of the Earth in [rad]
//------------------------------------------------------------------------------
void Rotation ( const Vec3D& r, const Mat3D& E, RotationType Sense,
                double f, double& lon, double& lat_g, double& lat_c )
{
  Vec3D s;
```

```
    // Planetocentric latitude and longitude
    s = E*r;  lat_c = s[theta];   lon   = s[phi];
    // Planetographic latitude and longitude
    if (Sense==Direct) lon=-lon;  lon = Modulo(lon,2*pi);
    lat_g = atan ( tan(lat_c) / ((1.0-f)*(1.0-f)) );
}
```

7.2 Illumination Conditions

Apart from the description of the actual rotational phase, planetary observers are also interested in the conditions of illumination. These are described by the phase, the position angle of the Sun, and the planet's magnitude.

7.2.1 Phase and Elongation

Because the planets in the Solar System do not shine with their own light, but are illuminated by the Sun, generally only a portion of the planet's disk visible from the Earth is illuminated. A measure of the illuminated portion is given by the angle between the planetocentric direction vectors of the Sun and Earth, which is known as the *phase angle* i. A phase angle of $i = 0°$ corresponds to a fully illuminated planetary disk, and one of $i = 90°$ to half phase.

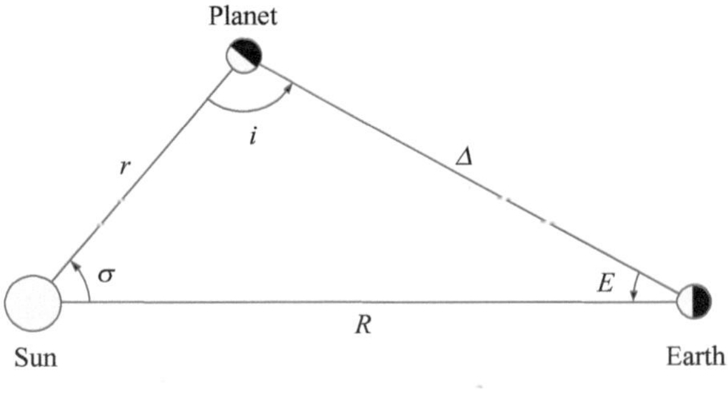

Fig. 7.4. Phase angle i and elongation E

In the Sun–Earth–Planet triangle (Fig. 7.4) let the distance of the planet from the Sun be r, that of the Earth R, and the distance between the planet and the Earth Δ. The phase angle may be obtained from the cosine law

$$\cos i = \frac{\Delta^2 + r^2 - R^2}{2r\Delta} \quad . \tag{7.12}$$

The illuminated fraction of the planetary disk, known as the *phase* is obtained from

$$k = \frac{1 + \cos i}{2} \quad . \tag{7.13}$$

Note that k does not apply to the spherical surface of the planet, but to the surface of the planet's apparent disk.

As with the phase angle, the cosine law may also be applied to the Sun–Earth–Planet triangle to give us

$$\cos E = \frac{\Delta^2 + R^2 - r^2}{2R\Delta} \quad , \tag{7.14}$$

where E is the *elongation* of the Sun, i.e., the angular distance between the Sun and the planet as seen from Earth. The elongation is an easy way of estimating whether the planet may be observed, or whether it is too close to the Sun in the sky to be visible.

The equations required to calculate the elongation and phase of a planet are combined in the `Illum` function. The values to be entered are the heliocentric coordinates of the planet and the Earth. They must be given in a common coordinate system and for the same equinox.

```
//------------------------------------------------------------------------------
// Illum: Calculates illumination parameters for the planets
//   r           Heliocentric position of body
//   r_Earth     Heliocentric position of the Earth
//   Elong       Elongation in [rad]
//   phi         Phase angle in [rad]
//   k           Phase
//------------------------------------------------------------------------------
void Illum ( const Vec3D& r, const Vec3D& r_Earth,
             double& Elong, double& phi, double& k )
{
  Vec3D   r_geoc;
  double  R,RE,D,c_phi;
  r_geoc = r - r_Earth;  // Geocentric position of the planet
  R  = Norm(r);          // Sun-planet distance
  RE = Norm(r_Earth);    // Sun-Earth  distance
  D  = Norm(r_geoc);     // Earth-planet distance
  // Elongation, phase angle and phase
  Elong = acos ( (D*D + RE*RE - R*R) / (2.0*D*RE) );
  c_phi = (D*D + R*R - RE*RE) / (2.0*D*R);
  phi   = acos (c_phi);
  k     = 0.5*(1.0+c_phi);
}
```

7.2.2 The Position Angle of the Sun

Because the planets do not orbit precisely in the plane of the ecliptic, it is rare for sunlight to fall on their surfaces from exactly (our) east or west. In general, therefore, the dividing line between the illuminated and dark hemispheres of a planet (the terminator) does not pass through the two poles of the apparent disk.

The orientation of the illuminated phase relative to the observer's north is unambiguously described by the position angle $\vartheta_\odot$ of the Sun, which is occasionally described as the illumination angle (Fig. 7.5). Using the previously determined relationship (7.4) for the position angle of the rotational axis, we can also calculate

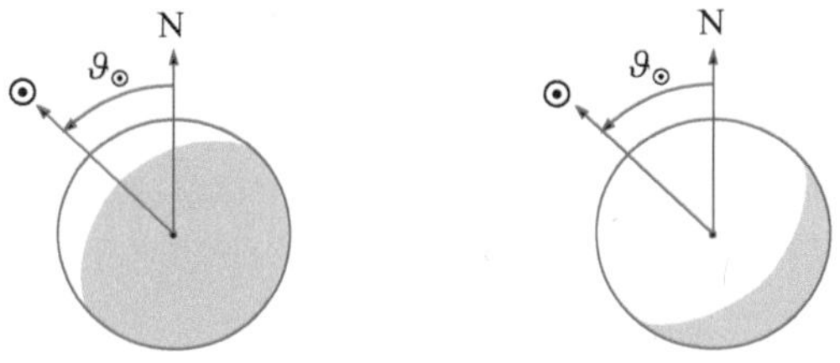

Fig. 7.5. Position angle of the Sun at various phase angles

the position angle of the Sun. Instead of the direction vector of the axis, we substitute the direction vector directed from the planet towards the Sun.

We can therefore employ the `PosAng` procedure to calculate the illumination angle. We thus enter the geocentric equatorial coordinates of the planet for vector $\boldsymbol{r}$ and the planetocentric equatorial coordinates of the Sun for the direction vector $\boldsymbol{d}$. These may be obtained by altering the sign of the heliocentric planetary coordinates that are obtained in any case during the calculation. Because position angles are normally referred to the actual direction of North, the solar and planetary coordinates should also be referred to the equinox of date.

7.2.3 Apparent Magnitude

By the apparent magnitude of a planet we mean the total brightness, expressed in magnitudes, of the illuminated portion visible from Earth. It depends on the planet's distance from the Sun and from the Earth, and also on the size and composition of its surface or atmosphere. Consequently, the apparent magnitude $V(1,0)$, of a planet at a distance of $r = \Delta = 1$ AU from the Sun and the Earth, and at a phase angle of $i = 0^\circ$, would, in accordance with its diameter and reflectivity, range between magnitude $-0\overset{\mathrm{m}}{.}4$ (Mercury) and $-9\overset{\mathrm{m}}{.}4$ (Jupiter).

The strength of the illumination of a planet is inversely proportional to the square of its distance from the Sun. Similarly, the observed radiation flux for a planet at a given distance from the Sun decreases as the square of its distance from the Earth. Bearing in mind that a decrease in the radiation flux by a factor of 100 corresponds to a change of +5 magnitudes in the brightness, we find that the apparent brightness of a planet is given by

$$m = V(1,0) + 5\log\left(\frac{r \cdot \Delta}{\mathrm{AU}^2}\right) + \Delta m(i) \quad . \tag{7.15}$$

Here, the $\Delta m(i)$ term represents the magnitude's phase-dependence. This involves both the decrease in illuminated area with decreasing phase, and also the dependence of scattering and reflection upon the angle of illumination. In general, $\Delta m(i)$ is expressed in the form of a power law, determined from observational data.

In the case of Saturn, the ring system's contribution to the planet's brightness also needs to be taken into account. This may be described as a function of the

Table 7.3. Apparent magnitudes of the planets

Planet	$V(1,0)$	$\Delta m(i)$
Mercury	$-0\overset{\mathrm{m}}{.}42$	$+3\overset{\mathrm{m}}{.}80 \cdot \left(\frac{i}{100^\circ}\right) - 2\overset{\mathrm{m}}{.}73 \cdot \left(\frac{i}{100^\circ}\right)^2 + 2\overset{\mathrm{m}}{.}00 \cdot \left(\frac{i}{100^\circ}\right)^3$
Venus	$-4\overset{\mathrm{m}}{.}40$	$+0\overset{\mathrm{m}}{.}09 \cdot \left(\frac{i}{100^\circ}\right) + 2\overset{\mathrm{m}}{.}39 \cdot \left(\frac{i}{100^\circ}\right)^2 - 0\overset{\mathrm{m}}{.}65 \cdot \left(\frac{i}{100^\circ}\right)^3$
Earth	$-3\overset{\mathrm{m}}{.}86$	
Mars	$-1\overset{\mathrm{m}}{.}52$	$+1\overset{\mathrm{m}}{.}60 \cdot \left(\frac{i}{100^\circ}\right)$
Jupiter	$-9\overset{\mathrm{m}}{.}40$	$+0\overset{\mathrm{m}}{.}50 \cdot \left(\frac{i}{100^\circ}\right)$
Saturn	$-8\overset{\mathrm{m}}{.}88$	$-2\overset{\mathrm{m}}{.}60 \cdot \lvert\sin\varphi'_\oplus\rvert + 1\overset{\mathrm{m}}{.}25 \cdot \lvert\sin\varphi'_\oplus\rvert^2 + 4\overset{\mathrm{m}}{.}40 \cdot \lvert\lambda'_\odot - \lambda'_\oplus\rvert/100^\circ$
Uranus	$-7\overset{\mathrm{m}}{.}19$	
Neptune	$-6\overset{\mathrm{m}}{.}87$	
Pluto	$-1\overset{\mathrm{m}}{.}0$	

planetocentric latitude $\varphi'_\oplus$ of the Earth, and the difference in the planetocentric longitudes of the Sun and the Earth, $\lambda'_\odot - \lambda'_\oplus$.

Expressions for the absolute magnitude $V(1,0)$ and the phase-dependent factor $\Delta m(i)$ for the planets are given in Table 7.3. The function `Bright` uses these values to determine the apparent magnitudes of the planets.

```
//------------------------------------------------------------------------------
// Bright: Calculates the visual magnitude of a planet
//   Planet    Identifies the planet
//   r         Heliocentric distance of the planet  in [AU]
//   Delta     Distance of the planet from the Earth  in [AU]
//   phi       Phase angle in [rad]
//   lat       Planetocentric latitude of the Earth in [rad]
//   Dlong     Difference of planetocentric longitudes of the Sun and the
//             Earth in [rad]
//   <return>  Brightness of the planet in stellar magnitudes [mag]
// Notes: lat and Dlong must be given for Saturn only and may be omitted
//        otherwise
//------------------------------------------------------------------------------
double Bright ( PlanetType Planet, double r, double Delta, double phi,
                double lat, double Dlong )
{
  // Variables
  double  p, sd, dl, mag;
  // Normalized phase angle
  p = Deg*phi/100.0;
  // Brightness at unit distance
  switch ( Planet ) {
    case Sun:       return 0.0;
    case Mercury:   mag = -0.42+(3.80-(2.73-2.00*p)*p)*p; break;
    case Venus:     mag = -4.40+(0.09+(2.39-0.65*p)*p)*p; break;
    case Earth:     mag = -3.86;           break;
    case Mars:      mag = -1.52+1.6*p; break;
    case Jupiter:   mag = -9.40+0.5*p; break;
    case Saturn:
      sd = fabs(sin(lat));
      dl = fabs(Modulo(Deg*Dlong+180.0,360.0)-180.0)/100.0;
      mag = -8.88 - 2.60*sd + 1.25*sd*sd + 4.40*dl;
```

```
      break;
    case Uranus:   mag = -7.19; break;
    case Neptune:  mag = -6.87; break;
    case Pluto:    mag = -1.0;
  }
  // Apparent brightness
  return ( mag + 5.0*log10(r*Delta) );
}
```

7.2.4 Apparent Diameter

The apparent diameter of a planet is the angle that the planet subtends at the Earth. It is obtained from the equatorial radius R_{Eq} of the planet at a distance Δ as

$$\varnothing_{\mathrm{Eq}} = 2 \cdot \arcsin\left(\frac{R_{\mathrm{Eq}}}{\Delta}\right) \quad . \tag{7.16}$$

With planets that have polar flattening, we have to differentiate between the equatorial and polar diameters. The apparent polar diameter is always smaller than the apparent equatorial diameter and depends both on the flattening and on the planetocentric latitude $\varphi'_{\oplus}$ of the Earth. For small values of flattening, the apparent polar diameter may be obtained, to a first approximation, from

$$\varnothing_{\mathrm{Pol}} = (1 - f\cos^2\varphi'_{\oplus})\,\varnothing_{\mathrm{Eq}} \quad . \tag{7.17}$$

7.3 The PHYS Program

The `Phys` program calculates a table of physical ephemerides for the major planets and the Sun. It uses the functions previously described to determine the planetographic coordinates of the Earth, the illumination conditions and the position angle of the north pole of the rotational axis.

The required bodies coordinates are calculated by the `PertPosition` function for the equinox of date using analytical series. The effects of nutation and aberration are small enough for them to be ignored in the context of the limited accuracy that is required for practical observational purposes. On the other hand, the light-time between the planet and the Earth does have to be taken into account. Consequently, the first step is to determine the geometrical distance between the Earth and the body, from which the light-time is derived. Later in the calculation, the coordinates and rotation angle that are derived are based upon the time at which the light is emitted by the body.

We do not need to take account of precession to the current equinox of date in determining the planetographic or heliographic coordinates of the Earth. These values are determined by the relative positions of the celestial bodies and are thus independent of the coordinate system used. Only in calculating the position angle, which is with respect to the actual direction of North, and thus to the equinox of date, do we need to carry out an appropriate coordinate transformation.

After the physical ephemerides for the planets have been calculated, the Sun's apparent diameter, the position angle of the axis and the heliographic coordinates of the Earth are determined. The orientation of the planet-fixed coordinate system relative to the Earth's equator and equinox is determined within `Orient` and passed directly to `Rotation`. The geocentric coordinates of the Sun (which are also required for the calculation of the position angle) are obtained by inverting the signs of the heliocentric coordinates of the Earth.

In addition, it must be noted that heliographic longitude is measured in the opposite direction to the way in which planetographic longitudes are defined. The Earth's heliographic longitude always increases. In addition, it is the normal convention to give the position angle of the Sun's axis as lying within the range $-180° \ldots 180°$, and this is taken into account by special instructions within the program.

```
//------------------------------------------------------------------------
// File:    Phys.cpp
// Purpose: Physical ephemerides of the major planets and the Sun
// (c) 1999 Oliver Montenbruck, Thomas Pfleger
//------------------------------------------------------------------------

#include <cmath>
#include <iomanip>
#include <iostream>
#include "APC_Const.h"
#include "APC_Math.h"
#include "APC_Phys.h"
#include "APC_Planets.h"
#include "APC_PrecNut.h"
#include "APC_Spheric.h"
#include "APC_Sun.h"
#include "APC_Time.h"
#include "APC_VecMat3D.h"
using namespace std;

//------------------------------------------------------------------------
// Main program
//------------------------------------------------------------------------
void main()
{
  // Body names
  const char* Name[] =
    { "Sun", "Mercury", "Venus", "Earth", "Mars",
      "Jupiter", "Saturn", "Uranus", "Neptune", "Pluto" };
  // Variables
  PlanetType    Planet;
  int           iPlanet, year, month, day;
  double        hour, T, T0;
  Vec3D         r, r_Earth, r_geoc;
  double        Delta, R_equ, D_equ, f;
  Mat3D         E_I, E_II, E_III,  P;
  RotationType  Sense;
  double        long_I, long_II, long_III, lat_g, lat_c;
```

```
double          long_Sun, latg_Sun, latc_Sun;
double          Elong, phi, k, mag, PosAng_Ax, PosAng_Sun;

// Title
cout << endl
     << " PHYS: physical ephemerides of the planets and the Sun" << endl
     << "        (c) 1999 Oliver Montenbruck, Thomas Pfleger    " << endl
     << endl;
// Prompt user for input
cout << " Date (yyyy mm dd hh.hhh) ... ";
cin >> year >> month >> day >> hour; cin.ignore(81,'\n');
T = (Mjd(year,month,day)+hour/24.0-MJD_J2000)/36525.0;
// Header
cout << endl
     << "             D      V      i    PA(S)   PA(A)"
     << "    L(I)  L(II) L(III)     B" << endl
     << "            '  \"     mag     o      o       o  "
     << "     o      o      o      o" << endl;

// Precession (mean-of-date to EME2000)
P = PrecMatrix_Equ(T,T_J2000);
// Heliocentric position of the Earth, mean equator and equinox of date
r_Earth = Ecl2EquMatrix(T) * PertPosition(Earth,T);
// Loop over bodies
for (iPlanet=Sun; iPlanet<=Pluto; iPlanet++) {
  Planet = (PlanetType) iPlanet;
  if (Planet==Earth) continue;  // Skip Earth
  // Geometric geocentric position and light time
  r = Ecl2EquMatrix(T) * PertPosition(Planet,T);
  Delta = Norm(r-r_Earth);
  T0 = T - (Delta/c_light)/36525.0,
  // Antedated planetary position at emission of light time
  r = Ecl2EquMatrix(T) * PertPosition(Planet,T0);
  // Geocentric position
  r_geoc = r - r_Earth;
  // Apparent semi-diameter (in ["])
  Shape ( Planet, R_equ,f );
  D_equ = Arcs*2.0*asin(R_equ/(Delta*AU));
  // Transformation from mean equator and equinox of date
  // to body-fixed frame
  Orient ( Planet, Sys_I  ,T0, E_I  , Sense );  E_I   = E_I   * P;
  Orient ( Planet, Sys_II ,T0, E_II , Sense );  E_II  = E_II  * P;
  Orient ( Planet, Sys_III,T0, E_III, Sense );  E_III = E_III * P;
  // Planetocentric longitude and latitude of the Earth
  Rotation ( -r_geoc, E_I,   Sense, f, long_I,  lat_g,lat_c );
  Rotation ( -r_geoc, E_II,  Sense, f, long_II, lat_g,lat_c );
  Rotation ( -r_geoc, E_III, Sense, f, long_III,lat_g,lat_c );
  // Position angle of the rotation axis
  PosAng_Ax  = PosAng ( r_geoc, Row(E_I,z) );
  // Specific computations for Sun and planets
  if (Planet==Sun) {
    // Sun specific convention for longitude and position angle
    // of the rotation axis
```

```
      long_I = 2.0*pi-long_I;
      if (PosAng_Ax>pi) PosAng_Ax-=2.0*pi;
    }
    else  {
      // Ilumination and apparent magnitude
      Illum ( r, r_Earth, Elong,phi,k );
      if (Planet==Saturn) {
        // Planetocentric longitude and latitude of the Sun
        Rotation ( -r, E_I, Sense, f, long_Sun, latg_Sun, latc_Sun );
        mag = Bright ( Planet, Norm(r), Norm(r_geoc), phi,
                       lat_c, long_Sun-long_I );
      }
      else
        mag = Bright ( Planet, Norm(r),Norm(r_geoc), phi);
      // Position angle of the Sun
      PosAng_Sun = PosAng ( r_geoc, -r );
    };

    // Output
    if (Planet==Sun)
      cout << " " << left << setw(6) << Name[Planet] << right << fixed
           << setprecision(2) << setw(8) << Angle(D_equ/60.0,DMMm)
           << setprecision(2) << setw(29) << Deg*PosAng_Ax;
    else
      cout << " " << left << setw(9) << Name[Planet] << right << fixed
           << setprecision(2) << setw(5) << D_equ
           << setprecision(1) << setw(6) << mag
           << setprecision(1) << setw(7) << Deg*phi
           << setprecision(2) << setw(8) << Deg*PosAng_Sun
           << setprecision(2) << setw(8) << Deg*PosAng_Ax;
    // Print planetographic longitude(s) of the Earth
    switch (Planet) {
      // Bodies with System I only
      case Sun: case Mercury: case Venus: case Mars: case Pluto:
        cout << setprecision(2) << setw(8) << Deg*long_I
             << setw(14) << " ";
        break;
      case Jupiter:
        cout << setprecision(2) << setw(8) << Deg*long_I
             << setprecision(2) << setw(7) << Deg*long_II
             << setprecision(2) << setw(7) << Deg*long_III;
        break;
      case Saturn:
        cout << setprecision(2) << setw(8)  << Deg*long_I
             << setprecision(2) << setw(14) << Deg*long_III;
        break;
      // Bodies with System III only
      case Uranus: case Neptune:
        cout << setprecision(2) << setw(22) << Deg*long_III;
    }
    cout << setprecision(2) << setw(8)  << Deg*lat_g << endl;
  } // End Planet loop
}
```

As an example we will calculate the physical ephemerides for 1999 August 30 at 0^h Ephemeris Time. As usual, the input data is shown in italic.

The following are displayed: the apparent equatorial diameter `D` in arc-seconds, the visual magnitude `V`, the phase angle `i`, as well as the position angle of the Sun `PA(S)` and of the north pole `PA(A)`. In addition the planetographic coordinates of the Earth are given in the rotation systems appropriate for each planet.

```
PHYS: physical ephemerides of the planets and the Sun
      (c) 1999 Oliver Montenbruck, Thomas Pfleger

Date (yyyy mm dd hh.hhh) ... 1999 08 30 00.0
```

	D	V	i	PA(S)	PA(A)	L(I)	L(II)	L(III)	B
	' "	mag	o	o	o	o	o	o	o
Sun	31 40.64				20.43	203.33			7.15
Mercury	5.30	-1.4	31.6	119.79	21.10	328.45			5.31
Venus	55.62	-4.2	156.5	79.44	17.13	334.31			8.68
Mars	7.95	0.3	44.2	283.48	37.85	29.70			12.81
Jupiter	45.22	-2.7	10.0	71.42	339.34	329.80	176.12	24.39	3.78
Saturn	18.73	0.2	5.9	74.63	359.41	93.75		118.66	-25.77
Uranus	3.69	5.7	1.1	251.46	262.90			155.83	-34.84
Neptune	2.29	7.9	1.1	257.45	355.90			256.25	-28.71
Pluto	0.14	13.8	1.9	279.11	75.60	311.55			-22.30

Venus is close to inferior conjunction, a few degrees south of the Sun, and shows the very large, narrow crescent ($i = 156^\circ\!.5$) phase that may sometimes be seen, even in daylight, under very favourable conditions. The direction from which the sunlight is falling is almost due east ($\vartheta_\odot = 79^\circ\!.4$).

Jupiter is conspicuous because of its large apparent diameter and its brightness. At this time, Jupiter is close to the turning point of its opposition loop and appoaches opposition which is about to take place two months later.

Mars, on the other hand, is already past opposition and exhibits a distinct phase ($i = 44^\circ\!.2$). In addition, its small diameter of slightly less than eight arc-seconds means that recognizing any features on its surface is very difficult.

8. The Orbit of the Moon

8.1 General Description of the Lunar Orbit

The motion of the Moon is principally determined by two bodies, the Earth and the Sun. If we consider the gravitational forces that affect the Moon, we find that it is not its nearest neighbour, the Earth, that has the greatest effect, but the more distant Sun. Although gravitational force decreases as the square of the distance, the Sun's gravity exceeds that of the Earth, because of its far greater mass. Putting figures to this (Earth-Moon distance $r \approx 380\,000$ km, Sun-Moon distance $R \approx 150$ million km, Sun/Earth mass ratio $M/m \approx 330\,000$), we find that the Sun's gravitational attraction is about twice as great as that of the Earth:

$$\frac{F_\odot}{F_\oplus} = \frac{M}{R^2}\Big/\frac{m}{r^2} \approx 2 \quad .$$

Regardless of the relative positions of the Sun, Earth, and Moon, the sum of the two forces therefore always has a component directed towards the Sun, and never away from it. The Moon may thus be described as following an elliptical orbit around the Sun at a distance of some 150 million km, superimposed upon which there are small monthly oscillations. Because the gravitational force is never directed away from the Sun, the orbit, despite these oscillations, is always concave towards the Sun.

Although the Moon's orbit is primarily determined by the Sun's gravity, it does not seem to be particularly advisable to describe it in heliocentric coordinates. We are, after all, primarily interested in its motion around the Earth. Because the Earth and the Moon are relatively close neighbours, they are subject to almost identical gravitational forces exerted by the Sun. If these forces were exactly equal, then the Sun would have no overall influence on the *relative* positions of the Earth and the Moon. It would only be responsible for their common yearly orbit around the Sun, while the Earth's gravity would determine the Moon's monthly motion. The difference between the forces exerted by the Sun on the Moon and on the Earth is about r/R times the amount of the force itself, and is thus about 200 times smaller than the force that the Earth exerts on the Moon. The most useful description of the Moon's orbit is therefore one expressed in a geocentric reference frame. In it, the Moon moves around the Earth in a monthly Keplerian orbit, which is perturbed to a greater or lesser degree by the Sun.

The mean orbital ellipse has an eccentricity of $e = 0.055$ and is inclined to the ecliptic at about $5\overset{\circ}{.}1$. The orbital period from perigee to perigee (known as the *anomalistic* month) has an average value of 27.55 days. This is almost exactly 2 days less than the time between the dates of two successive New or Full phases (the *synodic* month), 29.53 days. This difference arises from the apparent motion of the Sun against the background sky, which amounts to about 30° in one month. After one orbit around the Earth, the Moon requires roughly another two days to catch up with the Sun, and show the same phase.

Because of the eccentric orbit, the Moon varies by up to $6\overset{\circ}{.}3$ from its mean position. This most apparent irregularity is known as the *major inequality*. It has nothing to do with solar perturbations, but is solely caused by the elliptical shape of the orbit. The amount of the major inequality follows from the equation of the centre (6.1), which was described in connection with the series expansions of planetary orbits. Ptolemy pointed out in the *Almagest* that it alone was not sufficient to explain the Moon's orbit satisfactorily. He was the first to notice *evection*, a perturbation amounting to $1\overset{\circ}{.}3$, which depends on the relative location of the Sun and Moon. Considerably later, two further perturbations (*variation* and *annual inequality*) were discovered by Tycho Brahe from his more accurate observations, before Newton provided the basis for a theoretical understanding of lunar motion with his theory of universal gravitation. In 1770, Mayer published the first lunar tables that were sufficiently accurate for determining position and time at sea. Nowadays a total of more than one thousand individual periodic perturbation terms are known.

As well as the greater or lesser periodic oscillations in the Moon's orbit, the influence of the Sun also causes a series of secular effects, the most important of which is a slow rotation of the plane of the lunar orbit. The line of the nodes, i.e., the intersection between the lunar orbit and the ecliptic, rotates by a full 360° in 18.6 years, while the inclination remains almost fixed. The rotation of the line of nodes is *retrograde*, which means that the longitude of the ascending node *decreases* relative to the vernal equinox by about 20° per year. The reason for this shift is that the Sun's gravity constantly tries to pull the Moon down towards the ecliptic and to force the axis of symmetry of the orbit to lie at right-angles to that plane. Like the axis of a gyroscope, however, this is not aligned with the force, so it begins a slow precessional motion, which causes it to rotate once around the pole of the ecliptic in the 18.6 years already mentioned. The position of perigee, i.e., the orientation of the semi-major axis of the orbit in space, also varies. Perigee has a direct (or prograde) shift along the orbit of more than 40° per year. A complete rotation takes about 8.5 years.

The various parameters for the mean motion of the Sun and the Moon that have to be taken into account in theories of lunar motion (with the customary designations) and with

$$T = (\mathrm{JD} - 2451545)/36525$$

are:

- the mean longitude of the Moon (L_0)

$$\begin{aligned} L_0 &= 218\overset{\circ}{.}31617 + 481267\overset{\circ}{.}88088 \cdot T - 4\overset{\prime\prime}{.}06 \cdot T^2 \\ &= 0\overset{\mathrm{r}}{.}60643382 + 1336\overset{\mathrm{r}}{.}85522467 \cdot T - 0\overset{\mathrm{r}}{.}00000313 \cdot T^2 \quad , \qquad (8.1) \end{aligned}$$

- the Moon's mean anomaly (l)

$$\begin{aligned} l &= 134^\circ\!.96292 + 477198^\circ\!.86753 \cdot T + 33''\!.25 \cdot T^2 \\ &= 0^r\!.37489701 + 1325^r\!.55240982 \cdot T + 0^r\!.00002565 \cdot T^2 \quad , \end{aligned} \tag{8.2}$$

- the Sun's mean anomaly (l')

$$\begin{aligned} l' &= 357^\circ\!.52543 + \ 35999^\circ\!.04944 \cdot T - 0''\!.58 \cdot T^2 \\ &= 0^r\!.99312619 + \ \ 99^r\!.99735956 \cdot T - 0^r\!.00000044 \cdot T^2 \quad , \end{aligned} \tag{8.3}$$

- the mean distance of the Moon from the ascending node (F)

$$\begin{aligned} F &= 93^\circ\!.27283 + 483202^\circ\!.01873 \cdot T - 11''\!.56 \cdot T^2 \\ &= 0^r\!.25909118 + 1342^r\!.22782980 \cdot T - 0^r\!.00000892 \cdot T^2 \quad , \end{aligned} \tag{8.4}$$

- and the mean elongation of the Moon (D), i.e., the difference between the mean longitudes of the Sun and the Moon,

$$\begin{aligned} D &= 297^\circ\!.85027 + 445267^\circ\!.11135 \cdot T - 5''\!.15 \cdot T^2 \\ &= 0^r\!.82736186 + 1236^r\!.85308708 \cdot T - 0^r\!.00000397 \cdot T^2 \quad . \end{aligned} \tag{8.5}$$

The individual values are given in degrees and units of one revolution ($1^r = 360^\circ$). L_0 refers to the mean equinox of date whereas the remaining quantities are angles, which are not affected by precession. The longitude of the ascending node Ω is not explicitly employed. It is obtained from the difference $\Omega = L_0 - F$.

The true ecliptic longitude of the Moon (λ) differs from the mean longitude by a series of periodic terms, which have l, l', F and D as arguments:

$$\lambda = L_0 + \Delta\lambda$$

where

$$\begin{aligned} \Delta\lambda = \; & +22640''\cdot\sin(l) && \text{(major inequality)} \\ & -4586''\cdot\sin(l - 2D) && \text{(evection)} \\ & +2370''\cdot\sin(2D) && \text{(variation)} \\ & +769''\cdot\sin(2l) && \text{(major inequality)} \\ & -668''\cdot\sin(l') && \text{(annual inequality)} \\ & -412''\cdot\sin(2F) && \text{(reduction to the ecliptic)} \\ & -212''\cdot\sin(2l - 2D) && \\ & -206''\cdot\sin(l + l' - 2D) && \\ & +192''\cdot\sin(l + 2D) && \\ & -165''\cdot\sin(l' - 2D) && \\ & +148''\cdot\sin(l - l') && \\ & -125''\cdot\sin(D) && \text{(parallactic inequality)} \\ & -110''\cdot\sin(l + l') && \\ & -55''\cdot\sin(2F - 2D) \quad . && \end{aligned}$$

For an unperturbed orbit, the ecliptic latitude β depends on

$$\begin{aligned} \sin\beta &= \sin i \cdot \sin u \\ \beta &\approx i \cdot \sin u \end{aligned}$$

that is, on the orbital inclination i and the position in the orbit. Here u denotes the argument of latitude, i.e. the angle between the position vector and the line of nodes. Taking perturbations by the Sun into account, the latitude of the Moon may be expressed, to a first approximation, by a slightly altered equation, namely:

$$\beta \approx 18520'' \sin(S) + N$$

where $S = F + \Delta S$. ΔS consists of a series of terms, which are mainly the same as in the expansion for $\Delta\lambda$. If we consider the most important differences only, we have:

$$\Delta S \approx \Delta\lambda + 412'' \sin(2F) + 541'' \sin(l') \quad .$$

The quantity N incorporates a small number of additional latitude variations, which are caused by an oscillation of the inclination of the orbit:

$$N = -526'' \sin(F-2D) + 44'' \sin(l+F-2D) - 31'' \sin(-l+F-2D) \ldots \quad (8.6)$$

This greatly simplified description will serve to give a rough idea of the way lunar theory is expressed and how the Moon's ecliptic longitude and latitude are calculated. The equations just described are evaluated in the `MiniMoon` function, which was given in Chap. 3 without detailed discussion. A glance at this program should clarify the individual steps taken in the calculation.

A point that should be mentioned is that we have knowingly avoided using a series expansion of β itself to calculate the ecliptic latitude. Such a series, whose most important terms are

$$\begin{aligned} \beta = {} & 18461'' \sin(F) + 1010'' \sin(l+F) + 1000'' \sin(l-F) - 624'' \sin(F-2D) \\ & -199'' \sin(l-F-2D) - 167'' \sin(l+F-2D) + \ldots \quad , \end{aligned}$$

is certainly easier to employ, but requires, an unnecessarily high expenditure of computational effort. This is because the individual terms contain just odd multiples of F, whereas the terms in the expansions of $\Delta\lambda$ and ΔS always incorporate only even multiples of F. For each term in $\Delta\lambda$, there is a corresponding term in ΔS, requiring the same angular function. If the program is appropriately arranged, therefore, no additional trigonometric terms have to be evaluated in calculating ΔS. Conversely, the series for $\Delta\lambda$ and $\Delta\beta$ have no terms that may be evaluated together.

8.2 Brown's Lunar Theory

The lunar theory developed by E.W. Brown at the beginning of the 20th century is one of the best-known analytical expressions of lunar motion. The version presented here is the Nautical Almanac Office's *Improved Lunar Ephemeris* of 1954. Only a fraction of the theory's more than one thousand perturbation terms are used in the `MoonPos` program. The accuracy thus obtainable is about one second of arc.

The calculation begins by determining the mean arguments L, l, l', F and D from (8.1) to (8.5). The time T is expressed as Julian centuries *Ephemeris Time* since epoch J2000. The mean arguments are also subject to small long-period oscillations, which are to be added to the values given above as corrections:

$$\begin{array}{rrrrrl}
\Delta L_0 & \Delta l & \Delta l' & \Delta F & \Delta D & \\
+0''\!.84 & +2''\!.94 & -6''\!.40 & +0''\!.21 & +7''\!.24 & \cdot\ \sin(2\pi(0.19833+0.05611T)) \\
+0''\!.31 & +0''\!.31 & 0''\!.0 & +0''\!.31 & +0''\!.31 & \cdot\ \sin(2\pi(0.27869+0.04508T)) \\
+14''\!.27 & +14''\!.27 & 0''\!.0 & +14''\!.27 & +14''\!.27 & \cdot\ \sin(2\pi(0.16827-0.36903T)) \\
+7''\!.26 & +9''\!.34 & 0''\!.0 & -88''\!.70 & +7''\!.26 & \cdot\ \sin(2\pi(0.34734-5.37261T)) \\
+0''\!.28 & +1''\!.12 & 0''\!.0 & -15''\!.30 & +0''\!.28 & \cdot\ \sin(2\pi(0.10498-5.37899T)) \\
+0''\!.24 & +0''\!.83 & -1''\!.89 & +0''\!.24 & +2''\!.13 & \cdot\ \sin(2\pi(0.42681-0.41855T)) \\
0''\!.00 & 0''\!.00 & 0''\!.00 & -1''\!.86 & 0''\!.00 & \cdot\ \sin(2\pi(0.14943-5.37511T))
\end{array} \qquad (8.7)$$

Using the improved mean arguments thus obtained, five series of perturbation terms are calculated:

$\Delta\lambda$	perturbations in ecliptic longitude,
$\Delta S, N, \gamma_1 C$	perturbations in ecliptic latitude and
$\Delta \sin \Pi$	perturbations in parallax .

All these series have a common structure:

$$\left\{\begin{array}{c}\Delta\lambda,\ \Delta S,\ N \\ \gamma_1 C,\ \Delta\sin\Pi\end{array}\right\} = \sum_n \left\{\begin{array}{c}a_n,\ b_n,\ c_n \\ d_n,\ e_n\end{array}\right\} \cdot \left\{\begin{array}{c}\sin \\ \cos\end{array}\right\} (p_n l + q_n l' + r_n F + s_n D) \quad .$$

The array (p, q, r, s) describes how an individual summand depends on l, l', F and D and hence the periodicity of the term. An effective way of carrying out the calculation is to combine all the terms that have similar characteristics. The following is just a small portion of the whole table of perturbation terms:

$\Delta\lambda$	ΔS	$\gamma_1 C$	$\Delta \sin \Pi$	p	q	r	s
$+22639''\!.500$	$+22609''\!.07$	$+0''\!.079$	$+186''\!.5398$	+1	0	0	0
$-4586''\!.465$	$-4578''\!.13$	$-0''\!.077$	$+34''\!.3117$	+1	0	0	−2
$+2369''\!.912$	$+2373''\!.36$	$+0''\!.601$	$+28''\!.2333$	0	0	0	+2
$+769''\!.016$	$+767''\!.96$	$+0''\!.107$	$+10''\!.1657$	+2	0	0	0
$-668''\!.146$	$-126''\!.98$	$-1''\!.302$	$-0''\!.3997$	0	+1	0	0
$-411''\!.608$	$-0''\!.20$	$+0''\!.000$	$-0''\!.0124$	0	0	+2	0

Because of its different coefficients, the series for the quantity N cannot be calculated at the same time as the other series. However, it contains only a few terms.

Because of the large number of perturbation terms, it is worthwhile taking some pains over the calculation of the various angular functions that occur. The calculation of a sine or cosine function – when compared with elementary operations like the four basic functions – requires a relatively large amount of effort and time. It is not, however, necessary to calculate every angular function explicitly, if use is made of the addition theorems

$$\begin{aligned}\cos(\alpha_1+\alpha_2) &= \cos\alpha_1\cos\alpha_2 - \sin\alpha_1\sin\alpha_2\\ \sin(\alpha_1+\alpha_2) &= \sin\alpha_1\cos\alpha_2 + \cos\alpha_1\sin\alpha_2 \quad .\end{aligned}$$

Using the short `AddThe` function

```
void AddThe(double c1, double s1, double c2, double s2,
            double& c, double& s)
  { c=c1*c2-s1*s2; s=s1*c2+c1*s2; }
```

we can first calculate sines and cosines for multiples of the mean arguments:

$$\texttt{Cos[j+o][i]} = \begin{cases}\cos(jl) & (i=0)\\ \cos(jl') & (i=1)\\ \cos(jF) & (i=2)\\ \cos(jD) & (i=3)\end{cases} \qquad \texttt{Sin[j+o][i]} = \begin{cases}\sin(jl) & (i=0)\\ \sin(jl') & (i=1)\\ \sin(jF) & (i=2)\\ \sin(jD) & (i=3)\end{cases} \quad .$$

For l (i=0) and using

```
const int o=6;              // Index offset

Cos[o][i]=1.0;  Cos[o+1][i]=cos(arg);  Cos[o-1][i]=+Cos[o+1][i];
Sin[o][i]=0.0;  Sin[o+1][i]=sin(arg);  Sin[o-1][i]=-Sin[o+1][i];
for (int j=2;j<=max;j++) {
  AddThe ( Cos[o+j-1][i],Sin[o+j-1][i], Cos[o+1][i],Sin[o+1][i],
           Cos[o+j][i],Sin[o+j][i] );
  Cos[o-j][i]=+Cos[o+j][i];
  Sin[o-j][i]=-Sin[o+j][i];
};
```

we can obtain all the values of $\cos(jl)$ and $\sin(jl)$ for `j=-max...+max` with only two calls of the sine or cosine functions. For a C++ implementation of the ILE series it is advisable to define a dedicated class `ILE_Pert`, which, among others, contains the members `Cos` and `Sin`.

```
// Constants
const int o   = 6;          // Index offset
const int dim = 2*o+1;      // Work array dimension
// Definition of ILE_Pert class for summing up Brown's lunar
// perturbation series
class ILE_Pert
{
  public:
    // Initialization (mean arguments, long-periodic corrections, etc.)
    void Init (double T);
```

```
    // Perturbation term
    void Term (int p, int q, int r, int s, double& x, double& y);
    // Summation of solar perturbations
    void AddSol ( double coeffl, double coeffS, double coeffg,
                  double coeffP, int p, int q, int r, int s );
    // Summation of perturbation in latitude
    void AddN (double coeffN, int p, int q, int r, int s);
    // Planetary perturbations
    void Planetary (double T);
    // Coordinates
    double lambda();  // ecliptic longitude in [rad]
    double beta();    // ecliptic latitude in [rad]
    double dist();    // geocentric distance in [km]
  private:
    double Dgam;                           // Longperiodic perturbation
    double Dlam, DS, gam1C, sinPi, N;      // Periodic perturbations
    double L0, l,ls,F,D;                   // Mean arguments of lunar orbit
    double Cos[dim][4], Sin[dim][4];       // (Co)sine of mean arguments
};
```

The `Term` method, which uses the precomputed values `Cos[][]` and `Sin[][]`, then allows

$$x = \cos(pl + ql' + rF + sD) \quad \text{und} \quad y = \sin(pl + ql' + rF + sD)$$

to be calculated simply for given coefficients (p, q, r, s).

```
// Term: calculates x=cos(p*l+q*ls+r*F+s*D) and y=sin(p*l+q*ls+r*F+s*D)
void ILE_Pert::Term (int p, int q, int r, int s, double& x, double& y)
{
  int i[4];
  i[0]=p; i[1]=q; i[2]=r; i[3]=s;  x=1.0; y=0.0;
  for (int k=0; k<=3; k++)
    if (i[k]!=0) AddThe(x,y,Cos[o+i[k]][k],Sin[o+i[k]][k],x,y);
}
```

The application of this function may be explained very clearly by taking the series for the latitude perturbation N as an example. The `AddN` method is designed to sum all the terms in the perturbation series mentioned into the variable `N`:

```
// AddN: Summation of perturbation in latitude
void ILE_Pert::AddN (double coeffN, int p, int q, int r, int s)  {
  double x,y;
  Term(p,q,r,s,x,y);  N += coeffN*y;
}
```

Using `Term`, the values of the associated angular functions are calculated first (N is made up of sine terms, only). The sine term is then multiplied by the coefficients and added to N. An individual term is evaluated in each call of `AddN`, which accepts the coefficient and the characteristic values (p, q, r, s) as input. Naturally, before `AddN` is called, all the values of the arrays `Sin[][]` and `Cos[][]` must be determined. This is accomplished by the `Init` method, which computes the required auxiliary data for given time T. The code segment for computing the perturbation in latitude N is then given by:

```
ILE_Pert Pert;
Pert.Init(T);
// Solar perturbations in latitude
Pert.AddN(-526.069, 0, 0,1,-2);   Pert.AddN(  -3.352, 0, 0,1,-4);
Pert.AddN( +44.297,+1, 0,1,-2);   Pert.AddN(  -6.000,+1, 0,1,-4);
Pert.AddN( +20.599,-1, 0,1, 0);   Pert.AddN( -30.598,-1, 0,1,-2);
Pert.AddN( -24.649,-2, 0,1, 0);   Pert.AddN(  -2.000,-2, 0,1,-2);
Pert.AddN( -22.571, 0,+1,1,-2);   Pert.AddN( +10.985, 0,-1,1,-2);
```

The `AddSol` method

```
// AddSol: Summation of solar perturbation terms
void ILE_Pert::AddSol (
  double coeffl, double coeffS, double coeffg, double coeffP,
  int p, int q, int r, int s )
{
  double x,y;
  Term (p,q,r,s,x,y);
  Dlam  += coeffl*y; DS    += coeffS*y;
  gam1C += coeffg*x; sinPi += coeffP*x;
}
```

which calculates the series $\Delta\lambda$, ΔS, $\gamma_1 C$, and $\Delta \sin \Pi$, functions in precisely the same way.

The treatment of the solar perturbation terms in the method just described is not completely comprehensive, however. The reason for this is that, strictly speaking, the coefficients of the perturbation terms are slightly different from the values quoted. Each coefficient in the series for $\Delta\lambda$, ΔS, N, $\gamma_1 C$, and $\Delta \sin \Pi$ is actually a function of various parameters – such as the eccentricity of the solar orbit, for example – for which Brown took specific numerical values. To take the correct values for these parameters into account, each term (which is characterized by the array (p, q, r, s)) must be multiplied by a factor

$$(1.000002208)^{|p|} \cdot (1.0 - 0.002495388(T+1))^{|q|} \cdot (1.000002708 + 139.978\Delta\gamma)^{|r|} \ .$$

Here $\Delta\gamma$ is given by the value of

$$\begin{aligned} \Delta\gamma &= -0.000003332 \cdot \sin(2\pi(0.59734 - 5.37261T)) \\ &\quad -0.000000539 \cdot \sin(2\pi(0.35498 - 5.37899T)) \\ &\quad -0.000000064 \cdot \sin(2\pi(0.39943 - 5.37511T)) \quad , \end{aligned}$$

which is calculated in `ILE_Pert::Init`. Anyway, these corrections may be taken into account very simply, if in calculating the arrays `Cos[][]` and `Sin[][]` as described above, we multiply the values

`Cos[o+1][0]` and `Sin[o+1][0]` by (1.000002208),
`Cos[o+1][1]` and `Sin[o+1][1]` by $(1.0 - 0.002495388(T+1))$ and
`Cos[o+1][2]` and `Sin[o+1][2]` by $(1.000002708 + 139.978\Delta\gamma)$.

The appropriate correction factors will already have been taken into account in all the previously calculated angular functions, therefore, when the repeated calls to `AddThe` are made. For example, `Sin[o+3,1]` contains the value of $(1.0 - 0.002495388(T+1))^{|3|} \cdot \sin(3l')$ instead of $\sin(3l')$. This small change enables all the terms to be automatically, and without great difficulty, multiplied by the right factors.

This concludes the calculation of the various lunar perturbations that are caused by the Sun. However, the Sun is not the only body in the Solar System that affects the orbit of the Moon. A complete theory of the lunar orbit cannot neglect planetary perturbations. In the `MoonPos` function the most important contributions by Venus and Jupiter are also taken into account:

$$\begin{aligned}
\Delta\lambda_{\text{plan}} = {} & +0\overset{\prime\prime}{.}82\sin(0\overset{\mathrm{r}}{.}7736 - 62\overset{\mathrm{r}}{.}5512T) + 0\overset{\prime\prime}{.}31\sin(0\overset{\mathrm{r}}{.}0466 - 125\overset{\mathrm{r}}{.}1025T) \\
& +0\overset{\prime\prime}{.}35\sin(0\overset{\mathrm{r}}{.}5785 - 25\overset{\mathrm{r}}{.}1042T) + 0\overset{\prime\prime}{.}66\sin(0\overset{\mathrm{r}}{.}4591 + 1335\overset{\mathrm{r}}{.}8075T) \\
& +0\overset{\prime\prime}{.}64\sin(0\overset{\mathrm{r}}{.}3130 - 91\overset{\mathrm{r}}{.}5680T) + 1\overset{\prime\prime}{.}14\sin(0\overset{\mathrm{r}}{.}1480 + 1331\overset{\mathrm{r}}{.}2898T) \\
& +0\overset{\prime\prime}{.}21\sin(0\overset{\mathrm{r}}{.}5918 + 1056\overset{\mathrm{r}}{.}5859T) + 0\overset{\prime\prime}{.}44\sin(0\overset{\mathrm{r}}{.}5784 + 1322\overset{\mathrm{r}}{.}8595T) \\
& +0\overset{\prime\prime}{.}24\sin(0\overset{\mathrm{r}}{.}2275 - 5\overset{\mathrm{r}}{.}7374T) + 0\overset{\prime\prime}{.}28\sin(0\overset{\mathrm{r}}{.}2965 + 2\overset{\mathrm{r}}{.}6929T) \\
& +0\overset{\prime\prime}{.}33\sin(0\overset{\mathrm{r}}{.}3132 + 6\overset{\mathrm{r}}{.}3368T) \quad .
\end{aligned}$$

These perturbations in longitude are calculated in `ILE_Pert::Planetary`.

Now all the necessary values are known. The ecliptic longitude of the Moon is obtained by adding the solar and planetary perturbations to the mean longitude:

$$\lambda = L_0 + \Delta\lambda + \Delta\lambda_{\text{plan}} \quad .$$

L_0 here is the value of the mean longitude already corrected as given in (8.7). To obtain the ecliptic latitude β, ΔS is first added to the mean distance F from the node:

$$S = F + \Delta S \quad .$$

We then have

$$\begin{aligned}
\beta = {} & (1.000002708 + 139.978\Delta\gamma) \cdot \{18519\overset{\prime\prime}{.}70 + \gamma_1 C\} \cdot \sin(S) \\
& - \{6\overset{\prime\prime}{.}24\} \cdot \sin(3S) + N \quad .
\end{aligned}$$

The distance of the Moon is calculated from the sine of the horizontal parallax:

$$r = (1/\sin\Pi)\, R_\oplus \quad \text{(radius of the Earth } R_\oplus \approx 6378.14 \text{ km)} \quad .$$

$\sin\Pi$ has the value

$$\sin\Pi = 0.999953253 \cdot (3422\overset{\prime\prime}{.}7 + \Delta\sin\Pi) \cdot \frac{\pi}{180 \cdot 3600''} \quad .$$

The factor

$$\frac{\pi}{180 \cdot 3600''} = \frac{1}{206264\overset{\prime\prime}{.}81}$$

here merely serves to convert seconds of arc into radians.

The complete MoonPos function follows, including those methods of ILE_Pert that have not been listed before. For a given time T, (expressed in Julian centuries Ephemeris Time since epoch J2000), MoonPos calculates the geocentric ecliptic coordinates of the Moon (in [km]) with respect to the mean equinox of date.

```
// Sine
double Sine (double x) { return sin(pi2*Frac(x)); }

//
// Initialization
//
void ILE_Pert::Init (double T)
{
  // Variables
  double dL0, dl, dls, dF, dD;           // Longperiodic perturbations
  double T2, arg, fac;                   // Auxiliary variables
  double S1, S2, S3, S4, S5, S6, S7;
  int    max;

  T2=T*T; // Time

  // Reset perturbations
  Dlam=0.0; DS=0.0; gam1C=0.0; sinPi=3422.7000; N=0.0;

  // Longperiodic perturbations
  S1 = Sine (0.19833+0.05611*T);  S2 = Sine (0.27869+0.04508*T);
  S3 = Sine (0.16827-0.36903*T);  S4 = Sine (0.34734-5.37261*T);
  S5 = Sine (0.10498-5.37899*T);  S6 = Sine (0.42681-0.41855*T);
  S7 = Sine (0.14943-5.37511*T);
  dL0 = 0.84*S1+0.31*S2+14.27*S3+ 7.26*S4+ 0.28*S5+0.24*S6;
  dl  = 2.94*S1+0.31*S2+14.27*S3+ 9.34*S4+ 1.12*S5+0.83*S6;
  dls =-6.40*S1                                   -1.89*S6;
  dF  = 0.21*S1+0.31*S2+14.27*S3-88.70*S4-15.30*S5+0.24*S6-1.86*S7;
  dD  = dL0-dls;
  Dgam   = -3332e-9 * Sine (0.59734-5.37261*T)
             -539e-9 * Sine (0.35498-5.37899*T)
              -64e-9 * Sine (0.39943-5.37511*T);

  // Mean arguments of the lunar orbit (incl. longperiodic corrections)
  // L0 mean longitude of the Moon
  // l  mean anomaly of the Moon     l' mean anomaly of the Sun
  // F  mean distance from the node  D  mean elongation from the Sun
  L0 = pi2*Frac(0.60643382+1336.85522467*T-0.00000313*T2) + dL0/Arcs;
  l  = pi2*Frac(0.37489701+1325.55240982*T+0.00002565*T2) + dl /Arcs;
  ls = pi2*Frac(0.99312619+  99.99735956*T-0.00000044*T2) + dls/Arcs;
  F  = pi2*Frac(0.25909118+1342.22782980*T-0.00000892*T2) + dF /Arcs;
  D  = pi2*Frac(0.82736186+1236.85308708*T-0.00000397*T2) + dD /Arcs;

  // Cosine and sine of multiples of mean arguments
  // incl. secular correction
  for (int i=0; i<=3; i++) {
    switch(i) {
      case 0: arg=l;  max=4; fac=1.000002208;                 break;
```

```
        case 1: arg=ls; max=3; fac=0.997504612-0.002495388*T; break;
        case 2: arg=F;  max=4; fac=1.000002708+139.978*Dgam;  break;
        case 3: arg=D;  max=6; fac=1.0;                       break;
      };
      Cos[o][i]=1.0;  Cos[o+1][i]=cos(arg)*fac;  Cos[o-1][i]=+Cos[o+1][i];
      Sin[o][i]=0.0;  Sin[o+1][i]=sin(arg)*fac;  Sin[o-1][i]=-Sin[o+1][i];
      for (int j=2;j<=max;j++) {
        AddThe ( Cos[o+j-1][i],Sin[o+j-1][i], Cos[o+1][i],Sin[o+1][i],
                 Cos[o+j][i],Sin[o+j][i] );
        Cos[o-j][i]=+Cos[o+j][i];
        Sin[o-j][i]=-Sin[o+j][i];
      };
    };
  }

  // Planetary: Perturbations of ecliptic latitude by Venus and Jupiter
  void ILE_Pert::Planetary (double T)
  {
    Dlam +=
         +0.82*Sine(0.7736  -62.5512*T)+0.31*Sine(0.0466 -125.1025*T)
         +0.35*Sine(0.5785  -25.1042*T)+0.66*Sine(0.4591+1335.8075*T)
         +0.64*Sine(0.3130  -91.5680*T)+1.14*Sine(0.1480+1331.2898*T)
         +0.21*Sine(0.5918+1056.5859*T)+0.44*Sine(0.5784+1322.8595*T)
         +0.24*Sine(0.2275   -5.7374*T)+0.28*Sine(0.2965   +2.6929*T)
         +0.33*Sine(0.3132   +6.3368*T);
  }
  // lambda, beta, dist

  double ILE_Pert::lambda() {
    return Modulo ( L0+Dlam/Arcs, pi2 );
  }
  double ILE_Pert::beta() {
    double S   = F + DS/Arcs;
    double fac = 1.000002708+139.978*Dgam;
    return (fac*(18518.511+1.189+gam1C)*sin(S)-6.24*sin(3*S)+N) / Arcs;
  }
  double ILE_Pert::dist() {
    return R_Earth * Arcs / (sinPi * 0.999953253);
  }

//------------------------------------------------------------------------------
// MoonPos: Computes the Moon's ecliptic position using Brown's theory
//          (Improved Lunar Ephemeris)
//   T          Time in Julian centuries since J2000
//   <return>   Geocentric position of the Moon (in [km]) referred to the
//              ecliptic and equinox of date
//------------------------------------------------------------------------------
Vec3D MoonPos (double T)
{
  ILE_Pert Pert;

  // Initialization
  Pert.Init(T);
```

```
// Solar perturbations
Pert.AddSol (   13.902,   14.06,-0.001,   0.2607,0, 0, 0, 4);
Pert.AddSol (    0.403,   -4.01,+0.394,   0.0023,0, 0, 0, 3);
Pert.AddSol ( 2369.912, 2373.36,+0.601,  28.2333,0, 0, 0, 2);
Pert.AddSol ( -125.154, -112.79,-0.725,  -0.9781,0, 0, 0, 1);
Pert.AddSol (    1.979,    6.98,-0.445,   0.0433,1, 0, 0, 4);
Pert.AddSol (  191.953,  192.72,+0.029,   3.0861,1, 0, 0, 2);
Pert.AddSol (   -8.466,  -13.51,+0.455,  -0.1093,1, 0, 0, 1);
Pert.AddSol (22639.500,22609.07,+0.079, 186.5398,1, 0, 0, 0);
Pert.AddSol (   18.609,    3.59,-0.094,   0.0118,1, 0, 0,-1);
Pert.AddSol (-4586.465,-4578.13,-0.077,  34.3117,1, 0, 0,-2);
Pert.AddSol (   +3.215,    5.44,+0.192,  -0.0386,1, 0, 0,-3);
Pert.AddSol (  -38.428,  -38.64,+0.001,   0.6008,1, 0, 0,-4);
Pert.AddSol (   -0.393,   -1.43,-0.092,   0.0086,1, 0, 0,-6);
Pert.AddSol (   -0.289,   -1.59,+0.123,  -0.0053,0, 1, 0, 4);
Pert.AddSol (  -24.420,  -25.10,+0.040,  -0.3000,0, 1, 0, 2);
Pert.AddSol (   18.023,   17.93,+0.007,   0.1494,0, 1, 0, 1);
Pert.AddSol ( -668.146, -126.98,-1.302,  -0.3997,0, 1, 0, 0);
Pert.AddSol (    0.560,    0.32,-0.001,  -0.0037,0, 1, 0,-1);
Pert.AddSol ( -165.145, -165.06,+0.054,   1.9178,0, 1, 0,-2);
Pert.AddSol (   -1.877,   -6.46,-0.416,   0.0339,0, 1, 0,-4);
Pert.AddSol (    0.213,    1.02,-0.074,   0.0054,2, 0, 0, 4);
Pert.AddSol (   14.387,   14.78,-0.017,   0.2833,2, 0, 0, 2);
Pert.AddSol (   -0.586,   -1.20,+0.054,  -0.0100,2, 0, 0, 1);
Pert.AddSol (  769.016,  767.96,+0.107,  10.1657,2, 0, 0, 0);
Pert.AddSol (   +1.750,    2.01,-0.018,   0.0155,2, 0, 0,-1);
Pert.AddSol ( -211.656, -152.53,+5.679,  -0.3039,2, 0, 0,-2);
Pert.AddSol (   +1.225,    0.91,-0.030,  -0.0088,2, 0, 0,-3);
Pert.AddSol (  -30.773,  -34.07,-0.308,   0.3722,2, 0, 0,-4);
Pert.AddSol (   -0.570,   -1.40,-0.074,   0.0109,2, 0, 0,-6);
Pert.AddSol (   -2.921,  -11.75,+0.787,  -0.0484,1, 1, 0, 2);
Pert.AddSol (   +1.267,    1.52,-0.022,   0.0164,1, 1, 0, 1);
Pert.AddSol ( -109.673, -115.18,+0.461,  -0.9490,1, 1, 0, 0);
Pert.AddSol ( -205.962, -182.36,+2.056,  +1.4437,1, 1, 0,-2);
Pert.AddSol (    0.233,    0.36, 0.012,  -0.0025,1, 1, 0,-3);
Pert.AddSol (   -4.391,   -9.66,-0.471,   0.0673,1, 1, 0,-4);
Pert.AddSol (    0.283,    1.53,-0.111,  +0.0060,1,-1, 0,+4);
Pert.AddSol (   14.577,   31.70,-1.540,  +0.2302,1,-1, 0, 2);
Pert.AddSol (  147.687,  138.76,+0.679,  +1.1528,1,-1, 0, 0);
Pert.AddSol (   -1.089,    0.55,+0.021,   0.0   ,1,-1, 0,-1);
Pert.AddSol (   28.475,   23.59,-0.443,  -0.2257,1,-1, 0,-2);
Pert.AddSol (   -0.276,   -0.38,-0.006,  -0.0036,1,-1, 0,-3);
Pert.AddSol (    0.636,    2.27,+0.146,  -0.0102,1,-1, 0,-4);
Pert.AddSol (   -0.189,   -1.68,+0.131,  -0.0028,0, 2, 0, 2);
Pert.AddSol (   -7.486,   -0.66,-0.037,  -0.0086,0, 2, 0, 0);
Pert.AddSol (   -8.096,  -16.35,-0.740,   0.0918,0, 2, 0,-2);
Pert.AddSol (   -5.741,   -0.04, 0.0  ,  -0.0009,0, 0, 2, 2);
Pert.AddSol (    0.255,    0.0 , 0.0  ,   0.0   ,0, 0, 2, 1);
Pert.AddSol ( -411.608,   -0.20, 0.0  ,  -0.0124,0, 0, 2, 0);
Pert.AddSol (    0.584,    0.84, 0.0  ,  +0.0071,0, 0, 2,-1);
Pert.AddSol (  -55.173,  -52.14, 0.0  ,  -0.1052,0, 0, 2,-2);
Pert.AddSol (    0.254,    0.25, 0.0  ,  -0.0017,0, 0, 2,-3);
Pert.AddSol (   +0.025,   -1.67, 0.0  ,  +0.0031,0, 0, 2,-4);
```

```
Pert.AddSol (    1.060,     2.96,-0.166,   0.0243,3, 0, 0,+2);
Pert.AddSol (   36.124,    50.64,-1.300,   0.6215,3, 0, 0, 0);
Pert.AddSol (  -13.193,   -16.40,+0.258,  -0.1187,3, 0, 0,-2);
Pert.AddSol (   -1.187,    -0.74,+0.042,   0.0074,3, 0, 0,-4);
Pert.AddSol (   -0.293,    -0.31,-0.002,   0.0046,3, 0, 0,-6);
Pert.AddSol (   -0.290,    -1.45,+0.116,  -0.0051,2, 1, 0, 2);
Pert.AddSol (   -7.649,   -10.56,+0.259,  -0.1038,2, 1, 0, 0);
Pert.AddSol (   -8.627,    -7.59,+0.078,  -0.0192,2, 1, 0,-2);
Pert.AddSol (   -2.740,    -2.54,+0.022,   0.0324,2, 1, 0,-4);
Pert.AddSol (    1.181,     3.32,-0.212,   0.0213,2,-1, 0,+2);
Pert.AddSol (    9.703,    11.67,-0.151,   0.1268,2,-1, 0, 0);
Pert.AddSol (   -0.352,    -0.37,+0.001,  -0.0028,2,-1, 0,-1);
Pert.AddSol (   -2.494,    -1.17,-0.003,  -0.0017,2,-1, 0,-2);
Pert.AddSol (    0.360,     0.20,-0.012,  -0.0043,2,-1, 0,-4);
Pert.AddSol (   -1.167,    -1.25,+0.008,  -0.0106,1, 2, 0, 0);
Pert.AddSol (   -7.412,    -6.12,+0.117,   0.0484,1, 2, 0,-2);
Pert.AddSol (   -0.311,    -0.65,-0.032,   0.0044,1, 2, 0,-4);
Pert.AddSol (   +0.757,     1.82,-0.105,   0.0112,1,-2, 0, 2);
Pert.AddSol (   +2.580,     2.32,+0.027,   0.0196,1,-2, 0, 0);
Pert.AddSol (   +2.533,     2.40,-0.014,  -0.0212,1,-2, 0,-2);
Pert.AddSol (   -0.344,    -0.57,-0.025,  +0.0036,0, 3, 0,-2);
Pert.AddSol (   -0.992,    -0.02, 0.0  ,   0.0   ,1, 0, 2, 2);
Pert.AddSol (  -45.099,    -0.02, 0.0  ,  -0.0010,1, 0, 2, 0);
Pert.AddSol (   -0.179,    -9.52, 0.0  ,  -0.0833,1, 0, 2,-2);
Pert.AddSol (   -0.301,    -0.33, 0.0  ,   0.0014,1, 0, 2,-4);
Pert.AddSol (   -6.382,    -3.37, 0.0  ,  -0.0481,1, 0,-2, 2);
Pert.AddSol (   39.528,    85.13, 0.0  ,  -0.7136,1, 0,-2, 0);
Pert.AddSol (    9.366,     0.71, 0.0  ,  -0.0112,1, 0,-2,-2);
Pert.AddSol (    0.202,     0.02, 0.0  ,   0.0   ,1, 0,-2,-4);
Pert.AddSol (    0.415,     0.10, 0.0  ,   0.0013,0, 1, 2, 0);
Pert.AddSol (   -2.152,    -2.26, 0.0  ,  -0.0066,0, 1, 2,-2);
Pert.AddSol (   -1.440,    -1.30, 0.0  ,  +0.0014,0, 1,-2, 2);
Pert.AddSol (    0.384,    -0.04, 0.0  ,   0.0   ,0, 1,-2,-2);
Pert.AddSol (   +1.938,    +3.60,-0.145,  +0.0401,4, 0, 0, 0);
Pert.AddSol (   -0.952,    -1.58,+0.052,  -0.0130,4, 0, 0,-2);
Pert.AddSol (   -0.551,    -0.94,+0.032,  -0.0097,3, 1, 0, 0);
Pert.AddSol (   -0.482,    -0.57,+0.005,  -0.0045,3, 1, 0,-2);
Pert.AddSol (    0.681,     0.96,-0.026,   0.0115,3,-1, 0, 0);
Pert.AddSol (   -0.297,    -0.27, 0.002,  -0.0009,2, 2, 0,-2);
Pert.AddSol (    0.254,    +0.21,-0.003,   0.0   ,2,-2, 0,-2);
Pert.AddSol (   -0.250,    -0.22, 0.004,   0.0014,1, 3, 0,-2);
Pert.AddSol (   -3.996,     0.0 , 0.0  ,  +0.0004,2, 0, 2, 0);
Pert.AddSol (    0.557,    -0.75, 0.0  ,  -0.0090,2, 0, 2,-2);
Pert.AddSol (   -0.459,    -0.38, 0.0  ,  -0.0053,2, 0,-2, 2);
Pert.AddSol (   -1.298,     0.74, 0.0  ,  +0.0004,2, 0,-2, 0);
Pert.AddSol (    0.538,     1.14, 0.0  ,  -0.0141,2, 0,-2,-2);
Pert.AddSol (    0.263,     0.02, 0.0  ,   0.0   ,1, 1, 2, 0);
Pert.AddSol (    0.426,    +0.07, 0.0  ,  -0.0006,1, 1,-2,-2);
Pert.AddSol (   -0.304,    +0.03, 0.0  ,  +0.0003,1,-1, 2, 0);
Pert.AddSol (   -0.372,    -0.19, 0.0  ,  -0.0027,1,-1,-2, 2);
Pert.AddSol (   +0.418,     0.0 , 0.0  ,   0.0   ,0, 0, 4, 0);
Pert.AddSol (   -0.330,    -0.04, 0.0  ,   0.0   ,3, 0, 2, 0);
```

```
  // Solar perturbations in latitude
  Pert.AddN(-526.069, 0, 0,1,-2);    Pert.AddN(  -3.352, 0, 0,1,-4);
  Pert.AddN( +44.297,+1, 0,1,-2);    Pert.AddN(  -6.000,+1, 0,1,-4);
  Pert.AddN( +20.599,-1, 0,1, 0);    Pert.AddN( -30.598,-1, 0,1,-2);
  Pert.AddN( -24.649,-2, 0,1, 0);    Pert.AddN(  -2.000,-2, 0,1,-2);
  Pert.AddN( -22.571, 0,+1,1,-2);    Pert.AddN( +10.985, 0,-1,1,-2);
  // Planetary perturbations
  Pert.Planetary(T);

  // Position vector
  return Vec3D ( Polar(Pert.lambda(),Pert.beta(),Pert.dist()) );
}
```

The MoonPos function may be supplemented by MoonEqu, which calculates the true equatorial coordinates, instead of mean ecliptic coordinates.

```
//------------------------------------------------------------------------------
// MoonEqu: Computes the Moon's equatorial position using Brown's theory
//          (Improved Lunar Ephemeris)
//   T         Time in Julian centuries since J2000
//   <return>  Geocentric equatorial position of the Moon (in [km])
//             referred to the true equinox of date
//------------------------------------------------------------------------------
Vec3D MoonEqu (double T) {
  return NutMatrix(T) * Ecl2EquMatrix(T) * MoonPos(T);
}
```

8.3 The Chebyshev Approximation

Despite all the tricks used in evaluating the perturbation series, the calculation of lunar coordinates using the MoonPos function is still relatively costly in time and effort. This is particularly noticeable when numerous positions of the Moon are required at any one time. In the next chapter, for example, we want to predict occultations of stars by the Moon and we shall use an iterative procedure to determine the time of conjunction of the Moon and a specific star. To be able to solve such computationally intensive calculations efficiently, we will now develop a dedicated C++ class by which the coordinates of a body may be approximated in a form that is suitable for practical use. This will then enable us to save computing time, when more coordinate values are required, than we expend in obtaining the approximation.

One simple example of an approximation has already been discussed in Chap. 3, when we represented the altitude of the Sun or the Moon by a parabola. There, for every three points on the function, we determined a second-order polynomial, which approximately represented the change in the actual function over that interval. Similarly, we can take n different points (x_i, y_i) where $i = 1 \dots n$, and unambiguously determine a polynomial $P_n(x)$ of order $n-1$, that fits the given points exactly. There is a whole series of procedures that may be used to ascertain what this polynomial is. Here we will mention only Lagrange's and Newton's methods.

Unfortunately, in many practical cases a polynomial determined in such a way proves to be unsuitable. The given values of the function are always correctly represented, but between these fixed points the polynomial shows variations, which are not related to the actual curve of the function that is being approximated, and these variations grow in an uncontrolled manner at each end of the interval covered. This effect is greater, the higher the order of the polynomial chosen. Usable results are therefore generally obtained with Lagrange's interpolation method for low orders only (up to about $n = 5$). We require a procedure that guarantees a smooth approximation even with high-order polynomials. We shall now show how such an approximation may be obtained. The key to this are the so-called *Chebyshev polynomials* (Fig. 8.1).

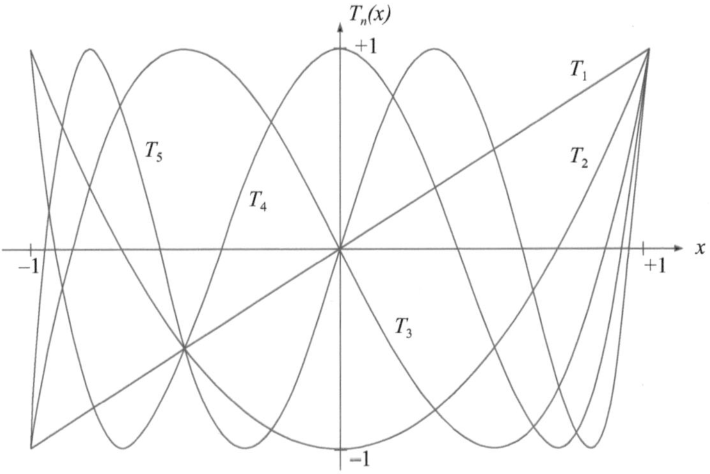

Fig. 8.1. The Chebyshev polynomials T_1 to T_5

The n-th order Chebyshev polynomial (for $|x| \leq 1$) is defined as

$$T_n(x) = \cos(n \cdot \arccos x) \quad . \tag{8.8}$$

It is not immediately obvious from this trigonometric representation that we are actually dealing with a polynomial. This is seen easily only for $n = 0$ ($T_0 = 1$) and $n = 1$ ($T_1 = x$). By using the addition theorem for the cosine function, however, we can rapidly derive a recursive relation for T_n, from which it is quite obvious that (8.8) actually defines polynomials. In general,

$$\cos(\alpha + \beta) + \cos(\alpha - \beta) \quad = \quad 2\cos(\alpha)\cos(\beta)$$

is valid, as is

$$\cos((n+1)\varphi) \quad = \quad 2\cos(n\varphi)\cos(\varphi) - \cos((n-1)\varphi) \quad . \tag{8.9}$$

If we substitute $\varphi = \arccos x$, then we have

$$T_{n+1}(x) \quad = \quad 2xT_n(x) - T_{n-1}(x) \quad \text{for} \quad n \geq 1 \quad . \tag{8.10}$$

We shall later make detailed use of this recursive expression. In any case, this relation means that $T_n(x)$ is definitely an n-th order polynomial of x. Written in full, the first Chebyshev polynomials are:

$$\begin{aligned} T_0(x) &= 1 & T_3(x) &= 4x^3 - 3x \\ T_1(x) &= x & T_4(x) &= 8x^4 - 8x^2 + 1 \\ T_2(x) &= 2x^2 - 1 & T_5(x) &= 16x^5 - 20x^3 + 5x \quad . \end{aligned}$$

We now need to look at the properties of Chebyshev polynomials in the interval $[-1, +1]$ in more detail. As may be seen easily by substitution in the definition (8.8), $T_n(x)$ becomes zero at exactly n points within this interval:

$$T_n(x) = 0 \quad \text{for} \quad x = \cos(\pi \cdot (k + 1/2)/n) \quad \text{where } k = 0, \ldots, n-1 \,. \tag{8.11}$$

For $|x| \leq 1$, the values of $|T_n(x)|$ cannot exceed 1.

To approximate an arbitrary function $f(x)$ in the interval $[a, b]$, we now replace the independent variable x by the normalized variable $\hat{x}$ for the interval $[-1, +1]$. For this we require the conversions

$$\hat{x} = \frac{x - \frac{1}{2}(a+b)}{\frac{1}{2}(b-a)} \qquad \text{for } x \in [a, b] \rightarrow \hat{x} \in [-1, +1]$$

and

$$x = \hat{x} \cdot \frac{1}{2}(b-a) + \frac{1}{2}(a+b) \quad \text{for } \hat{x} \in [-1, +1] \rightarrow x \in [a, b] \quad .$$

A function $f(x)$ may then be expressed in the form

$$f(x) \approx f^*(x) = \sum_{j=0}^{n} c_j T_j(\hat{x}) \; - \; c_0/2 \tag{8.12}$$

by Chebyshev polynomials up to order n. The coefficients c_j in this sum are calculated from

$$c_j = \frac{2}{n+1} \sum_{k=0}^{n} f(x_k^{n+1}) T_j(\hat{x}_k^{n+1}) \quad , \tag{8.13}$$

where $\hat{x}_k^{n+1}$ represents the k-th root of T_{n+1}. Written in full, the individual terms in this equation, because of (8.8) and (8.11), are:

$$\begin{aligned} x_k^{n+1} &= \frac{(b-a)}{2} \cdot \cos\left(\pi \frac{2k+1}{2n+2}\right) + \frac{(a+b)}{2} \qquad k = 0, 1, \ldots, n \\ T_j(\hat{x}_k^{n+1}) &= \cos\left(j\pi \frac{2k+1}{2n+2}\right) \quad . \end{aligned}$$

We can see that to calculate the coefficient c_j, the given function f must be evaluated at $(n+1)$ points. These are fixed as the points at which T_{n+1} is zero, and cannot be chosen freely as in the Lagrangian interpolation. This is the secret of the Chebyshev approximation's success. f^* is not just an n-th order polynomial that agrees with f at $(n+1)$ points $(f^*(x_k^{n+1}) = f(x_k^{n+1}))$. By choosing points that are closer spaced at the ends of the interpolation interval than they are in the centre, a far smoother, overall approximation error is guaranteed. This is particularly well seen if, in calculating f^*, we do not sum all the terms in (8.12). If we neglect (say) the highest term $c_n T_n$, then we can definitely say that, because $|T_n| \leq 1$ for all values in $[a, b]$, the resulting error is smaller than $|c_n|$. Such an estimate is impossible in other interpolation procedures.

Naturally the Chebyshev approximation may be used in this form only if we can also calculate the required values of the function for all the given points. However, we want to approximate the coordinates of celestial bodies that, in principle, are available for any arbitrary time. Because of this, this restricted form is suitable for our application.

Generally, in what follows we assume that there is a time-dependent function that delivers a three-dimensional vector of type Vec3D:

```
// Prototype of function to fit: Vec3D f(double t)
typedef Vec3D (*C3Dfunct)(double t);
```

Examples are the functions MoonPos and MoonEqu, but also the functions for calculating heliocentric planetary positions described in Chap. 6. For functions of the specified type, Chebyshev approximations may be calculated and evaluated using the class Cheb3D.

```
// Cheb3D: three dimensional Chebyshev approximation

class Cheb3D
{
  public:
    // Constructor, destructor
    Cheb3D (
       C3Dfunct f,    // Pointer to function to fit
       int n,         // Degree of fit polynomial
       double dt      // Interval size
    );
    ~Cheb3D();
    // Fit: approximates function m_f over the interval [ta, tb]
    void Fit (double ta, double tb);
    // Value: evaluation of approximation at argument t;
    //        performs a new fit if necessary
    Vec3D Value (double t);

  private:
    C3Dfunct m_f;              // Function
    int      m_n;              // Degree
    bool     m_Valid;          // Status flag for validity of coefficients
    double   m_dt;             // Interval size
    double   m_ta,m_tb;        // Interval
    double   *Cx,*Cy,*Cz;      // Chebyshev coefficients
};
```

The constructor

```
Cheb3D::Cheb3D (C3Dfunct f, int n, double dt)
  : m_n(n), m_f(f), m_dt(dt), m_Valid(false)
{
  Cx = new double[n+1];  Cy = new double[n+1];  Cz = new double[n+1];
};
```

therefore first performs the task of allocating the required memory for the polynomial coefficients Cx, Cy and Cz of the x-, y- and z-coordinates. In addition the degree of the polynomial, the desired standard interval size, and a pointer to the approximation function are stored in the appropriate attributes. Finally, the Fit method is used to calculate the coefficients of the Chebyshev polynomial for all three coordinates over a predetermined interval $[t_a, t_b]$.

```
// Fit: approximates function m_f over the interval [ta, tb]
void Cheb3D::Fit (double ta, double tb)
{
  int     j, k;
  double  tau, t, fac;
  double* T=new double[m_n+1];
  Vec3D   f;
  // Clear all coefficients
  for (j=0; j<=m_n; j++) Cx[j]=Cy[j]=Cz[j]=0.0;
  // Loop over roots of T^(n+1)
  for (k=0; k<=m_n; k++) {
    tau = cos((2*k+1)*pi/(2*m_n+2));             // Grid point tau_k
    t = ((tb-ta)/2.0) * tau + ((tb+ta)/2.0); // Time t
    f = m_f(t);                                  // Function value at time t
    // Calculate coefficients C_j using recurrence relation
    for (j=0; j<=m_n; j++) {
      switch (j) {
        case  0: T[j]=1.0; break;
        case  1: T[j]=tau; break;
        default: T[j]=2.0*tau*T[j-1]-T[j-2];
      };
      // Increment coefficient C_j by f(t)*T_j(tau)
      Cx[j]+=f[x]*T[j]; Cy[j]+=f[y]*T[j]; Cz[j]+=f[z]*T[j];
    };
  };
  // Scaling
  fac = 2.0/(m_n+1);
  for (j=0; j<=m_n; j++) { Cx[j]*=fac; Cy[j]*=fac; Cz[j]*=fac; }
  // Update some members
  m_ta = ta;
  m_tb = tb;
  m_Valid = true;
  delete[] T;
}
```

Within this method the function values $f(t)$ and – by recursion – the values $T_j(\tau)$ of the Chebyshev polynomials of orders $j = 0, \ldots, n$ are calculated for the individual roots $\tau = \hat{x}_k^{n+1}$ of the Chebyshev polynomial of order $n+1$. From here the desired coefficients are derived in accordance with (8.13).

In evaluating a given expansion with Chebyshev polynomials, it is not necessary to calculate the polynomials explicitly as well. An algorithm given by Clenshaw, when applied to the recursive relation from (8.10) yields the following rule for evaluating a Chebyshev expansion (8.12) of order n with coefficients $(c_0, c_1, \ldots, c_n)$:

- Set $f_{n+1} = 0$ and $f_{n+2} = 0$.
- Calculate with the normalized argument $\hat{x}$ the series
  $$f_i = 2\hat{x} f_{i+1} - f_{i+2} + c_i \quad \text{for } i = n, n-1, \ldots, 0 \quad .$$
- The required value of the function is then
  $$f(x) = (f_0 - f_2)/2 = \hat{x} f_1 - f_2 + c_0/2 \quad .$$

The `Value` methods evaluates a Chebyshev-expansion using this method:

```
// Value: evaluation of approximation at argument t
Vec3D Cheb3D::Value (double t) {
  const double eps=0.01; // Fractional overlap
  Vec3D  f1, f2, old_f1;
  double tau, k;
  // Generate new coefficients as required
  if ( !m_Valid || (t<m_ta) || (m_tb<t) ) {
    k = floor(t/m_dt);
    Fit ( (k-eps)*m_dt, (k+1+eps)*m_dt );
  }
  // Evaluate approximation
  tau = (2.0*t-m_ta-m_tb)/(m_tb-m_ta);
  for (int i=m_n; i>=1; i--) {
    old_f1 = f1;
    f1 = 2.0*tau*f1-f2+Vec3D(Cx[i],Cy[i],Cz[i]);
    f2 = old_f1;
  };
  return tau*f1-f2+0.5*Vec3D(Cx[0],Cy[0],Cz[0]);
}
```

8.4 The LUNA Program

The `Luna` program combines the various routines given in this chapter in one small application. It may be used to calculate a lunar ephemeris – i.e., a table of lunar positions – like those found in many yearbooks. The output consists of the apparent equatorial coordinates of the Moon (referred to the equinox of date), its distance in Earth radii, and its horizontal parallax. `Luna` expands the lunar coordinates as a series of Chebyshev polynomials over an interval of ten days at a time. This series-expansion may then be easily and rapidly evaluated. Eleven lunar positions are calculated so that the approximation given by this series may be determined to the required accuracy. If we therefore require lunar coordinates at intervals of only about one day or more, the series expansion causes a certain penalty in increased computation. On the other hand, ephemerides at smaller intervals (as are required for navigation, for example), may be prepared with a considerable gain in the amount of time required.

```
//------------------------------------------------------------------------------
// Module:  Luna program (Luna.cpp)
// Purpose: Computes a lunar ephemeris
// (c) 1999 Oliver Montenbruck, Thomas Pfleger
//------------------------------------------------------------------------------

#include <cmath>
#include <fstream>
#include <iomanip>
#include <iostream>
#include "APC_Cheb.h"
#include "APC_Const.h"
#include "APC_Math.h"
#include "APC_Moon.h"
#include "APC_Time.h"
#include "APC_VecMat3D.h"
using namespace std;

//------------------------------------------------------------------------------
// GetEph: Prompts user for date of start, step size and end of ephemeris
//   MjdStart  Start time of ephemeris as Modified Julian Date
//   Step      Step size of ephemeris in days
//   MjdEnd    End time of ephemeris as Modified Julian Date
//------------------------------------------------------------------------------
void GetEph (double& MjdStart, double& Step, double& MjdEnd)
{
  int    year, month, day;
  double hour;
  cout << endl
       << " Begin and end of the ephemeris: " << endl
       << endl;
  cout << "  first date (yyyy mm dd hh.hhh) ... ";
  cin >> year >> month >> day >> hour; cin.ignore(81,'\n');
  MjdStart = Mjd(year,month,day) + hour/24.0 ;
  cout << "  final date (yyyy mm dd hh.hhh) ... ";
  cin >> year >> month >> day >> hour; cin.ignore(81,'\n');
  MjdEnd = Mjd(year,month,day) + hour/24.0 ;
  cout << "  step size (dd hh.hh)           ... ";
  cin >> day >> hour; cin.ignore(81,'\n');
  Step  = day + hour/24.0;
}

//------------------------------------------------------------------------------
//
// Main program
//
//------------------------------------------------------------------------------
void main()
{
```

```
  // Constants
  const int    Degree    = 10;
  const double Interval  = 10.0/36525.0;  // 10d in [cy]

  // Variables
  int    n_line = 0;
  double MjdStart, Step, MjdEnd;
  double Date, T;
  double Dist, Parallax;
  Vec3D  r_Moon;
  Cheb3D ChebMoonEqu (MoonEqu, Degree, Interval);
  // Title
  cout << endl
       << "                        LUNA: lunar ephemeris            " << endl
       << "         (c) 1999 Oliver Montenbruck, Thomas Pfleger  " << endl
       << endl;

  // Prompt user for date of start, step size and end of ephemeris
  GetEph (MjdStart, Step, MjdEnd);
  // Header
  cout << endl << endl
       << "    Date       ET         RA              Dec       Distance  "
       << " Parallax " << endl
       << "                         h  m  s        o  '  \"  Earth radii"
       << "    '  \"   " << endl;

  // Compute ephemeris
  Date = MjdStart;
  // Time loop
  while ( Date < MjdEnd + Step/2 ) {
    // Geocentric coordinates of the Moon from Chebyshev approximation
    // (true equator and equinox of date)
    T = ( Date - MJD_J2000 ) / 36525.0;
    r_Moon = ChebMoonEqu.Value(T);
    // Distance and parallax
    Dist     = r_Moon[r];
    Parallax = asin(R_Earth/Dist);
    // Output
    cout << " " << DateTime(Date,HHMM)
         << fixed
         << setprecision(2) << setw(14) << Angle(Deg*r_Moon[phi]/15.0,DMMSSs)
         << "  " << showpos
         << setprecision(1) << setw(11) << Angle(Deg*r_Moon[theta],DMMSSs)
         << noshowpos
         << setprecision(3) << setw(10) << Dist/R_Earth
         << setprecision(2) << setw(12) << Angle(60.0*Deg*Parallax,DMMm)
         << endl;
    ++n_line; if ( (n_line % 5) ==0 ) cout << endl;
    Date += Step; // Next time step
  };  // End time loop

}
```

When Luna is run, the expansion of the lunar coordinates is first calculated. The coordinates are then displayed in rapid sequence, until a new series expansion (for the subsequent interval) is required.

To illustrate the operation of the program, we will calculate a lunar ephemeris for 1989 January, with a step of two days. Luna requires only the limiting dates for the ephemeris, in the form: Year, Month, Day, and Hours, together with the step interval in the form: Days and Hours. It must be noted that here (as with calculations of planetary positions), all times are in Ephemeris Time. The input data are shown here in italics.

```
                LUNA: lunar ephemeris
      (c) 1999 Oliver Montenbruck, Thomas Pfleger


 Begin and end of the ephemeris:

  first date (yyyy mm dd hh.hhh) ... 1989 01 01 00.0
  final date (yyyy mm dd hh.hhh) ... 1989 01 31 00.0
  step size (dd hh.hh)           ... 2 00.0

    Date        ET         RA           Dec       Distance    Parallax
                         h  m  s      o  '  "   Earth radii     '  "
 1989/01/01 00:00   13 05 26.21  -10 42 58.7     63.053     54 31.41
 1989/01/03 00:00   14 38 15.31  -20 23 19.3     61.961     55 29.10
 1989/01/05 00:00   16 26 11.87  -26 51 42.9     60.407     56 54.71
 1989/01/07 00:00   18 27 52.77  -27 43 03.3     58.861     58 24.44
 1989/01/09 00:00   20 29 55.18  -21 44 30.5     57.789     59 29.44

 1989/01/11 00:00   22 21 11.90  -10 29 51.2     57.442     59 51.00
 1989/01/13 00:00    0 03 44.05  + 2 54 49.4     57.757     59 31.44
 1989/01/15 00:00    1 46 21.72  +15 30 04.5     58.481     58 47.18
 1989/01/17 00:00    3 36 54.09  +24 39 14.7     59.389     57 53.29
 1989/01/19 00:00    5 35 18.56  +28 13 06.1     60.375     56 56.57

 1989/01/21 00:00    7 31 01.72  +25 29 44.7     61.406     55 59.16
 1989/01/23 00:00    9 14 03.91  +17 51 26.1     62.408     55 05.27
 1989/01/25 00:00   10 44 27.57  + 7 31 36.2     63.194     54 24.13
 1989/01/27 00:00   12 08 35.18  - 3 38 59.4     63.512     54 07.79
 1989/01/29 00:00   13 34 26.38  -14 17 56.9     63.141     54 26.88

 1989/01/31 00:00   15 10 02.69  -23 01 46.5     62.014     55 26.26
```

What is striking here are the high declinations that the Moon reached on approximately 1989 January 7 and 19. In that year the lines of nodes of the lunar orbit lay in such a position that the obliquity of the ecliptic (ca. $23\overset{\circ}{.}5$) and the inclination of the lunar orbit (ca. $5\overset{\circ}{.}1$) were superimposed with the same sign, so the declination could exceed 28°. At specific times during that year, therefore, the Moon was particularly high or low in the sky, and the azimuths of its rising and setting points reached extreme values.

9. Solar Eclipses

Approximately every 30 days the Moon turns its unilluminated face towards the Earth at New Moon. An observer looking down on the north pole of the ecliptic would then see the Sun, Moon and Earth in a straight line. Despite this, the shadow cast by the Moon rarely touches the Earth. Because of the inclination of its orbit, at the time of New Moon the Moon is usually above or below the Earth's orbital plane, so its shadow misses the Earth. The Moon crosses the ecliptic on only two days each month, but if one of these dates coincides with New Moon, then the Sun, Moon and Earth are in line, and the Moon's shadow falls on part of the Earth. Overall, such a solar eclipse occurs twice a year. Anyone within the umbra, which is about 100 km across, sees a total eclipse of the Sun. This is a relatively small area, so few persons ever have the opportunity of witnessing a solar eclipse.

If the plane of the Moon's orbit remained fixed in space, all eclipses would take place in two specific months. Because of the regression of the line of nodes of the lunar orbit, the dates of eclipses shift by an average of three weeks each year. The period of 18.6 years taken for one complete regression of the line of nodes is therefore directly reflected in the pattern of solar eclipses (see Fig. 9.1).

9.1 Phases of the Moon and Eclipses

To obtain an idea of possible solar eclipses in a year, we can first determine the dates of New Moon for every month and then investigate the position of the Moon relative to the ecliptic. The time of New Moon is the moment when the ecliptic longitudes of the Sun and Moon ($\lambda_\odot$ and λ_M, respectively) agree, i.e., when the difference $\lambda_M - \lambda_\odot$ disappears. At First Quarter the difference then amounts to 90° ($\pi/2$), at Full Moon it is 180° (π) and 270° ($3\pi/2$) at the Last Quarter.

The determination of the phases of the Moon is therefore synonymous with the task of determining the zero points of the function

$$\Delta\lambda = \lambda_M - \lambda_\odot - k \cdot \pi/2 \quad , \tag{9.1}$$

where the k term identifies the respective lunar phase: New Moon ($k = 0$), First Quarter ($k = 1$), Full Moon ($k = 2$), or Last Quarter ($k = 3$). In calculating $\Delta\lambda$, if necessary, multiples of 2π need to be added or subtracted, because the result of determining the zero points must lie in the interval $[-\pi, +\pi]$. In the following function `PhasesFunc`, the functions `SunPos` and `MoonPos` are used to calculate the ecliptic longitude of the Sun and Moon. In the case of the Sun, the light-time correction of around eight minutes is also taken into account.

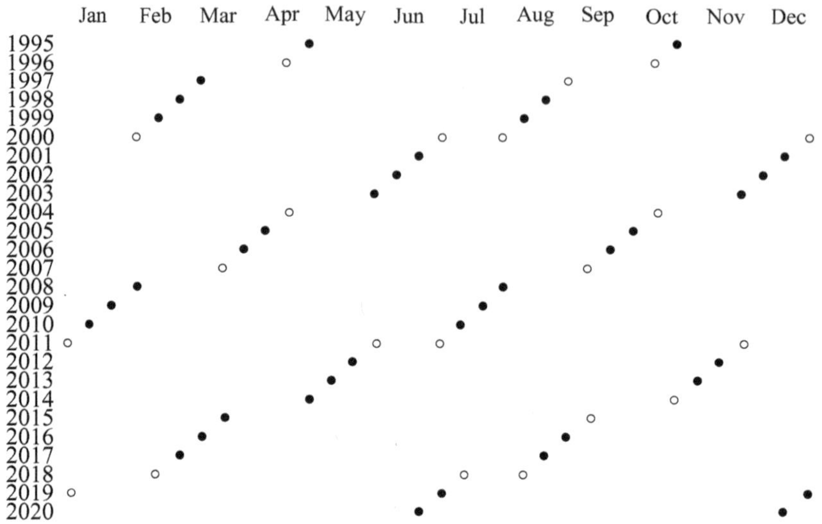

Fig. 9.1. Solar eclipses between 1995 and 2020 (• =total/annular, ○ =partial)

```
enum enLunarPhase { NewMoon, FirstQuarter, FullMoon, LastQuarter };
enLunarPhase Phase = NewMoon;
//------------------------------------------------------------------------
// PhasesFunc: Goal function for search of phase events [-pi, pi]
//   T          Ephemeris Time (Julian centuries since J2000)
//   <return>   Difference between the longitude of the Moon from the Sun
//              and the nominal value for a given phase (New Moon 0, First
//              Quarter pi/2, etc.) (in [rad])
//------------------------------------------------------------------------
double PhasesFunc (double T) {
  const double tau_Sun  = 8.32 / (1440.0*36525.0);    // 8.32 min  [cy]
  const double LongDiff = MoonPos(T)[phi] - SunPos(T-tau_Sun)[phi];
  return  Modulo ( LongDiff - Phase*pi/2.0 + pi, pi2 ) - pi;
}
```

By carrying out a search with a step value of seven days, a time interval $[t_0, t_1]$ is first identified, within which $\Delta\lambda$ changes to a positive sign. Afterwards the exact zero point may be found by iteration using well-known numerical procedures. In the Phases program we use the so-called Pegasus method, which is related to the *regula falsi* method, but which provides better convergence (see Sect. 10.2).

To identify possible eclipses it essentially suffices to examine the ecliptic latitude β of the Moon at New Moon (for solar eclipses) and Full Moon (for lunar eclipses). Corresponding to the inclination of the Moon's orbit to the ecliptic, β varies in the course of a year between $-5°$ and $+5°$. Solar eclipses usually take place at the two New Moons, where the ecliptic latitude exhibits the smallest (absolute) values. In a single year, however, even more than two eclipses may occur. Modern examples of this are the years 1982 and 2000, each of which has four partial eclip-

Table 9.1. The Phases program's criteria for the occurrence of solar and lunar eclipses, depending on the ecliptic latitude β of the Moon at New or Full Moon

New Moon		
$\|\beta\| < 0°52'20''$	Central solar eclipse certain	c
$\|\beta\| < 1°02'36''$	Central solar eclipse possible	c?
$\|\beta\| < 1°24'33''$	Partial solar eclipse certain	p
$\|\beta\| < 1°34'50''$	Partial solar eclipse possible	p?
$\|\beta\| > 1°34'50''$	No solar eclipse possible	
Full Moon		
$\|\beta\| < 0°21'50''$	Total lunar eclipse certain	t
$\|\beta\| < 0°32'14''$	Total lunar eclipse possible	t?
$\|\beta\| < 0°53'24''$	Partial lunar eclipse certain	p
$\|\beta\| < 1°03'50''$	Partial lunar eclipse possible	p?
$\|\beta\| < 1°26'15''$	Penumbral lunar eclipse certain	P
$\|\beta\| < 1°36'43''$	Penumbral lunar eclipse possible	P?
$\|\beta\| > 1°36'43''$	No lunar eclipse possible	

ses. The highest value of ecliptic latitude at which a total eclipse is possible is about one degree. A partial solar eclipse may, on the other hand, occur with $\beta \approx 1.5°$. In addition, it is already possible to say something about where a specific eclipse is visible. If, for example, the Moon is north of the ecliptic ($\beta > 0$), then the area of the shadow will principally lie on the northern hemisphere of the Earth.

The necessary steps have been brought together in a single program Phases, but which is not reproduced here in full for reasons of space. After entry of the desired year, the times of the four phases of the Moon are determined, and using the criteria in Table 9.1, the times are marked at which solar or lunar eclipses are possible or probable. An example of the use of this program is given at the end of this chapter.

9.2 Geometry of an Eclipse

The rays of light leaving the Sun produce two cone-shaped regions, which are known as the umbra and penumbra (Fig. 9.2). For an observer situated in the umbra, the Moon appears larger than the Sun and therefore completely hides the latter.

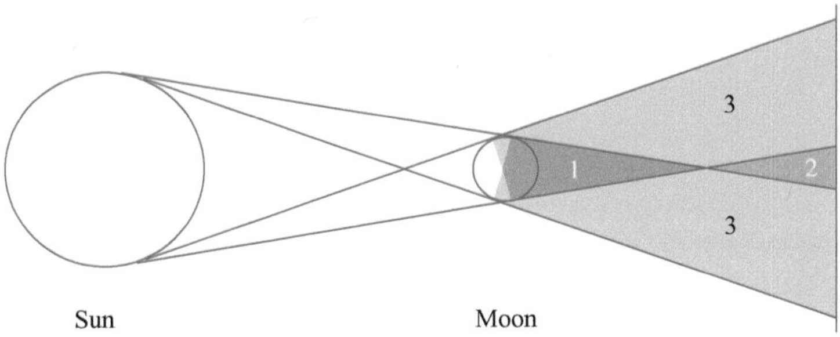

Fig. 9.2. Umbra (1=total eclipse, 2=annular eclipse) and penumbra (3)

The point of the umbral cone lies at an average distance of about 375 000 km from the Moon. Farther away, it defines a region in which the solar eclipse appears annular. From here the Moon appears as a dark disk in front of the Sun, which it is unable to cover fully. Part of the Sun remains visible as a bright ring around the Moon. An example of this type of annular solar eclipse is the one that occurred on 1976 April 29, which was observable from part of the Mediterranean Sea. Within the penumbra, the Sun is only partially eclipsed, and appears as a crescent of greater or lesser width. The percentage that is covered is greater the closer one is to the umbral shadow.

The distance of the Moon from the Earth is, on average, 380 000 km, but because of the eccentricity of the lunar orbit, it varies by more than 25 000 km. As a result, however, the Earth is always close to the point of the umbral shadow cone during an eclipse. Even at perigee, when the Earth and Moon are closest, the diameter of the umbral shadow on the surface of the Earth is rarely more than 200 km across. At apogee the eclipse is annular. The border-line case occurs with eclipses that are total for a short period, but which are otherwise annular. An example was the eclipse of 1987 March 29.

Measured in a plane perpendicular to the line joining the Sun and the Moon, and at a distance s behind the Moon, the diameters of the umbra (d) and of the penumbra (D) are:

$$\begin{aligned} d(s) &= D_\odot\left(\frac{s}{r_{\odot M}}\right) - D_M\left(1+\frac{s}{r_{\odot M}}\right) \quad \text{and} \\ D(s) &= D_\odot\left(\frac{s}{r_{\odot M}}\right) + D_M\left(1+\frac{s}{r_{\odot M}}\right) \quad . \end{aligned} \tag{9.2}$$

Here $D_\odot$ and D_M are the diameters of the Sun and Moon respectively:

$$\begin{aligned} D_\odot &= 1\,392\,000\ \text{km} &= 218.25\ R_\oplus \\ D_M &= 3\,476\ \text{km} &= 0.5450\ R_\oplus \quad . \end{aligned}$$

$r_{\odot M}$ is the distance between the Sun and the Moon.

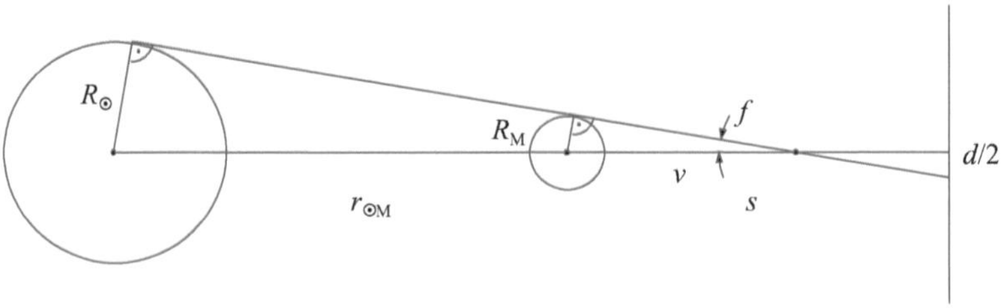

Fig. 9.3. Calculating the diameter of the shadow

The derivation of the equations for the diameter of the shadow is illustrated in Fig. 9.3. There f is half of the apex angle of the shadow cone, and v the distance of the apex from the centre of the Moon. If $R_\odot$ and R_M are the radii of the Sun and Moon, respectively, then

$$\sin(f)\cdot v = R_\mathrm{M} \quad \text{and} \quad \sin(f)\cdot(v + r_{\odot\mathrm{M}}) = R_\odot$$

or rearranged

$$\sin(f) = \frac{R_\odot - R_\mathrm{M}}{r_{\odot\mathrm{M}}} \quad \text{and} \quad v = \frac{R_\mathrm{M}\cdot r_{\odot\mathrm{M}}}{R_\odot - R_\mathrm{M}} \quad .$$

The diameter of the shadow at a distance s from the Moon is therefore

$$\begin{aligned} d &= 2(s-v)\cdot\tan(f) \\ &\approx 2(s-v)\cdot\sin(f) \\ &= 2R_\odot\left(\frac{s}{r_{\odot\mathrm{M}}}\right) - 2R_\mathrm{M}\left(1+\frac{s}{r_{\odot\mathrm{M}}}\right) \quad . \end{aligned}$$

The equation for the semi-diameter D of the shadow may be derived in a similar way. The sign of d in the above equations is chosen such that in the true shadow, i.e., for a total eclipse, d is *negative*, and is positive for a partial eclipse. Whereas the semi-diameter of the shadow D close to the Earth is approximately half the size of the Earth itself, the diameter d of the umbra varies between $-0.04\,R_\oplus$ (250 km) and $+0.06\,R_\oplus$ (350 km).

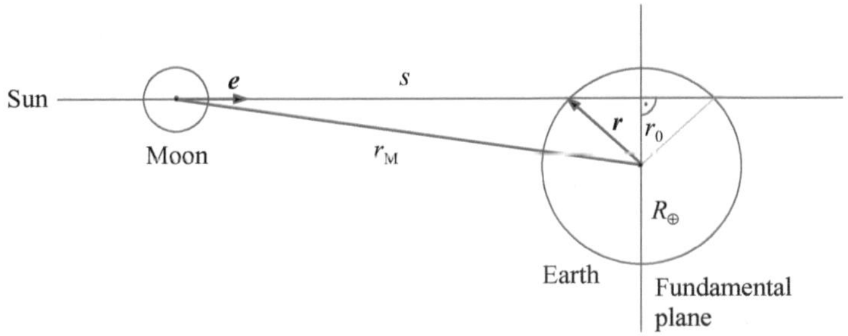

Fig. 9.4. The intersection of the axis of the shadow with the Earth

The position of the umbra on the Earth is obtained if we draw a straight line through the centres of the Sun and Moon, and determine where this intersects the surface of the Earth (see Fig. 9.4). The direction of this straight line is described by the vector

$$\boldsymbol{e} = \frac{\boldsymbol{r}_\mathrm{M} - \boldsymbol{r}_\odot}{|\boldsymbol{r}_\mathrm{M} - \boldsymbol{r}_\odot|} \tag{9.3}$$

of unit length, which may be calculated from the coordinates $\boldsymbol{r}_\odot$ and $\boldsymbol{r}_\mathrm{M}$ of the Sun and Moon. For a point $\boldsymbol{r}$ of the shadow we then have

$$\boldsymbol{r} = \boldsymbol{r}_\mathrm{M} + s\boldsymbol{e} \tag{9.4}$$

or, expressing this as individual components

$$\begin{pmatrix} x \\ y \\ z \end{pmatrix} = \begin{pmatrix} x_M \\ y_M \\ z_M \end{pmatrix} + s \cdot \begin{pmatrix} e_x \\ e_y \\ e_z \end{pmatrix} \quad .$$

Here s is the distance between the point and the centre of the Moon. Because the tip of $\boldsymbol{r}$ lies on the surface of the Earth, and if we initially assume the Earth to be a true sphere with radius $R_\oplus$ we have:

$$\boldsymbol{r}^2 = x^2 + y^2 + z^2 = R_\oplus^2 \quad . \tag{9.5}$$

Combining, we have the quadratic equation

$$s^2 + 2(\boldsymbol{r}_M \cdot \boldsymbol{e})s + (r_M^2 - R_\oplus^2) = 0 \quad .$$

This has two solutions

$$s = \begin{cases} s_0 - \sqrt{\Delta} & \text{day side of the Earth} \\ s_0 + \sqrt{\Delta} & \text{night side of the Earth} \end{cases}$$

where

$$\begin{aligned} s_0 &= -\boldsymbol{r}_M \cdot \boldsymbol{e} = -(x_M e_x + y_M e_y + z_M e_z) \qquad \text{and} \\ \Delta &= s_0^2 + R_\oplus^2 - r_M^2 \quad . \end{aligned} \tag{9.6}$$

s_0 is clearly the distance of the Moon from what is known as the fundamental plane, which is perpendicular to the axis of the shadow and passes through the centre of the Earth. The required point of intersection of the axis of the shadow with the surface of the Earth lies at a distance $\sqrt{\Delta}$ away from this plane on the side that is closest to the Sun and the Moon. The second point of intersection lies on the night side of the Earth, and is of no further relevance for eclipses. Combining the equations, the coordinates of the umbra on the surface of the Earth are given by

$$\boldsymbol{r} = \boldsymbol{r}_M + (s_0 - \sqrt{\Delta}) \cdot \boldsymbol{e} \quad . \tag{9.7}$$

We naturally take the discriminant Δ as positive, as otherwise the axis of the shadow would miss the Earth. Nevertheless, a solar eclipse begins before this central phase. As soon as the penumbral shadow touches the Earth, a partial eclipse is observed from somewhere on Earth. If

$$r_0 = \sqrt{r_M^2 - s_0^2}$$

is the distance between the axis of the shadow and the centre of the Earth, and d_0 and D_0 are the diameters of the umbra and the penumbra on the fundamental plane, then we can distinguish between the individual phases of the eclipse by using the following relationships:

$$\begin{array}{lclcll} R_\oplus + D_0/2 & < & r_0 & & & \text{no eclipse} \\ R_\oplus + |d_0|/2 & < & r_0 & < & R_\oplus + D_0/2 & \text{partial phase} \\ R_\oplus & < & r_0 & < & R_\oplus + |d_0|/2 & \text{non-central phase} \\ & & r_0 & < & R_\oplus & \text{central phase} \quad . \end{array}$$

In the non-central phase, the umbral shadow cone brushes the Earth, so that a total or annular eclipse is visible from a small area of the Earth. The axis of the shadow itself, however, does not intersect the surface of the Earth. There is thus no point on the Earth, from which the Moon appears directly in front of the Sun, such that their centres coincide.

These conclusions are based on the assumption that the Earth is a true sphere. To take the slight flattening of the Earth into account in calculating eclipses, some small corrections are required. The surface of the Earth may be envisaged as a sphere of radius $R_\oplus$, that is compressed in a direction parallel to its axis by the factor

$$1 - f = 0.996647 \quad .$$

Any point (x, y, z) on this rotational ellipsoid obeys the equation

$$\boldsymbol{r}^2 = x^2 + y^2 + z^2/(1-f)^2 = R_\oplus^2 \quad , \tag{9.8}$$

which is obtained from (9.5), if we replace z by $z/(1-f)$. The z-axis of the coordinate system used here is parallel to the Earth's axis (i.e., we are using equatorial coordinates).

Because of the fact that the Earth may be represented as a sphere that is correspondingly stretched, there is a simple method of calculating the position of the umbral shadow correctly. The z coordinates of the Sun and the Moon are multiplied by the factor $1/(1-f)$, a straight line is laid through these modified positions, and then the point where this intersects a sphere with the radius of the Earth is determined. The only difference between the point of intersection thus calculated and the true point of intersection of the axis of the shadow and the surface of the Earth is that the z coordinate is too great by the factor $1/(1-f)$. Using this procedure, the relationships determined above may be employed without any alteration.

9.3 Geographic Coordinates and the Flattening of the Earth

Geographic longitude (λ) and latitude (φ) are generally used to describe the coordinates of a point on the surface of the Earth. These are closely related to the equatorial coordinates right ascension (α) and declination (δ), which are employed in astronomy. Both coordinate systems are orientated parallel to the plane of the equator and to the Earth's axis of rotation.

Conversion between declination and geographic latitude would be superfluous if the Earth were a true sphere, because both coordinates would then be identical. Because the Earth is a flattened ellipsoid of rotation – even if only slightly – the concept of geographic latitude must first be refined. φ denotes the angle between the local horizon at any one point and the Earth's axis. This is shown in Fig. 9.5. A cross-section of the Earth through the North and South Poles is an ellipse, to which the horizon is tangent. As may be seen, φ is not identical with geocentric latitude

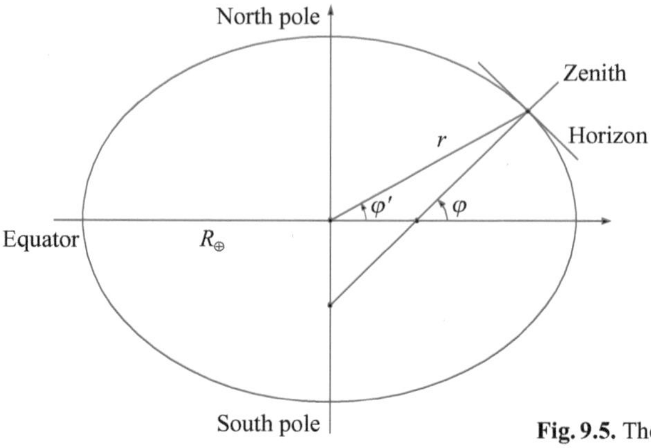

Fig. 9.5. The flattening of the Earth

φ' or declination δ, both of which indicate the angle between the position vector and the equatorial plane. φ and φ' are linked by the equations

$$\begin{aligned} r \cdot \cos(\varphi') &= \frac{R_\oplus}{\sqrt{1 - e^2 \sin^2 \varphi}} \cos \varphi \\ r \cdot \sin(\varphi') &= \frac{(1 - e^2) R_\oplus}{\sqrt{1 - e^2 \sin^2 \varphi}} \sin \varphi \\ \tan(\varphi') &= (1 - e^2) \tan \varphi \end{aligned} \tag{9.9}$$

Here $R_\oplus = 6378.14$ km is the Earth's equatorial radius. e is the eccentricity of the terrestrial ellipsoid. Because of the centrifugal force produced by the Earth's rotation, the Earth is deformed such that the equatorial radius is about 20 km larger than the distance R_{Pol} between the centre and the poles. The relationship between these two values defines the flattening

$$f = \frac{R_\oplus - R_{\mathrm{Pol}}}{R_\oplus} = \frac{21.385 \text{ km}}{6378.14 \text{ km}} = \frac{1}{298.257} = 0.003353 \quad . \tag{9.10}$$

From this the eccentricity of the solid ellipsoid may be derived as

$$e = \sqrt{1 - (1 - f)^2} = \sqrt{2f - f^2} \quad . \tag{9.11}$$

Instead of the exact equations (9.9), we can use the perfectly adequate approximation

$$\varphi = \varphi' + 0\overset{\circ}{.}1924 \cdot \sin(2\varphi') \tag{9.12}$$

to determine geographic latitude. This clearly indicates that the difference between φ and φ' is a maximum for intermediate latitudes where it amounts to about twelve arc-minutes. At the poles and the equator, the difference vanishes.

The second coordinate used to describe the position of a point on the Earth is geographic longitude. It essentially corresponds to right ascension in an equatorial coordinate system. But whereas right ascension is reckoned from the vernal equinox, a fixed point in space, geographic longitude is measured from the (Earth-fixed) Greenwich meridian. The two coordinates therefore differ essentially by an angle $\Theta_0(t)$, which is dependent on the rotation of the Earth in time t:

$$\alpha \;=\; \Theta_0(t) + \lambda \quad . \tag{9.13}$$

For the position vectors we likewise have

$$\begin{pmatrix} r\cos(\varphi')\cos(\lambda) \\ r\cos(\varphi')\sin(\lambda) \\ r\sin(\varphi') \end{pmatrix} = \boldsymbol{R}_z(\Theta_0(t)) \begin{pmatrix} r\cos(\varphi')\cos(\alpha) \\ r\cos(\varphi')\sin(\alpha) \\ r\sin(\varphi') \end{pmatrix} \quad . \tag{9.14}$$

$\Theta_0(t)$ is the corresponding right ascension of the zero meridian, and is therefore simply Greenwich Sidereal Time. The calculation of Sidereal Time has already been described in Section 3.3, where a suitable procedure (`GMST`) has been given.

An important difficulty that arises here is the necessity for careful discrimination between Universal Time (UT) and Ephemeris Time (ET or TDB, TT). Although the coordinates of the Sun and Moon must always be calculated only as a function of uniform Ephemeris Time, Universal Time is required as an argument for calculating Sidereal Time. Universal Time cannot, however, be measured with a clock, but, because of its definition, can only be determined from observations. As a result it is only possible to take account retrospectively of the correct difference ΔT = ET − UT between the two time systems. Some of the measured values for this difference are given in Table (3.1) for the twentieth century. To avoid having to enter the values every time, we can also resort to using a polynomial approximation of the tabular values.

Table 9.2. Polynomial approximations of the difference ΔT = ET − UT between Ephemeris Time and Universal Time (T=(JD-2451545)/36525)

Period	ΔT = ET − UT	t
1825 − 1850	$10\overset{s}{.}4 - 80\overset{s}{.}8t + 413\overset{s}{.}9t^2 - 572\overset{s}{.}3t^3$	$T + 1.75$
1850 − 1875	$6\overset{s}{.}6 + 46\overset{s}{.}3t - 358\overset{s}{.}4t^2 + 18\overset{s}{.}8t^3$	$T + 1.50$
1875 − 1900	$-3\overset{s}{.}9 - 10\overset{s}{.}8t - 166\overset{s}{.}2t^2 + 867\overset{s}{.}4t^3$	$T + 1.25$
1900 − 1925	$-2\overset{s}{.}6 + 114\overset{s}{.}1t + 327\overset{s}{.}5t^2 - 1467\overset{s}{.}4t^3$	$T + 1.00$
1925 − 1950	$24\overset{s}{.}2 - 6\overset{s}{.}3t - 8\overset{s}{.}2t^2 + 483\overset{s}{.}4t^3$	$T + 0.75$
1950 − 1975	$29\overset{s}{.}3 + 32\overset{s}{.}5t - 3\overset{s}{.}8t^2 + 550\overset{s}{.}7t^3$	$T + 0.50$
1975 − 2000	$45\overset{s}{.}3 + 130\overset{s}{.}5t - 570\overset{s}{.}5t^2 + 1516\overset{s}{.}7t^3$	$T + 0.25$

For times between the years 1825 and 2000 a set of polynomials is given in Table 9.2, each of which covers an interval of 25 years with a typical accuracy of 1 s. Because the accuracy degrades notably outside the specified time, the `ETminUT` function checks the respective interval boundaries and if they are not met returns a value of `false` in the variable `valid`. Excepted from this is the last polynomial, which may also be used for extrapolation to the year 2005. Because the value of ET–UT must be determined from observations, it is unfortunately not possible to give meaningful long-term predictions. The current data are, however, published in various yearbooks, where they may be found.

```
//------------------------------------------------------------------------------
// ETminUT: Difference ET-UT of ephemeris time and universal time
//   T          Time in Julian centuries since J2000
//   DTsec      ET-UT in [s]
//   valid      Flag indicating T in domain of approximation
// Notes: The approximation spans the years from 1825 to 2005
//------------------------------------------------------------------------------
void ETminUT (double T, double& DTsec, bool& valid)
{
  int    i = (int) floor(T/0.25);
  double t = T-i*0.25;

  if ( (T<-1.75) || (0.05<T) ) {
    valid = false; DTsec = 0.0;
  }
  else {
    valid = true;
    switch (i) {
      case -7: DTsec=10.4+t*(-80.8+t*( 413.9+t*( -572.3))); break; // 1825-
      case -6: DTsec= 6.6+t*( 46.3+t*(-358.4+t*(   18.8))); break; // 1850-
      case -5: DTsec=-3.9+t*(-10.8+t*(-166.2+t*(  867.4))); break; // 1875-
      case -4: DTsec=-2.6+t*(114.1+t*( 327.5+t*(-1467.4))); break; // 1900-
      case -3: DTsec=24.2+t*( -6.3+t*(  -8.2+t*(  483.4))); break; // 1925-
      case -2: DTsec=29.3+t*( 32.5+t*(  -3.8+t*(  550.7))); break; // 1950-
      case -1: DTsec=45.3+t*(130.5+t*(-570.5+t*( 1516.7))); break; // 1975-
      case  0: t+=0.25;
               DTsec=45.3+t*(130.5+t*(-570.5+t*( 1516.7)));        // 2000-
    }                                                              // 2005
  }
}
```

The value of the difference between ET and UT varies very slowly and may, therefore, be regarded as constant for the duration of a solar eclipse.

9.4 Duration of an Eclipse

The duration of the total or annular phase of an eclipse is not the same for every point on the central line. Apart from the diameter of the umbra, it largely depends on the velocity with which the shadow moves across the surface of the Earth. An accurate calculation of the duration of totality requires an iterative determination of the times at which the total phase of the eclipse begins and ends, and is thus computation-intensive. To assess the course of the eclipse and choose a favourable observing site on the central line, however, the following method is completely adequate.

The equatorial coordinates

$$\boldsymbol{r} = (x, y, z) \quad \text{and} \quad \boldsymbol{r}' = (x', y', z')$$

of the points at which the axis of the shadow intersects the Earth at times t and $t + \Delta t$, separated by a short interval, may be calculated from equations (9.6) and (9.7), which have already been discussed[1]. The difference between these two vectors is a measure of the velocity at which the umbral shadow is moving. Nevertheless, $\boldsymbol{r}' - \boldsymbol{r}$ does not give us the required motion of the shadow relative to the surface of the Earth.

The angle w through which the Earth rotates in a time Δt is derived from the duration of a complete rotation, $23^{\mathrm{h}}56^{\mathrm{m}}$, and is

$$w = 2\pi \cdot \frac{\Delta t}{1436^{\mathrm{m}}} \qquad \text{(radians)} \quad .$$

For a point on the surface of the Earth that has the equatorial coordinates $\boldsymbol{r}' = (x', y', z')$ at time $t + \Delta t$, the coordinates at time t are

$$\boldsymbol{r}'' = \begin{pmatrix} +x' \cdot \cos w + y' \cdot \sin w \\ -x' \cdot \sin w + y' \cdot \cos w \\ +z' \end{pmatrix} \approx \begin{pmatrix} x' + wy' \\ y' - wx' \\ z' \end{pmatrix} ,$$

which are derived from $\boldsymbol{r}'$ by rotation around the z-axis through angle w. The path that the centre of the umbra traces across the surface of the Earth during the time Δt is therefore

$$\Delta \boldsymbol{r} = \boldsymbol{r}'' - \boldsymbol{r}$$

or

$$\begin{pmatrix} \Delta x \\ \Delta y \\ \Delta z \end{pmatrix} = \begin{pmatrix} x' + wy' & - & x \\ y' - wx' & - & y \\ z' & - & z \end{pmatrix} .$$

If $\boldsymbol{e} = (e_x, e_y, e_z)$ describes the direction of the axis of the shadow, as in (9.4), then $\Delta \boldsymbol{r}$ may be resolved into one component $\Delta \boldsymbol{r}_{\parallel}$ parallel to $\boldsymbol{e}$ and another $\Delta \boldsymbol{r}_{\perp}$ perpendicular to the axis of the shadow. The lengths of these two components are

$$\Delta r_{\parallel} = \Delta \boldsymbol{r} \cdot \boldsymbol{e} = \Delta x e_x + \Delta y e_y + \Delta z e_z$$

and

$$\Delta r_{\perp} = \sqrt{(\Delta r)^2 - (\Delta r_{\parallel})^2} \quad .$$

$\Delta r_{\perp}$ is the length of the path traced out, perpendicular to the direction of incidence of the shadow, by the umbra's point of contact in time Δt. The duration τ of totality or of the annular phase, for a diameter $|d|$ of the umbral cone, thus equals

$$\tau = \frac{|d|}{\Delta r_{\perp}} \cdot \Delta t \quad .$$

[1] Because of other approximations that are made here, the flattening of the Earth is ignored in this section.

9.5 Solar and Lunar Coordinates

Accurate coordinates of the Sun and the Moon are naturally important prerequisites for calculating solar eclipses. To be able to determine the central line to an accuracy of about 10 km, for example, the position of the Moon in its orbit must be known with comparable accuracy. For an average distance of 380 000 km, this corresponds to an angle of 5″. The geocentric solar coordinates must be known with similar accuracy. These requirements are fully met by the `SunPos` and `MoonPos` procedures, which have already been discussed. The bases of the perturbation series that they use need not be examined here. They may be found in Chap. 6 and Chap. 8.

For our purposes, however, it is desirable to use procedures that are not only accurate, but also fast, because of the large number of solar and lunar positions that are required in calculating the course of an eclipse. So the first step is to derive a Chebyshev expansion of the individual coordinates. By this we mean obtaining specific polynomials that express the orbits of the Sun and Moon over a short period of time with sufficient accuracy, but which are significantly simpler to evaluate than complete, universally applicable, series expansions. This method was adopted in calculating lunar ephemerides (Chap. 8). The basis for Chebyshev expansions, and the corresponding class `Cheb3D` were discussed there, so the reader should already be familiar with it.

The expansion of the lunar coordinates merely requires declaration of an object of class `Cheb3D`. In the constructor the function to be approximated, the required order, and the approximation interval must be specified:

```
// Constants
const double Interval = 1.0/36525.0;    // 1d (in centuries)

// Global lunar ephemeris data
Cheb3D ChebMoonEqu(MoonEqu,10,Interval);
```

Because solar eclipses never last more than a few hours, it is sufficient to use polynomials of order ten, covering a one-day interval for the expansion. Furthermore, it is obvious that function `MoonEqu` may be employed instead of `MoonPos`, because the calculation of solar eclipses as described takes place in an equatorial coordinate system. By this method, no additional transformations between the ecliptic and equatorial systems are necessary later. The polynomials for the lunar coordinates may be evaluated for any required time using the `ChebMoonEqu.Value(T)` method:

```
r_Moon = ChebMoonEqu.Value(T_ET);  // [km]
```

It should be noted that the actual calculation of the Chebyshev approximation is independently carried out on the first call to `Value`, in case the user has not previously explicitly called the `Fit` method.

To express the solar coordinates in an analogous way, we require another small function. `SunEqu` calculates the true equatorial coordinates of the Sun for a given time. This requires that nutation (see Sect. 6.4.2) is taken into account after conversion of the ecliptic solar coordinates into equatorial coordinates. The position vector thus obtained expresses the position of the Sun in a coordinate system that is aligned with the actual (and not to the mean) orientation of the Earth's axis and equator. For the lunar orbit nutation is already taken into account in a corresponding manner in the `MoonEqu` function.

```
//------------------------------------------------------------------------------
// SunEqu: Computes the Sun's equatorial position using analytical series
//   T          Time in Julian centuries since J2000
//   <return>   Geocentric position of the Sun (in [AU]), referred to the
//              equator and equinox of date
//------------------------------------------------------------------------------
Vec3D SunEqu ( double T )
{
  return NutMatrix(T) * Ecl2EquMatrix(T) * SunPos(T);
}
```

The Chebyshev expansion of solar coordinates again takes place by declaring a corresponding `Cheb3D` object:

```
  Cheb3D ChebSunEqu(SunEqu,5,Interval);
```

Because of the slower motion of the Sun, however, a lower degree of expansion suffices here.

To properly compare lunar and solar positions, light-time must, in addition, be considered when computing the solar coordinates with `SunEqu` (and likewise `ChebSunEqu.Value`). It amounts to about 8.32 minutes between the Sun and the Earth, during which interval the Sun moves about 20″ along its orbit. As the *observed* position of the Sun at a specific instant t corresponds to its geometrical position at an earlier time $t - 8^{\text{m}}\!.32$ at which the light was emitted, the calculation of the solar coordinates must be correspondingly adjusted.

```
  const double tau_Sun = 8.32 / ( 1440.0*36525.0);      // 8.32 min  [cy]
  r_Sun = ChebSunEqu.Value (T_ET-tau_Sun)*AU;           // [km]
```

With the Moon, the effect has already been incorporated in Brown's theory of lunar motion, and therefore does not require to be specifically calculated.

9.6 The ECLIPSE Program

The program `Eclipse` calculates the central line of a solar eclipse and the duration of the total or annular phases, starting with the date of the next New Moon.

The core of the program is the `Intersect` function, in which the intersection of the axis of the shadow with the surface of the Earth is determined from the geocentric coordinates of the Sun and the Moon. In addition, the diameters of the two shadow cones near the Earth are calculated, from which the respective phases of the eclipse are determined. The `Central` function converts the equatorial coordinates of the umbra provided by `Intersect` into geographic coordinates, and also determines the duration of totality.

The program requests the date of New Moon, the difference between Universal Time and Ephemeris Time, and the output step size to be entered. The value ΔT=ET–UT is, however, only known for earlier dates, and may otherwise be only estimated. Normally, in predicting eclipses, a value ET–UT=0 is therefore assumed. The coordinates of a point on the central line thus calculated are shifted in geographic longitude relative to the correct ones. This difference corresponds to the rotation of the Earth during the short interval ΔT. If λ' is the calculated longitude, then for the convention adopted here ($\lambda > 0$ towards the East), the actual geographic longitude is:

$$\lambda = \lambda' + 0.^\circ 25/^{\mathrm{m}} \cdot \Delta T \quad .$$

For the present-day value of $\Delta T \approx 1^{\mathrm{m}}$, the coordinates predicted by using $\Delta T = 0$ are thus about $1/4°$ too far West. At the equator this amounts to about 28 km.

```
//------------------------------------------------------------------------------
// Module:  Eclipse program (Eclipse.cpp)
// Purpose: Computes the centre line of a solar eclipse
// (c) 1999 Oliver Montenbruck, Thomas Pfleger
//------------------------------------------------------------------------------

#include <cmath>
#include <fstream>
#include <iomanip>
#include <iostream>

#include "APC_Cheb.h"
#include "APC_Const.h"
#include "APC_IO.h"
#include "APC_Math.h"
#include "APC_Moon.h"
#include "APC_PrecNut.h"
#include "APC_Spheric.h"
#include "APC_Sun.h"
#include "APC_Time.h"
#include "APC_VecMat3D.h"

using namespace std;

// Types
enum enPhase { NoEclipse, partial, NonCenAnn, NonCenTot, annular, total };
// Constants
const double Interval = 1.0/36525.0;    // 1d [cy]
// Global lunar and solar ephemeris data
Cheb3D ChebMoonEqu(MoonEqu,10,Interval);
Cheb3D ChebSunEqu (SunEqu,  5,Interval);

//------------------------------------------------------------------------------
// Intersect: Computes intersection point of shadow's axis and the Earth,
//            the shadow axis unit vector, umbral diameter and phase
//   T_ET      Ephemeris time in Julian centuries since J2000
```

```
//   r          Intersection point of shadow's axis and surface of the Earth
//              (geocentric coordinates in [km])
//   e          Shadow axis unit vector
//   D_umbra    Shadow diameter on the fundamental plane in [km]
//   Phase      Phase of the eclipse
//------------------------------------------------------------------------------
void Intersect ( double T_ET,
                 Vec3D& r, Vec3D& e, double& D_umbra, enPhase& Phase )
{
  const double tau_Sun  = 8.32 / ( 1440.0*36525.0);    // 8.32 min  [cy]
  const double fac      = 0.996647;     // Ratio polar/equat. Earth radius
  double r_MS,s0,Delta,r0,s;
  double D_penumbra;
  Vec3D  r_Moon,r_Sun;
  // Solar and lunar position w.r.t. true equator and equinox of date
  // (solar coordinates corrected for light time)
  r_Moon = ChebMoonEqu.Value(T_ET);                // [km]
  r_Sun  = ChebSunEqu.Value (T_ET-tau_Sun)*AU;     // [km]
  // Scale z-coordinate to compensate Earth flattening
  r_Moon = Vec3D ( r_Moon[x], r_Moon[y], r_Moon[z]/fac );
  r_Sun  = Vec3D ( r_Sun[x] , r_Sun[y] , r_Sun[z]/fac  );
  // Shadow axis (Sun->Moon unit vector)
  r_MS = Norm(r_Moon-r_Sun);
  e    = (r_Moon-r_Sun)/r_MS;
  // Distance of the Moon from the fundamental plane
  s0 = -Dot(r_Moon,e);
  // Distance of the shadow axis from the centre of the Earth
  Delta = s0*s0 + R_Earth*R_Earth - Dot(r_Moon,r_Moon);
  r0 = sqrt(R_Earth*R_Earth-Delta);
  // Umbra and penumbra diameter on the fundamental plane
  D_umbra    = 2.0*((R_Sun-R_Moon)*(s0/r_MS)-R_Moon);
  D_penumbra = 2.0*((R_Sun-R_Moon)*(s0/r_MS)+R_Moon);
  // Determine phase and shadow coordinates if required
  if (r0<R_Earth) {
    // Shadow axis intersects the Earth; total or annular eclipse:
    // Intersection of the shadow axis and the surface of the Earth
    s = s0-sqrt(Delta);     r = r_Moon + s*e;
    // Re-scale z-component
    r = Vec3D(r[x],r[y],fac*r[z]);
    // Umbra diameter at the surface of the Earth
    D_umbra = 2.0*((R_Sun-R_Moon)*(s/r_MS)-R_Moon);
    Phase = ( (D_umbra>0.0)? annular : total );
  }
  else {
    if ( r0 < R_Earth+0.5*fabs(D_umbra) ) {  // Non-central eclipse
      Phase = ( (D_umbra>0.0)? NonCenAnn : NonCenTot );
    }
    else {  // Partial or no eclipse
      Phase = ( (r0<R_Earth+0.5*fabs(D_penumbra))? partial : NoEclipse );
    };
    r = Vec3D(); // Dummy value (Null vector)
  };
};
```

```
//------------------------------------------------------------------------------
// Central: Computes centre-line coordinates, duration and phase
//   MjdUT     Universal time as Modified Julian Date
//   ET_UT     Difference of ephemeris time and universal time in [s]
//   Lambda    Geographic longitude (east pos.) of centre of shadow [rad]
//   Phi       Geographic latitude of centre of shadow [rad]
//   Durat     Duration of central phase at location (Lambda, Phi) in [min]
//   Phase     Phase of the eclipse
//------------------------------------------------------------------------------
void Central ( double MjdUT, double ET_UT,
               double &Lambda, double& Phi, double& Durat, enPhase& Phase )
{
  const double dt    = 0.1  / (1440.0*36525.0); // 0.1 min [cy]
  const double omega = pi2 * 1.002738*36525.0;  // [rad/cy]

  double    T_ET;               // Ephemeris time
  Vec3D     r,r_G,rr,dr;        // Shadow coordinates
  Vec3D     e,ee;               // Shadow axis
  double    D_umbra, DU;        // Umbra diameter
  enPhase   Ph=NoEclipse;       // Phase
  double    w;                  // Earth rotation angle
  double    drp;                // Shadow displacement

  // Ephemeris Time
  T_ET = ( MjdUT + ET_UT/86400.0 - MJD_J2000 ) / 36525.0;

  // Intersection of shadow axis with the surface of the Earth
  Intersect ( T_ET, r, e, D_umbra, Phase );

  // For central phase only: geogr. coord. and duration of central phase
  Lambda = Phi = Durat = 0.0;
  if ( Phase >= annular ) {
    // Geographic shadow coordinates
    r_G    = R_z(GMST(MjdUT))*r;                 // Greenwich coordinates
    Lambda = Modulo(r_G[phi]+pi,2*pi)-pi;        // East longitude
    Phi    = r_G[theta];                         // Geocentric latitude
    Phi    = Phi + 0.1924*Rad*sin(2*Phi);        // Geographic latitude
    // Duration of central phase for this place:
    // (a) shadow coordinates at time T+dt (or T-dt)
    if (Ph<annular) {
      Intersect ( T_ET-dt, rr, ee, DU, Ph ); w = -dt*omega;
    }
    else {
      Intersect ( T_ET+dt, rr, ee, DU, Ph ); w = +dt*omega;
    }
    // (b) displacement dr of the shadow on Earth and
    //     fraction drp perpendicular to the shadow axis
    dr  = Vec3D ( rr[x]-r[x]+w*r[y], rr[y]-r[y]-w*r[x], rr[z]-r[z] );
    drp = Norm ( dr - Dot(dr,e)*e );
    Durat = (36525.0*1440.0)*dt * fabs(D_umbra) / drp;  // [min]
  };
};
```

```
//------------------------------------------------------------------------------
// GetInput: Prompts user for approximate date and time of New Moon,
//           desired step size and ET-UT
//   MjdStart  Start time for calculation of eclipse
//   Step      Step size in [d]
//   MjdEnd    End time for calculation of eclipse
//   ET_UT     Difference ephemeris time - universal time in [s]
//------------------------------------------------------------------------------
void GetInput(double& MjdStart,double& Step, double& MjdEnd, double& ET_UT)
{
  int    year, month, day, step;
  double hour, MJD;
  bool   valid;
  // Date and step size
  cout << endl
       << " Date of New Moon (yyyy mm dd hh.h)      ... ";
  cin >> year >> month >> day >> hour; cin.ignore(81,'\n');
  cout << " Output step size (min)" << setw(21) << right << "... ";
  cin >> step; cin.ignore(81,'\n');
  // Round date to step size
  Step = step/1440.0;  // [d]
  MJD = Mjd(year,month,day) + floor((hour/24.0)/Step+0.5)*Step;
  // Start and end date (date of New Moon +/- 6 hours)
  MjdStart = MJD - 0.25;
  MjdEnd   = MJD + 0.25;
  // Difference between ephemeris time and universal time
  ETminUT ( (MJD-MJD_J2000)/36525.0, ET_UT, valid );
  if ( valid ) {
    cout << " Difference ET-UT (proposal:" << fixed << setw(6)
         << setprecision (1) << ET_UT << " sec) ... ";
    cin >> ET_UT; cin.ignore(81,'\n');
  }
  else {
    cout << " Difference ET-UT (sec)" << setw(21) << right << "... ";
    cin >> ET_UT; cin.ignore(81,'\n');
  }
}

//------------------------------------------------------------------------------
//
// Main program
//
//------------------------------------------------------------------------------
void main(int argc, char* argv[]) {

  // Variables
  double    MjdStart, Step, MjdEnd;
  double    MjdUT, ET_UT;
  double    Lambda, Phi, Durat;
  enPhase   Phase;
  char      OutputFile[APC_MaxFilename] = "";
  ofstream  OutFile;
```

```
  // Title
  cout << endl
    << "     ECLIPSE: central line and duration of solar eclipses " << endl
    << "          (c) 1999 Oliver Montenbruck, Thomas Pfleger     " << endl
    << endl;
  // Prompt user for input
  GetInput ( MjdStart, Step, MjdEnd, ET_UT );
  // Check for output file and eventually redirect output
  if (argc==2) {
    strncpy(OutputFile, argv[1], APC_MaxFilename);
    OutFile.open(OutputFile);
    if (OutFile.is_open()) cout = OutFile;
  }
  // Header
  cout << endl << endl
       << "    Date       UT          Phi      Lambda   Durat  Phase " << endl
       << "               h  m         o  '      o  '    min         " << endl
       << endl;

  // Phase and central line of the eclipse
  MjdUT = MjdStart;
  // Time loop
  while ( MjdUT < MjdEnd + Step/2 ) {
    // Phase and location of shadow point
    Central ( MjdUT, ET_UT, Lambda, Phi, Durat, Phase );
    // Output
    if ( Phase != NoEclipse ) {
      cout << " " << DateTime(MjdUT,HHMM);
      if ( Phase < annular )
        cout << "       -- --     -- --    ---";
      else {
        cout << showpos
             << "      " << setw(6) << Angle(Deg*Phi,DMM)
             << "   "    << setw(7) << Angle(Deg*Lambda,DMM)
             << noshowpos
             << fixed << setprecision(1) << setw(7) << Durat;
      };
      switch(Phase) {
        case partial  : cout << "   partial                "; break;
        case NonCenAnn: cout << "   annular (non-central) "; break;
        case NonCenTot: cout << "   total (non-central)   "; break;
        case annular  : cout << "   annular                "; break;
        case total    : cout << "   total                  ";
      };
      cout << endl;
    };
    MjdUT += Step;  // Next time step
  };  // End time loop
  if (OutFile.is_open()) OutFile.close();
}
```

As an example we will calculate the path of the prominent total eclipse that can be seen from the United States of America in 2017. The central line of this eclipse crosses the states of Oregon, Idaho, Wyoming, Nebraska, Missouri, Tennessee and Carolina. First we call Phases, in order to determine the dates of New Moon in 2017 and to check for possible eclipses.

```
        PHASES: Phases of the Moon and Check for Eclipses
           (c) 1999 Oliver Montenbruck, Thomas Pfleger

 Dates of Lunar Phases for the year ... 2017

 New Moon            First Quarter        Full Moon            Last Quarter
 2016/12/29 06:54    2017/01/05 19:48     2017/01/12 11:35     2017/01/19 22:15
 2017/01/28 00:08    2017/02/04 04:20     2017/02/11 00:34p?   2017/02/18 19:34
 2017/02/26 15:00c   2017/03/05 11:33     2017/03/12 14:55     2017/03/20 15:59
 2017/03/28 02:58    2017/04/03 18:41     2017/04/11 06:09     2017/04/19 09:58
 2017/04/26 12:17    2017/05/03 02:48     2017/05/10 21:44     2017/05/19 00:34
 2017/05/25 19:46    2017/06/01 12:43     2017/06/09 13:11     2017/06/17 11:34
 2017/06/24 02:32    2017/07/01 00:52     2017/07/09 04:08     2017/07/16 19:27
 2017/07/23 09:47    2017/07/30 15:24     2017/08/07 18:12p    2017/08/15 01:16
 2017/08/21 18:31c   2017/08/29 08:14     2017/09/06 07:04     2017/09/13 06:26
 2017/09/20 05:31    2017/09/28 02:55     2017/10/05 18:41     2017/10/12 12:27
 2017/10/19 19:13    2017/10/27 22:23     2017/11/04 05:24     2017/11/10 20:38
 2017/11/18 11:43    2017/11/26 17:04     2017/12/03 15:48     2017/12/10 07:52
 2017/12/18 06:32    2017/12/26 09:21     2018/01/02 02:25     2018/01/08 22:26
 2018/01/17 02:18    2018/01/24 22:21     2018/01/31 13:28t    2018/02/07 15:55

 All times in Ephemeris Time (ET).
```

The label 'c' on February 26 and August 21 of this year indicates that the ecliptic latitude of the Moon is small enough on these dates to reliably predict the occurence of central eclipses (cf. Table 9.1). Closer investigation with Eclipse shows that the regions of visibility are:

2017 Feb. 26: South Pacific, southern Chile and Argentina, South Atlantic, Angola (annular)

2017 Aug. 21: North Pacific, central USA, North Atlantic (total)

The path of the total eclipse of 2017 August 21 should now be determined with Eclipse. To do this we enter the date of New Moon just determined, the desired step value in minutes, and a figure for the difference between Ephemeris Time and Universal Time. For the desired eclipse we adopt a value of 80 s. All input is shown in italic in the example that follows. To avoid an unnecessarily long output, data for the partial phase have been truncated in the following print-out and the step size was reduced to 3 minutes during most of the total phase.

```
      ECLIPSE: central line and duration of solar eclipses
           (c) 1999 Oliver Montenbruck, Thomas Pfleger

 Date of New Moon (yyyy mm dd hh.h)       ... 2017 08 21 18.5
 Output step size (min)                   ...  1
 Difference ET-UT (sec)                   ... 80
```

Date	UT	Phi	Lambda	Durat	Phase
	h m	o '	o '	min	
2017/08/21	15:47	-- --	-- --	---	partial
2017/08/21	15:48	-- --	-- --	---	partial
. . . .	. .	.	.	.	.
2017/08/21	16:48	-- --	-- --	---	partial
2017/08/21	16:49	+40 21	-168 41	0.0	total
2017/08/21	16:50	+41 40	-161 54	1.1	total
2017/08/21	16:51	+42 18	-158 14	1.2	total
2017/08/21	16:54	+43 24	-150 52	1.3	total
2017/08/21	16:57	+44 02	-145 33	1.5	total
2017/08/21	17:00	+44 26	-141 11	1.6	total
2017/08/21	17:03	+44 42	-137 24	1.7	total
2017/08/21	17:06	+44 51	-134 01	1.8	total
2017/08/21	17:09	+44 56	-130 56	1.8	total
2017/08/21	17:12	+44 56	-128 06	1.9	total
2017/08/21	17:15	+44 54	-125 28	2.0	total
2017/08/21	17:18	+44 49	-123 00	2.1	total
2017/08/21	17:21	+44 41	-120 40	2.1	total
2017/08/21	17:24	+44 32	-118 28	2.2	total
2017/08/21	17:27	+44 20	-116 22	2.2	total
2017/08/21	17:30	+44 07	-114 23	2.3	total
2017/08/21	17:33	+43 53	-112 28	2.4	total
2017/08/21	17:36	+43 36	-110 38	2.4	total
2017/08/21	17:39	+43 19	-108 53	2.4	total
2017/08/21	17:42	+43 01	-107 11	2.5	total
2017/08/21	17:45	+42 41	-105 34	2.5	total
2017/08/21	17:48	+42 21	-103 59	2.6	total
2017/08/21	17:51	+41 59	-102 28	2.6	total
2017/08/21	17:54	+41 37	-100 59	2.6	total
2017/08/21	17:57	+41 13	- 99 33	2.6	total
2017/08/21	18:00	+40 49	- 98 10	2.7	total
2017/08/21	18:03	+40 24	- 96 48	2.7	total
2017/08/21	18:06	+39 59	- 95 29	2.7	total
2017/08/21	18:09	+39 33	- 94 12	2.7	total
2017/08/21	18:12	+39 06	- 92 56	2.7	total
2017/08/21	18:15	+38 38	- 91 42	2.7	total
2017/08/21	18:18	+38 10	- 90 29	2.7	total
2017/08/21	18:21	+37 41	- 89 18	2.7	total
2017/08/21	18:24	+37 12	- 88 08	2.7	total
2017/08/21	18:27	+36 42	- 86 59	2.7	total
2017/08/21	18:30	+36 11	- 85 51	2.7	total
2017/08/21	18:33	+35 40	- 84 43	2.7	total
2017/08/21	18:36	+35 08	- 83 36	2.7	total
2017/08/21	18:39	+34 36	- 82 30	2.7	total
2017/08/21	18:42	+34 03	- 81 24	2.7	total
2017/08/21	18:45	+33 30	- 80 19	2.7	total
2017/08/21	18:48	+32 56	- 79 14	2.6	total
2017/08/21	18:51	+32 22	- 78 08	2.6	total
2017/08/21	18:54	+31 46	- 77 03	2.6	total
2017/08/21	18:57	+31 11	- 75 57	2.5	total
2017/08/21	19:00	+30 34	- 74 50	2.5	total

```
2017/08/21 19:03      +29 57   - 73 43    2.5   total
2017/08/21 19:06      +29 20   - 72 36    2.4   total
2017/08/21 19:09      +28 41   - 71 27    2.4   total
2017/08/21 19:12      +28 02   - 70 17    2.3   total
2017/08/21 19:15      +27 22   - 69 05    2.3   total
2017/08/21 19:18      +26 41   - 67 51    2.3   total
2017/08/21 19:21      +25 59   - 66 36    2.2   total
2017/08/21 19:24      +25 17   - 65 17    2.1   total
2017/08/21 19:27      +24 33   - 63 56    2.1   total
2017/08/21 19:30      +23 48   - 62 31    2.0   total
2017/08/21 19:33      +23 01   - 61 01    2.0   total
2017/08/21 19:36      +22 13   - 59 27    1.9   total
2017/08/21 19:39      +21 23   - 57 46    1.8   total
2017/08/21 19:42      +20 31   - 55 56    1.7   total
2017/08/21 19:45      +19 37   - 53 57    1.7   total
2017/08/21 19:48      +18 39   - 51 45    1.6   total
2017/08/21 19:51      +17 36   - 49 14    1.5   total
2017/08/21 19:54      +16 27   - 46 17    1.4   total
2017/08/21 19:57      +15 08   - 42 33    1.2   total
2017/08/21 20:00      +13 27   - 37 07    1.1   total
2017/08/21 20:01      +12 38   - 34 12    1.0   total
2017/08/21 20:02      -- --    -- --      ---   total (non-central)
2017/08/21 20:03      -- --    -- --      ---   partial
  . . . . .              .        .        .    .
2017/08/21 21:03      -- --    -- --      ---   partial
2017/08/21 21:04      -- --    -- --      ---   partial
```

If the phase of the eclipse is total, but not central, the axis of the shadow misses the Earth. But there is still an area on the Earth that lies within the umbral cone. This generally occurs at the beginning and end of an eclipse. Eclipses that are total or annular, but do not appear central anywhere on Earth, because the shadow-cone grazes the Earth tangentially, are very rare. Eclipses of this sort are not worth observing, because of their short duration, and also because they generally occur at high geographic latitudes.

9.7 Local Circumstances

As well as the position of the central line, information about the local circumstances describing the course of a solar eclipse in greater detail for a given point on the Earth are of considerable significance in the observation of eclipses.

Apart from the maximum phase, the magnitude and the duration, the contact times, i.e., the times at which the limbs of the Sun and Moon touch, are particularly important. First and fourth contact denote the times of entering and leaving the penumbral shadow (i.e., of the beginning and end of the partial phases), and thus the duration of the whole eclipse. Similarly, second and third contact (which are also known as internal contacts) define the extent of the total or annular phase, provided the observer is sited at a point that is crossed by the umbra.

Generally the contact times are supplemented by stating the position angle. At first and fourth contacts, the position angle (which is measured from North through

East), indicates the point on the solar limb at which the Moon first touches or finally leaves the disk of the Sun. The position angles of the two internal contacts indicate the positions at which the disk of the Sun finally disappears or first reappears (in a total eclipse), or the points at which it begins and ends as an unbroken ring (in an annular eclipse).

The geometrical requirement for observing first or fourth contact at solar eclipses is that the observer shall be located somewhere on the curved boundary of the Moon's penumbra. Second and third contact, in contrast, occur when the observer is at the boundary of the umbral shadow.

As in our discussion of stellar occultations (see Sect. 10.3), we can best describe the situation by using a system of coordinates based on the fundamental plane, which lies perpendicular to the axis of the shadow, and passes through the centre of the Earth (Fig. 9.4). The x-axis and the y-axis lie in the fundamental plane, and the z-axis indicates the direction from the Moon to the Sun. The x-axis is additionally chosen such that it lies through the intersection of the Earth's equator and the fundamental plane. The unit vectors $\boldsymbol{i}$, $\boldsymbol{j}$ and $\boldsymbol{k}$ directed along the x-, y-, and z-axes of the coordinate system based on the fundamental plane may then be represented by

$$\boldsymbol{i} = \begin{pmatrix} -k_y/\sqrt{k_x^2+k_y^2} \\ +k_x/\sqrt{k_x^2+k_y^2} \\ 0 \end{pmatrix} \qquad \boldsymbol{j} = \boldsymbol{k} \times \boldsymbol{i} \qquad \boldsymbol{k} = \frac{\boldsymbol{r}_\odot - \boldsymbol{r}_\mathrm{M}}{|\boldsymbol{r}_\odot - \boldsymbol{r}_\mathrm{M}|} \quad , \tag{9.15}$$

where the vectors $\boldsymbol{r}_\odot$ and $\boldsymbol{r}_\mathrm{M}$ denote the equatorial coordinates of the Sun and Moon. By projecting $\boldsymbol{r}_\mathrm{M}$ onto the vector bases $\boldsymbol{i}$, $\boldsymbol{j}$ and $\boldsymbol{k}$, we obtain the coordinates

$$\begin{pmatrix} x_\mathrm{M} \\ y_\mathrm{M} \\ z_\mathrm{M} \end{pmatrix} = \begin{pmatrix} \boldsymbol{r}_\mathrm{M} \cdot \boldsymbol{i} \\ \boldsymbol{r}_\mathrm{M} \cdot \boldsymbol{j} \\ \boldsymbol{r}_\mathrm{M} \cdot \boldsymbol{k} \end{pmatrix}$$

of the Moon in the fundamental-plane system.

Because, by definition, the line joining the Sun and Moon is parallel to the z-axis, x_M and y_M are simultaneously the x- and y-coordinates of the shadow axis on the fundamental plane. It follows from the orientation of the fundamental plane, moreover, that both the umbral and the penumbral cones intersect the fundamental plane as circles, where, in accordance with (9.3), the radii are given by

$$l = z_\mathrm{M} \frac{R_\odot \mp R_\mathrm{M}}{r_{\odot\mathrm{M}}} \mp R_\mathrm{M} \quad . \tag{9.16}$$

Here, $R_\odot$ and R_M are the radii of the Sun and Moon respectively, and $r_{\odot\mathrm{M}} = |\boldsymbol{r}_\odot - \boldsymbol{r}_\mathrm{M}|$ is the distance between them. The negative sign applies to the umbra, and the positive sign to the penumbra.

To determine when the observer lies on the boundary of one or other of the shadow zones, the geocentric position of the observer is projected onto the fundamental plane. As shown in Sect. 9.3, the equatorial coordinates of an observer at geocentric longitude λ, geocentric latitude φ', and distance r from the centre of the Earth, are

$$\boldsymbol{r}_\mathrm{O} = \boldsymbol{R}_z(-\Theta_0(t)) \begin{pmatrix} r\cos(\varphi')\cos(\lambda) \\ r\cos(\varphi')\sin(\lambda) \\ r\sin(\varphi') \end{pmatrix} , \qquad (9.17)$$

where $\Theta_0(t)$ is the Greenwich Sidereal Time. The observer's coordinates (x_O, y_O, z_O) in the system based on the fundamental plane are again similarly obtained by projecting $\boldsymbol{r}_\mathrm{O}$ onto the vector bases $\boldsymbol{i}$, $\boldsymbol{j}$ and $\boldsymbol{k}$. From the distance z_O of the observer above the fundamental plane, together with a knowledge of the angular semi-diameter f of the shadow cone and its diameter l on the fundamental plane, we can calculate the radius

$$L = l - z_\mathrm{O}\tan f \qquad (9.18)$$

of the shadow cone at the observer's position. For partial and annular eclipses, L is always positive, whereas, by convention, the radius of the umbral region in a total eclipse is negative.

A requirement for the observer's position to lie on the shadow cone is therefore that the distance of the observer from the axis of the shadow should be equal to the radius of the shadow cone at the observer's position. This is equivalent to saying that the distance, as expressed by the function

$$f(t) = (x_\mathrm{M} - x_\mathrm{O})^2 + (y_\mathrm{M} - y_\mathrm{O})^2 - L^2 \qquad (9.19)$$

should become equal to zero. The contact times may be determined by iteratively determining the points at which $f(t)$ becomes zero. The time of maximum eclipse may be similarly derived from the point at which $f(t)$ reaches a minimum.

Because the relative positions of the observer and the shadow axis directly reflect the apparent positions of the Sun and Moon, as seen by the observer, the position angle of contacts may be obtained simply from the known coordinates in the fundamental plane. If the position angle of the point of contact is measured from the point due north of the centre of the Sun, positive towards the East, then, irrespective of the type of eclipse, internal and external contacts are given by

$$\begin{pmatrix} \cos P \\ \sin P \end{pmatrix} = \begin{pmatrix} (y_\mathrm{M} - y_\mathrm{O})/L \\ (x_\mathrm{M} - x_\mathrm{O})/L \end{pmatrix} . \qquad (9.20)$$

In solving these equations for P, it is important to remember that, as already mentioned, for total eclipses the radius L takes negative values.

Apart from the position angle P, relative to the direction of North, it is also useful to have, for solar eclipses, the position angle relative to the direction of the observer's zenith, $V = P - C$. The parallactic angle C thus corresponds to the position angle of the zenith, and may be calculated sufficiently accurately from the observer's coordinates on the fundamental plane, using

$$\begin{pmatrix} \cos C \\ \sin C \end{pmatrix} \approx \begin{pmatrix} y_\mathrm{O}/\sqrt{x_\mathrm{O}^2 + y_\mathrm{O}^2} \\ x_\mathrm{O}/\sqrt{x_\mathrm{O}^2 + y_\mathrm{O}^2} \end{pmatrix} . \qquad (9.21)$$

To give specific information about the maximum extent of a solar eclipse, we make use of a quantity, M, known as the magnitude of the eclipse. It is defined as the fraction of the apparent solar diameter that is covered by the Moon at the instant of maximum eclipse. M may be calculated directly from the radii L_1 and L_2 of the penumbral and umbral cones at the observer's position. For total or annular eclipses it is given by

$$M_2 = \frac{L_1 - L_2}{L_1 + L_2} \quad ,$$

the ratio of the apparent radii of the Sun and Moon. For a partial eclipse we have

$$M_1 = \frac{L_1 - m}{L_1 + L_2} \quad ,$$

which also depends on the observer's distance m from the axis of the shadow at the time of maximum. It is important to note that the magnitude M is not identical to the eclipsed fraction of the solar disk and, in particular, that at total solar eclipses it assumes values that are greater than one.

9.8 The ECLTIMER Program

Although the track of the central line for a solar eclipse is often known and published years in advance, yearbooks generally contain no details of the local circumstances of an eclipse for a specific observing site. As an extension to the `Eclipse` program we will describe the `EclTimer` program, which enables the local circumstances of an eclipse to be calculated for any given point.

For reasons of space, we will describe just the basic structure of the program, which follows the method of calculation described in the preceding section. The complete, fully commented source code is contained on the CD that accompanies this book.

To determine the contact times, the function `Contacts` is used to calculate the points at which the function, describing the observer's distance from the shadow, $f(t) = m^2(t) - L^2(t)$ (where m is the observer's distance from the axis of the shadow, and L is the radius of the shadow) becomes zero. The shadow distance is calculated by the function `ShadowDist`, which itself involves the two procedures `Bessel` and `Observer`, which determine the orientation of the fundamental plane, the parameters of the shadow cone, and the observer's coordinates.

By means of quadratic interpolation, we next determine the first and fourth contact, when the observer lies on the penumbral cone. Here we use the interpolation routine `Quad`, discussed in Chap. 3, which also enables us to determine when the distance between the observer and the shadow axis is a mimimum. Provided we have a partial eclipse (at the very least), `Contacts` calculates the magnitude of the eclipse for the time of maximum.

If, from the distance of the observer from the axis of the shadow and the shadow's diameter at the time of maximum, the eclipse proves to be total or annular, then the times of second and third contacts may be determined. Given that the duration of the total or annular phase can never exceed a value of approximately 13 minutes, second and third contacts must lie within a corresponding interval of time before and after maximum. We can, therefore, employ a procedure similar to *regula falsi*, which gives a faster result than quadratic interpolation. In `EclTimer` we use a variant of the *regula falsi*, known as the Pegasus method (cf. Sect. 10.2).

Finally, using the `PosAngles` procedure, the position angles of the internal and external contacts are calculated relative to the direction of North and of the zenith.

When the program is run, `EclTimer` first requests the date of New Moon, and the difference between ET and UT, $\Delta T = \mathrm{ET} - \mathrm{UT}$. The geographic coordinates of the observing site are then entered. A summary of possible eclipse times may be obtained by using the `Phases` program.

As an example, we will calculate the local circumstances for the annular eclipse of 1994 May 10. Let us take the observing site as being Rabat in Morocco, with geographical coordinates $\lambda = 6.8333°$ West and $\varphi = 33.95°$. Let us assume that the value for the time difference ΔT, was 60^{s}. All input data are shown in italics (as usual).

```
      ECLTIMER: local circumstances of solar eclipses
        (c) 1999 Oliver Montenbruck, Thomas Pfleger

 Date of new Moon (yyyy mm dd hh.hhh)            ... 1994 05 10 17.1
 Difference ET-UT (proposal:  60.2 sec)          ... 60.0
 Observer's coordinates: East longitude [deg] ... -6.8333
                         latitude [deg]          ... +33.95

 Annular eclipse with M=0.931 (0.87).
 Maximum at 1994/05/10 18:58:42 UT.

                            h  m  s [UT]    P [o]     V [o]
 1st contact: 1994/05/10 17:50:47            267       207
 2nd contact: 1994/05/10 18:56:40            235       179
 3rd contact: 1994/05/10 19:00:29            120        65
 4th contact: 1994/05/10 19:59:19             89        38
```

During the annular phase, which lasts about four minutes for the chosen observing site, the Moon's apparent diameter is about 7% smaller than that of the Sun. In total, only about 87% of the Sun's apparent area is covered by the Moon. Because the eclipse occurs during the evening, the position angles relative to the zenith (`V`) are about 50°–60° smaller than those relative to North (`P`). It should also be noted that the Sun sets at Rabat around $19^{\mathrm{h}}18^{\mathrm{m}}$ UT on the day of the eclipse, so that the end of the eclipse itself cannot be observed.

10. Stellar Occultations

In the course of a day the Moon covers about 13° of its orbit, moving from West to East across the sky. This motion is easiest to see when its orbit takes it close to bright stars. It is particularly striking when a stellar occultation occurs: a star suddenly vanishes behind the eastern limb of the Moon, and then reappears on the other side after a certain interval. In total, there are about one thousand stars visible to the naked eye that the Moon may occult. They lie in a narrow band to the north and south of the ecliptic, and are never more than 8° distant from it.

An occultation is like a solar eclipse in many respects. Because the rays of light from the distant star are parallel, however, the Moon's shadow is not conical, but cylindrical with a constant diameter. Similarly there is no difference between umbra and penumbra. The shadow of the Moon cast by starlight is as large as the Moon itself (1/4 Earth diameter), and is therefore unable to cover the whole Earth. As a result, any one stellar occultation may be seen from just part of the Earth's surface.

The lack of a lunar atmosphere means that a star disappears suddenly behind the Moon, rather than gradually, and that it reappears with equal abruptness. The times of disappearance and reappearance are therefore easy to measure, enabling the position of the Moon to be determined very accurately. The observation of stellar occultations has been used for a long time to improve our knowledge of the motion of the Moon and of the rotation of the Earth.

Basic data for planning the observation of occultations are published in various astronomical yearbooks and journals. The information is, however, restricted to certain specific, major sites. The times of disappearance and reappearance may, however, be calculated for any other observing sites by use of what are known as station coefficients. The calculation of such predictions does not generally require the highest degree of accuracy, so certain simplifications may be made rather than following a rigorous procedure. For example, the irregular profile of the lunar limb may be ignored. Nevertheless, the prediction of stellar occultations is computation-intensive, because a whole series of stars must first be checked to see whether an occultation may occur in the period under consideration.

The `Occult` program calculates possible occultations over consecutive intervals of one day for an arbitrary selection of stars. During this period of time a maximum of one appulse of the Moon with a specific star is possible. The minimum separation arises approximately when the geocentric right ascension of the Moon and the star's right ascension are the same. This may be determined with a simple iteration. If this shows that the difference in declination is not too great, then an occultation is

Table 10.1. Coordinates at epoch J2000 and proper motions for 100 years for various stars in the Pleiades based on the PPM Catalogue. In addition the *Zodiacal Catalog* number (ZC), the visual brightness and the common name of each star is provided.

ZC	α_{J2000}	μ_α	δ_{J2000}	μ_δ	m_{vis}	Name	
	h m s	s	° ′ ″	″	m		
536	3 44 48.180	+0.06	+24 17 21.44	-5.1	5.4	Celaeno	16 Tau
537	3 44 52.532	+0.14	+24 6 48.01	-4.6	3.8	Electra	17 Tau
539	3 45 12.469	+0.07	+24 28 1.36	-5.8	4.4	Taygeta	19 Tau
541	3 45 49.566	+0.05	+24 22 3.63	-5.0	4.0	Maia	20 Tau
542	3 45 54.423	+0.04	+24 33 16.02	-4.6	5.9	Asterope	21 Tau
545	3 46 19.537	+0.08	+23 56 53.42	-5.7	4.3	Merope	23 Tau
552	3 47 29.073	+0.14	+24 6 18.38	-4.6	3.0	Alcyone	η Tau
560	3 49 9.739	+0.13	+24 3 12.24	-4.7	3.8	Atlas	27 Tau
561	3 49 11.181	+0.09	+24 8 12.58	-4.3	5.2	Pleione	28 Tau

possible for at least part of the Earth. Finally, the motion of the lunar shadow around the time of conjunction may be examined to determine whether an occultation will be visible from a given observing site.

10.1 Apparent Positions

The coordinates of a star for which one wishes to calculate details of an occultation may be found in various catalogues. Examples are the PPM Catalogue (*Position and Proper Motion Catalogue*), the *SAO Catalog* published by the Smithsonian Astrophysical Observatory, and the *Zodiacal Catalog* (Catalog of 3539 Zodiacal Stars for the Equinox 1950.0), which was specially prepared for use in determining stellar occultations. A short extract from a star catalogue with the most significant data about various members of the Pleiades is given in Table 10.1. As well as the right ascension and declination, the individual proper motions for one hundred years are given. As can be seen, the spatial motions of stars relative to the Solar System produce marked changes in their coordinates with time. It is therefore necessary first to convert the positions (α_0, δ_0) at the catalogue epoch t_0 to those prevailing at the approximate time of the occultation by using the given proper motion data (μ_α, μ_δ):

$$\begin{array}{rclcl} \alpha(t) & = & \alpha_0 & + & \mu_\alpha(t-t_0) \\ \delta(t) & = & \delta_0 & + & \mu_\delta(t-t_0) \end{array} \quad . \tag{10.1}$$

Even after this correction, however, the positions given in the catalogues mentioned cannot be used directly in calculating occultations. To compare positions with that of the Moon we require what are known as the apparent places of the stars concerned. These are linked to the specific, actual position of the vernal equinox and celestial equator, as governed by precession and nutation, i.e., to the true equinox of date. The orientation of the coordinate system is always parallel to the instantaneous rotation axis of the Earth. The coordinate system in which positions are given in the various stellar catalogues, on the other hand, is defined in terms of the mean equator and vernal equinox for a fixed epoch. Catalogues in current use are generally referred to the epoch and equinox of the year 2000, but there are still some catalogues that are based on previously employed equinox 1950.

The precession matrix, which describes the mean, long-term changes in the position of the vernal equinox, may be calculated by Equations (2.14) and (2.15). They allow the unit vector

$$\boldsymbol{e} = \begin{pmatrix} \cos\delta\,\cos\alpha \\ \cos\delta\,\sin\alpha \\ \sin\delta \end{pmatrix} \tag{10.2}$$

of the stellar coordinates to be converted from the mean equinox at the catalogue epoch to the mean equinox of date.

Because of the varying gravitational attraction of the Sun and the Moon, an additional periodic variation — nutation — is superimposed on the precession of the Earth's axis. Nutation is described by the two angles $\Delta\varepsilon$ and $\Delta\psi$, whose maximum values are $9''$ and $17''$ (cf. (6.14)). A star's true coordinates differ by about this amount from its mean coordinates. The required conversion may be accomplished using the transformation matrix (6.15).

Another necessary correction has to be applied to stellar coordinates and this is *aberration*, which arises from the finite speed of light. To an observer moving round the Sun with the Earth, the light from a star appears to arrive from a slightly different direction than it would to an observer stationary with respect to the Sun. The observed stellar position $\boldsymbol{e}'$ may be determined to sufficient accuracy by adding the ratio $\boldsymbol{v}_{\oplus}/c$ of the Earth's velocity and the velocity of light to the rectangular stellar coordinate vector $\boldsymbol{e}$ and normalizing the result:

$$\boldsymbol{e}' = \frac{\boldsymbol{e} + \boldsymbol{v}_{\oplus}/c}{|\boldsymbol{e} + \boldsymbol{v}_{\oplus}/c|} \quad . \tag{10.3}$$

The right ascension and declination of the star, taking aberration into account, may then be obtained from these adjusted Cartesian coordinates.

Coordinates taken from a catalogue may be fully corrected taking precession, nutation, and aberration into account, by using the following program segment:

```
// Proper motion
double RA  = RA_Cat  + mu_RA *(T-T_CatEpoch);
double Dec = Dec_Cat + mu_Dec*(T-T_CatEpoch);
// Transformation from mean equator and equinox of catalogue
// to true equator and equinox of date
Mat3D PN = NutMatrix(T)*PrecMatrix_Equ(T_CatEquinox,T);
// Precession, nutation
Vec3D e = PN * Vec3D(Polar(RA, Dec));
// Earth velocity vector w.r.t. mean equator and equinox of date
// in units of speed of light
Vec3D v_Earth = Ecl2EquMatrix(T)*KepVelocity(Earth,T)/c_light;
// Aberration
e = e + v_Earth;  e = e / Norm(e);
```

To calculate precession and nutation, the PrecMatrix_Equ and NutMatrix functions already discussed earlier are employed. The heliocentric velocity of the Earth is calculated using KepVelocity, based on an approximate Keplerian orbit, and then subsequently transformed from ecliptic to equatorial coordinates. The result obtained is the vector of the apparent stellar coordinates.

To handle a star catalogue on the basis of which occultations by the Moon will be predicted by Occult, we make use of a separate class Star:

```
class Star {
  public:
    static void SetCatEquinox( double T );
    static void SetCatEpoch( double T );
    static void SetEpoch( double T );
    Vec3D  Apparent();              // Calculate apparent place of star
    inline char*  Name() { return name; };
    inline double Mag() { return mag; };
    friend istream& operator >> (istream& is, Star& aStar);
  private:
    double RA_Cat;                  // Catalogue right ascension in [rad]
    double Dec_Cat;                 // Catalogue declination in [rad]
    double mu_RA;                   // Proper motion RA in [rad/cy]
    double mu_Dec;                  // Proper motion DEC in [rad/cy]
    double mag;                     // Visual brightness in [mag]
    char   name[12];                // Name
    static double T_CatEquinox;     // Catalogue equinox
    static double T_CatEpoch;       // Catalogue epoch
    static double T_Epoch;          // Current epoch
    static Mat3D  PN;               // Precession/nutation matrix
    static Vec3D  v_Earth;          // Earth velocity vector
};
```

The class provides an input operator >>, using which the coordinates, proper motion, magnitude, and name of a star may be read from a text file, and stored in the appropriate attributes. Common to all objects of the class are the attributes that have been declared as static for the storage of the catalogue's epoch and equinox, which may be set by the appropriate methods

```
  void Star::SetCatEquinox (double T) { T_CatEquinox = T; }
  void Star::SetCatEpoch (double T) { T_CatEpoch = T; }
```

In a similar fashion, the current epoch may be determined by the method

```
void Star::SetEpoch (double T) {
  T_Epoch = T;  // Save Epoch
  PN = NutMatrix(T)*PrecMatrix_Equ(T_CatEquinox, T);
  v_Earth = Ecl2EquMatrix(T)*KepVelocity(Earth,T)/c_light;
}
```

By this means, the values of the precession/nutation matrix and of the heliocentric velocity of the Earth will simultaneously be established, which are the same for all the stars. The said values subsequently serve for the calculation of the apparent coordinates within the Apparent method:

```
Vec3D Star::Apparent() {
  double RA  = RA_Cat  + mu_RA *(T_Epoch-T_CatEpoch);
  double Dec = Dec_Cat + mu_Dec*(T_Epoch-T_CatEpoch);
  Vec3D e = PN * Vec3D(Polar(RA, Dec));
  e = e + v_Earth;  e = e / Norm(e);
  return e;
};
```

To store the whole catalogue, an array of objects of the class `Star` is employed in `Occult` that is dynamically allocated and assigned in function `ReadCatalogue`:

```
//------------------------------------------------------------------------------
// ReadCatalogue: Reads star catalogue from input file
//   Filename     Name of catalogue file to read
//   Stars        Array of Star objects to store catalogue
//   StarsRead    Number of stars read from catalogue file
//------------------------------------------------------------------------------
void ReadCatalogue (char* Filename, Star*& Stars, int& StarsRead)
{
  ifstream inp;
  double   Y_Epoch, Y_Eqx;
  int      i;
  inp.open(Filename);
  // Read and store epoch and equinox
  inp >> Y_Epoch >> Y_Eqx;
  Star::SetCatEpoch((Y_Epoch-2000.0)/100.0);
  Star::SetCatEquinox((Y_Eqx-2000.0)/100.0);
  // Read stars
  StarsRead = 0; inp >> StarsRead; inp.ignore(81,'\n');
  Stars = new Star[StarsRead];  // Allocate memory for catalogue array
  if (Stars != NULL) {
    for (i=0; i<StarsRead; i++)  inp >> Stars[i];
  }
  else cerr << "Insufficient memory to read star catalogue." << endl;
  inp.close();
  cout << endl << "  " << StarsRead
       << " stars read from catalogue" << endl << endl;
}
```

10.2 Geocentric Conjunction

To predict stellar occultations, we have to choose those stars that may be occulted by the Moon in a given period of time, from a large number of stars close to the ecliptic. To do this, we must first determine, for each star, a time at which its right ascension $\alpha_\star$ agrees with the geocentric right ascension of the Moon $\alpha_{\rm M}$. A comparison of the declinations of the Moon and the star at the time of conjunction allows us to see whether the star will be hidden for at least some portion of the Earth. The time of conjunction also serves as a starting point for the more precise calculation of possible occultations, which will be discussed in later sections.

Between two times t_1 and t_2 the right ascension of the Moon varies between $\alpha_{\rm M}(t_1)$ and $\alpha_{\rm M}(t_2)$. If the right ascension of a star $\alpha_\star$ lies between these two values,

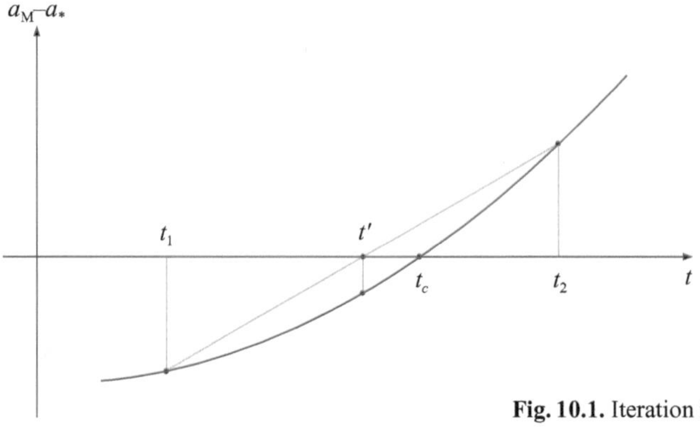

Fig. 10.1. Iteration for the time of conjunction

then we can determine the time of conjunction by means of a linear interpolation (Fig. 10.1). As the right ascension of the Moon increases very evenly, the following close approximation is valid:

$$t_c \approx t' = t_2 - \Delta\alpha_2 \cdot \frac{t_2 - t_1}{\Delta\alpha_2 - \Delta\alpha_1} \quad ,$$

where $\Delta\alpha_i = \alpha_M(t_i) - \alpha_\star$. According to whether the right ascension of the Moon at time t' is larger or smaller than that of the star, we replace t_1 or t_2 with t'. Then $\alpha_M(t_1) \leq \alpha_\star \leq \alpha_M(t_2)$ is again valid, so we can repeat the procedure to obtain a better time. Because of the regular motion of the Moon, this procedure, known as *regula falsi*, leads rapidly and definitely to the desired result. A small modification, however, allows the convergence to be significantly improved. To this end, after calculation of t' and $\Delta\alpha' = \alpha_M(t') - \alpha_\star$, the starting values for the subsequent iteration are chosen as follows:

- if $\Delta\alpha'\Delta\alpha_2 \leq 0$, replace $(t_1, \Delta\alpha_1)$ by $(t_2, \Delta\alpha_2)$;
- if $\Delta\alpha'\Delta\alpha_2 > 0$, replace $(t_1, \Delta\alpha_1)$ by $(t_1, \Delta\alpha_1\Delta\alpha_2/(\Delta\alpha_2 + \Delta\alpha'))$.

In addition, in both cases $(t_2, \Delta\alpha_2)$ is replaced by $(t', \Delta\alpha')$. This variant, known as the *Pegasus Procedure* of the *regula falsi* guarantees that no end point of the search interval is fixed during the course of the iteration, and thus obtains a significantly higher degree of convergence.

```
//------------------------------------------------------------------------------
// Pegasus: Root finder using the Pegasus method
//   PegasusFunct  Pointer to the function to be examined
//   LowerBound    Lower bound of search interval
//   UpperBound    Upper bound of search interval
//   Accuracy      Desired accuracy for the root
//   Root          Root found (valid only if Success is true)
//   Success       Flag indicating success of the routine
// Notes:
//   Pegasus assumes that the root to be found is bracketed in the interval
//   [LowerBound, UpperBound]. Ordinates for these abscissae must therefore
```

```
//   have different signs.
//------------------------------------------------------------------------------
void Pegasus ( PegasusFunct f,
               double LowerBound, double UpperBound, double  Accuracy,
               double& Root, bool& Success )
{
  const int MaxIterat = 30;
  double x1 = LowerBound; double f1 = f(x1);
  double x2 = UpperBound; double f2 = f(x2);
  double x3 = 0.0;        double f3 = 0.0;
  int Iterat = 0;
  // Initialization
  Success = false;
  Root    = x1;
  // Iteration
  if ( f1 * f2 < 0.0 )
    do {
      // Approximation of the root by interpolation
      x3 = x2 - f2/( (f2-f1)/(x2-x1) ); f3 = f(x3);
      // Replace (x1,f2) and (x2,f2) by new values, such that
      // the root is again within the interval [x1,x2]
      if ( f3 * f2 <= 0.0 ) {
        // Root in [x2,x3]
        x1 = x2; f1 = f2; // Replace (x1,f1) by (x2,f2)
        x2 = x3; f2 = f3; // Replace (x2,f2) by (x3,f3)
      }
      else {
        // Root in [x1,x3]
        f1 = f1 * f2/(f2+f3); // Replace (x1,f1) by (x1,f1')
        x2 = x3; f2 = f3;     // Replace (x2,f2) by (x3,f3)
      }
      if (fabs(f1) < fabs(f2)) Root=x1; else Root=x2;
      Success = (fabs(x2-x1) <= Accuracy);
      Iterat++;
    }
    while ( !Success && (Iterat<MaxIterat) );
}
```

With the Pegasus function shown here, the zero points of a scalar function $f(x)$ of type

```
// Pegasus function prototype: double f(double x);
typedef double (*PegasusFunct) (double x);
```

may generally be determined.

In searching for a time of conjunction, as well as in other stages of the computation, a whole series of lunar positions has to be determined. It is thus worthwhile employing the expansion of lunar coordinates using Chebyshev polynomials as discussed previously. From a few, precisely calculated positions, we are able to obtain a simple description of the lunar orbit as a polynomial, which, in just a few operations, may then be evaluated for any number of points in time. The maximum order (e.g., ten) of the polynomials employed are so chosen here that the lunar coordinates may be expressed over a period of around one day without a loss of accuracy. As described in Chap. 8, for this it suffices to declare an object of the class Cheb3D for the approximation of the MoonEqu function. Subsequently, the vector of the equatorial lunar coordinates may be determined through a call to the Value(T) method. The right ascension is obtained from this by the use of the array element operator [phi]. The coordinates thus determined are referred to the true equinox of date, that is, to the current position of the vernal equinox and celestial equator as determined from precession and nutation.

Because Pegasus expects a function as argument that depends on just one parameter, it is necessary to declare the Chebyshev approximation of the Moon as well as the right ascension of the star as global values. The goal function $\Delta\alpha(T)$ for seeking the zero points may be implemented as follows:

```
const double  Interval=1.0/36525.0;     // 1d [cy]
Cheb3D  ChebMoonEqu(MoonEqu, 10, Interval);
double  RA_Star;

//------------------------------------------------------------------------
// RA_Diff: Goal function for iteration of conjunction instant
//   T          Time in Julian centuries ET since J2000
//   <return>  Difference in RA between Moon and star in [rad]
//------------------------------------------------------------------------
double RA_Diff (double T) {
  Vec3D r_Moon = ChebMoonEqu.Value(T);   // Lunar position at time T
  return Modulo ( r_Moon[phi] - RA_Star + pi, pi2 ) - pi;
}
```

For a given star, the exact time of conjunction may thus be determined if, in the period of time under consideration $\Delta\alpha(T)$ undergoes any change in sign.

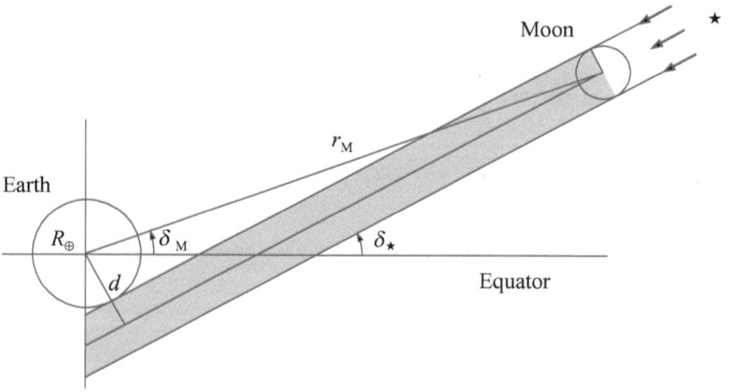

Fig. 10.2. The position of the Moon at the conjunction in right ascension

The possibility of an occultation may now be checked, using the lunar coordinates at the time of conjunction. As may be seen from Fig. 10.2, the shadow cast by the Moon touches the Earth only if the Moon's declination does not differ greatly from that of the star. The distance d between the centre of the Earth and the axis of the shadow should not be greater than the sum of the lunar radius R_{M} and the Earth's radius $R_{\oplus}$:

$$d \;<\; R_{\oplus} + R_{\mathrm{M}} \;\approx\; 1.3 R_{\oplus} \quad .$$

At the time of conjunction in right ascension, the condition

$$d(t_c) \;=\; r_{\mathrm{M}} \cdot |\sin(\delta_{\mathrm{M}} - \delta_{\star})| \;=\; |\boldsymbol{r}_{\mathrm{M}} - (\boldsymbol{e}\cdot\boldsymbol{r}_{\mathrm{M}})\boldsymbol{e}| \;<\; R_{\oplus} + R_{\mathrm{M}} \;\approx\; 1.3 R_{\oplus}$$

must therefore be met. Because of the inclination of the lunar orbit with respect to the celestial equator, $d(t_c)$ may nevertheless be as much as 0.2 Earth radii larger than the smallest limiting distance between the lunar shadow and the centre of the Earth that is otherwise attained. We can therefore select possible occultations by using the following test:

$$|\boldsymbol{r}_{\mathrm{M}} - (\boldsymbol{e}\cdot\boldsymbol{r}_{\mathrm{M}})\boldsymbol{e}| \left\{ \begin{array}{c} < \\ > \end{array} \right\} 1.5 R_{\oplus} \;\Rightarrow\; \left\{ \begin{array}{l} \text{occultation possible} \\ \text{no occultation} \end{array} \right\} . \qquad (10.4)$$

The first part of our stellar occultation program may now be written. `Conjunct` first determines whether the right ascension of the star actually lies within the range of right ascensions covered by the Moon between the two given times T_1 and T_2. If this is the case, then the time of conjunction is established and a test is carried out to see if the shadow of the Moon touches the Earth. If all the conditions are met, then the time of conjunction is returned to the calling program.

```
//------------------------------------------------------------------------------
// Conjunct: Checks for possibility of a lunar occultation
//   T1        Begin of search interval
//   T2        End of search interval
//   RA1       Right ascension of the Moon at T1 in [rad]
//   RA2       Right ascension of the Moon at T2 in [rad]
//   e         Direction of star to check (equatorial coordinates)
//   Conj      Flag indicating whether an occultation is at least likely
//   T_Conj    Time of conjunction
//------------------------------------------------------------------------------
void Conjunct ( double T1,  double T2, double RA1, double RA2, Vec3D& e,
                bool& Conj, double& T_Conj )
{
  const double eps = Rad*1.0e-4;        // Required accuracy in RA [rad]
  Vec3D r_Moon;
  T_Conj = 0.0;                         // Default value
  bool Success = false;
  if ( RA2<RA1 ) RA2+=pi2;              // Map to [RA1,RA1+2*pi[
  RA_Star = e[phi];
  if ( RA_Star<RA1 ) RA_Star+=pi2;      // Map to [RA1,RA1+2*pi[
  Conj = ( (RA1<=RA_Star) && (RA_Star<=RA2) );
  if ( Conj ) {
    Pegasus(RA_Diff, T1, T2, eps, T_Conj, Success);
    r_Moon = ChebMoonEqu.Value(T_Conj);
    Conj = ( Norm( r_Moon-e*Dot(r_Moon,e) ) < 1.5*R_Earth );
  }
}
```

10.3 The Fundamental Plane

The parallel light from a star casts a cylindrical shadow of the Moon onto the Earth. The shadow appears as a circle on any plane perpendicular to the axis of the shadow, and this has the same diameter as the Moon itself. Such a plane, which also passes through the centre of the Earth, is known as the *fundamental plane*. The relative positions of the Moon's shadow and an observer may be represented particularly well by projecting the positions onto the fundamental plane.

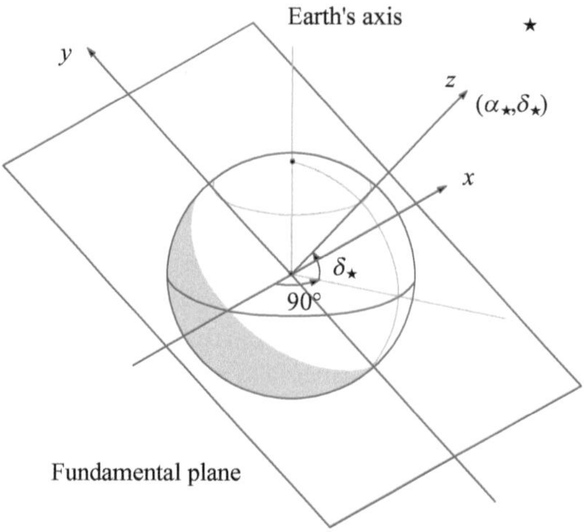

Fig. 10.3. The fundamental plane

If we lay a rectangular coordinate system through the centre of the Earth, as shown in Fig. 10.3, with the z-axis pointing in the direction of the star, which has the coordinates $(\alpha_\star, \delta_\star)$, and the x-axis in the Earth's equatorial plane, then the x- and y-axes define the fundamental plane. The corresponding unit vectors are given by

$$\boldsymbol{e}_x = \frac{\boldsymbol{e}_3 \times \boldsymbol{e}}{|\boldsymbol{e}_3 \times \boldsymbol{e}|} \qquad \boldsymbol{e}_y = \boldsymbol{e} \times \boldsymbol{e}_x \tag{10.5}$$

where $\boldsymbol{e} = \boldsymbol{e}_z$ is the unit vector in the direction of the star (cf. (10.2)) and $\boldsymbol{e}_3 = (0, 0, 1)^T$ denotes the unit vector in the direction of the celestial north pole. Upon projecting the equatorial coordinate vector $\boldsymbol{r}_\mathrm{M}$ of the Moon onto the base vectors, one obtains the coordinates

$$\begin{aligned} x_\mathrm{M} &= \boldsymbol{e}_x \cdot \boldsymbol{r}_\mathrm{M} \\ y_\mathrm{M} &= \boldsymbol{e}_y \cdot \boldsymbol{r}_\mathrm{M} \end{aligned} \tag{10.6}$$

of the centre of the lunar shadow on the fundamental plane.

In a similar manner, from the equatorial coordinates $\boldsymbol{r}_\mathrm{O}$ of the observer, we can obtain the point $(x_\mathrm{O}, y_\mathrm{O})$ on the fundamental plane that represents the projection of a ray of light from the star through the observer. The declination of the observing site describes the angle between the geocentric position vector and the plane containing the terrestrial and celestial equators. In this context it is also known as the *geocentric* latitude φ', which must not be confused with the *geographic* latitude φ. The latter value, which is far more frequently used, is the angle between the Earth's axis and the local horizon, and is, for example, a measure of the altitude of Polaris above the horizon. The flattening of the Earth is the reason for the difference between φ and φ' (cf. Fig. 9.5). Because of its rotation the Earth is not shaped like a sphere, but is instead an ellipsoid of rotation. The distance of the poles from the centre of the Earth is about 20 km less than the Earth's radius at the equator, $r_\oplus = 6378.14$ km. The ratio of this difference to the Earth's radius is the flattening[1] $f \approx 1/298$. In general the geocentric latitude of a point is not available, and it has to be calculated from the geographic latitude. The latter may be taken from current atlases or reference works.

The `Site` function uses (9.9) to determine the Cartesian coordinates

$$\boldsymbol{r}_\mathrm{O}^\mathrm{G} = \begin{pmatrix} r\cos\lambda\cos\varphi' \\ r\sin\lambda\cos\varphi' \\ r\sin\varphi' \end{pmatrix} = \frac{R_\oplus}{\sqrt{1-e^2\sin^2\varphi}} \begin{pmatrix} \cos\lambda\cos\varphi \\ \sin\lambda\cos\varphi \\ (1-e^2)\sin\varphi \end{pmatrix} \tag{10.7}$$

of the observer with respect to the Earth equator and the Greenwich meridian. Making use of the Greenwich sidereal time $\Theta_0(t)$ one then obtains the position vector

$$\boldsymbol{r}_\mathrm{O} = \boldsymbol{R}_z(-\Theta_0(t))\boldsymbol{r}_\mathrm{O}^\mathrm{G} \tag{10.8}$$

with respect to the equator and equinox. The sidereal time itself may be computed using the `GMST` function from Sect. 3.3.

```
//------------------------------------------------------------------------
// Site: Calculates the geocentric position of a site on the Earth
//   lambda     Geographical longitude (east positive) in [rad]
//   phi        Geographical latitude  in [rad]
//   <return>   Geocentric position in [km]
//------------------------------------------------------------------------
Vec3D Site (double lambda, double phi)
{
  // Constants
  const double f       = 1.0/298.257;    // Flattening
  const double e_sqr   = f*(2.0-f);      // Square of eccentricity
  const double cos_phi = cos(phi);       // (Co)sine of geographical latitude
  const double sin_phi = sin(phi);
  // Variables
  double   N = R_Earth / sqrt (1.0-e_sqr*(sin_phi*sin_phi));
  // Cartesian position vector [km]
  return Vec3D ( N*cos_phi*cos(lambda),
                 N*cos_phi*sin(lambda),
                 (1.0-e_sqr)*N*sin_phi   );
}
```

[1] f is traditionally used to denote the flattening as well as one of the two coordinates in the fundamental plane. This should not, however, give rise to confusion.

Here we need to consider the different forms of time measurement yet again. In calculating the position of the Moon we have tacitly assumed all the times to be in *Ephemeris Time* ET (= Dynamic Time TDB/TDT). This form of time-measurement, which is physically uniform, is always used in calculating the orbits of celestial bodies. This is basically what the name Ephemeris Time implies. But in calculating sidereal time (using the function `LMST` in Sect. 3.3) a knowledge of *Universal Time* is required, which essentially corresponds to ordinary clock time (if time zones are taken into account). The difference between ET and UT currently amounts to about one minute and may be taken from Table 3.1. In addition, a function `ETminUT` is given in Sect. 9.3 that calculates the desired value for the period between 1825 and 2005 using a polynomial approximation. For later years actual values are given in various astronomical almanacs. As the difference between Universal Time and Ephemeris Time alters very slowly, it has to be determined only once at the beginning of our program.

10.4 Disappearance and Reappearance

The actual circumstances of an occultation depend on the differences

$$\begin{aligned} f &= x_M - x_O \\ g &= y_M - y_O \end{aligned}$$

between the coordinates of the Moon and of the observer on the fundamental plane. If the observer's distance $\sqrt{f^2 + g^2}$ from the centre of the lunar shadow is less than the radius of the Moon R_M, then the star will be occulted by the Moon for that particular observer. The times of disappearance and reappearance are governed by the condition

$$f(t)^2 + g(t)^2 = R_M^2 = k^2 \cdot R_\oplus^2 \quad . \tag{10.9}$$

Here

$$k = R_M / R_\oplus = 0.2725$$

is the ratio between the radii of the Moon and the Earth.

In its orbit around the Earth, the Moon covers a distance equivalent to the diameter of the Earth in about four hours. During this time the occultation shadow moves across the Earth. In searching for times of contact it therefore suffices to begin about two-and-a-half hours before the time of conjunction. To determine the two times that fulfil condition (10.9), the value of

$$s(t) = f^2(t) + g^2(t) - k^2 R_\oplus^2$$

is calculated for intervals of 15^m. Initially s is positive, because the observing site is still outside the lunar shadow. When s becomes negative, the star disappears behind the limb of the Moon. From three consecutive values s_-, s_0, and s_+ a parabola may

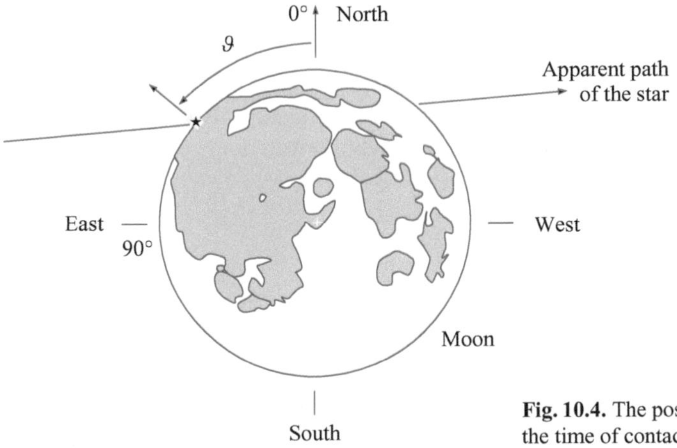

Fig. 10.4. The position angle of a star at the time of contact with the lunar limb

now be determined that closely approximates the changes in $s(t)$ over a period of half an hour. If s reaches zero once or twice during this period, then these points may be obtained by solving a quadratic equation. If not, the next three values for s are taken, and the same calculation is made. This procedure is precisely the same as that used to determine rising and setting times in Chap.3, where the function `Quad` was also introduced to provide quadratic interpolation and to determine the zero points of a function from three given points. This procedure is employed within `Examine` to determine the times of disappearance and reappearance. `Examine` is given in full at the end of this chapter.

In addition to the times of contact, other data of particular interest to observers are given in predictions of occultations: the position angle of the point of disappearance and reappearance, and the station coefficients. The definition of the position angle is shown in Fig. 10.4. It is the angle between North and the line from the centre of the Moon through the point of disappearance or reappearance on the lunar limb. Position angles are always measured anticlockwise from 0°–360°. The value of the position angle ϑ depends only on the fundamental-plane coordinates f and g at the times of disappearance and reappearance, and may be calculated using the relations

$$\begin{aligned} \cos\vartheta &= -g/\sqrt{f^2+g^2} \\ \sin\vartheta &= -f/\sqrt{f^2+g^2} \quad . \end{aligned} \tag{10.10}$$

The station coefficients are values that describe an approximate relationship between the contact times and the observing site. They are primarily used to convert the times of contact, predicted for one specific point, to those that will apply at a nearby site. The coefficient a describes how contact time t varies with geographic longitude λ; b is the corresponding value for geographic latitude φ:

$$a = \frac{dt}{d\lambda} \qquad b = \frac{dt}{d\varphi} \quad .$$

Typical values for these two coefficients are around $1^{\mathrm{m}}/^\circ$. For grazing occultations, when the times of disappearance and reappearance lie close together, the station coefficients increase rapidly, however. The equations required to calculate station coefficients mainly contain values that are already known. If one considers the condition (10.9) as an implicit relation between the contact time t and the observer's position $\boldsymbol{r}$, then differentiation yields the derivative

$$\frac{dt}{d\boldsymbol{r}} = \frac{f\boldsymbol{e}_x + g\boldsymbol{e}_y}{f\dot{f} + g\dot{g}} \tag{10.11}$$

of the time t of disappearance or reappearance with respect to the Cartesian coordinates of the observer. Here $\dot{f}$ and $\dot{g}$ are the time derivatives of f and g, which for the sake of simplicity may be expressed as the following differences:

$$\dot{f}(t) \approx \frac{f(t+0.25^{\mathrm{h}}) - f(t)}{0.25^{\mathrm{h}}} \qquad \dot{g}(t) \approx \frac{g(t+0.25^{\mathrm{h}}) - g(t)}{0.25^{\mathrm{h}}} \quad .$$

Making use of the derivatives

$$\frac{d\boldsymbol{r}}{d\lambda} = \begin{pmatrix} 0 \\ 0 \\ 1 \end{pmatrix} \times \boldsymbol{r} \qquad \frac{d\boldsymbol{r}}{d\varphi} = \boldsymbol{r} \times \frac{d\boldsymbol{r}}{d\lambda} \Big/ \left| \frac{d\boldsymbol{r}}{d\lambda} \right| \tag{10.12}$$

of the position vector with respect to the geographic longitude and latitude the desired station coefficients are obtained as

$$a = \frac{dt}{d\boldsymbol{r}} \frac{d\boldsymbol{r}}{d\lambda} \qquad b = \frac{dt}{d\boldsymbol{r}} \frac{d\boldsymbol{r}}{d\varphi} \quad . \tag{10.13}$$

The difference between geographic (φ) and geocentric (φ') latitudes may be neglected in calculating the station coefficients.

10.5 The OCCULT Program

The `Occult` program calculates stellar occultations by the Moon over the required period and for a given point. The times of the star's disappearance and reappearance at the lunar limb are calculated together with the corresponding position angles and the station coefficients that may be used to convert the times to those that apply at a nearby site.

The coordinates of the stars, whose occultations are being sought, must be entered into a data file called `Occult.dat` by default. The first line of this file contains the epoch and the equinox of the appropriate catalogue (generally 1950.0 or 2000.0) as well as the number of stars. These first two values allow for proper motion to be taken into account and for the stellar positions to be converted to the true equinox. Finally each star has a separate line containing

- right ascension (in $^{\mathrm{h\ m\ s}}$),
- proper motion in right ascension per hundred years (in ^{s}),

- declination (in ° ′ ″),
- proper motion in declination per hundred years (in ″) and
- the name

in the format

```
_hh_mm_ss.sss__ss.ss___+dd mm ss.ss__sss.s__mm.m___ccccccccccc
```

The length of the name is limited to a maximum of 11 letters, but may be increased at any time by altering class `Star`. The following example of the structure of the data file gives data for various stars of the Pleiades (cf. Table 10.1).

```
2000.0 2000.0 9
  3 44 48.180   0.06   +24 17 21.44   -5.1   5.4   Celaeno
  3 44 52.532   0.14   +24  6 48.01   -4.6   3.8   Electra
  3 45 12.469   0.07   +24 28  1.36   -5.8   4.4   Taygeta
  3 45 49.566   0.05   +24 22  3.63   -5.0   4.0   Maia
  3 45 54.423   0.04   +24 33 16.02   -4.6   5.9   Asterope
  3 46 19.537   0.08   +23 56 53.42   -5.7   4.3   Merope
  3 47 29.073   0.14   +24  6 18.38   -4.6   3.0   Alcyone
  3 49  9.739   0.13   +24  3 12.24   -4.7   3.8   Atlas
  3 49 11.181   0.09   +24  8 12.58   -4.3   5.2   Pleione
```

At the beginning of the program the geographic coordinates of the observing site are requested, together with the times between which occultations are to be determined. The overall interval is evaluated in steps of one day, for which the Chebyshev polynomials are first determined. For each star in the data file, after calculating the apparent coordinates, a check is carried out for the mid of the day in the `Examine` function to see if there is a possible occultation. The `Conjunct` routine, which has already been mentioned, next determines the time of conjunction of thc Moon and the star, and checks the distance of the axis of the shadow from the centre of the Earth. In case of a possible occultation `Examine` then determines the times of disappearance and reappearance for the given observing site. Subsequently, `Contact` computes the corresponding position angle and station coefficients for each of the contacts.

Out of the stellar occultations determined in this way, only those are output, which take place under favourable visibility conditions. To this end, `Contacts` checks the star's and the Sun's altitude above the horizon, the lunar phase and the illumination of the lunar limb as well as the star's brightness according to the following criteria:

- At the time of disappearance or reappearance, the star must at least be 5° above the horizon. For stars brighter than $1\overset{\mathrm{m}}{.}9$ this limit is reduced to 2°.
- The star's magnitude $7\overset{\mathrm{m}}{.}5$ or brighter.
- At the time of disappearance or reappearance, the Sun must at least be 6° below the horizon (civil twilight). For stars brighter that $5\overset{\mathrm{m}}{.}5$ a reduced limit of 3° applies. Occultations of stars with magnitudes $2\overset{\mathrm{m}}{.}0$ to $4\overset{\mathrm{m}}{.}0$ are considered

all the time from sunset to sunrise, in case of brighter stars even thoses events that take place during the day.

- For phase angles below 12°, i.e. about 24 hours before and after Full Moon, a limiting magnitude of $3^{\mathrm{m}}_{.}5$ is considered. Up to 24°, a value of $5^{\mathrm{m}}_{.}5$ holds, and a value of $6^{\mathrm{m}}_{.}5$, up to 36°.
- At the bright limb of the Moon, disappearances are predicted only down to magnitude $4^{\mathrm{m}}_{.}5$ and reappearances for stars brighter than $3^{\mathrm{m}}_{.}5$. For reappearances at the dark limb a limiting magnitude of $6^{\mathrm{m}}_{.}5$ applies.

It must also be mentioned that the output is not necessarily in chronological order of the individual occultations, but instead depends on the order of the entries in the data file.

```
//------------------------------------------------------------------------------
// Module:  Occult program (Occult.cpp)
// Purpose: Prediction of stellar occultations by the Moon
// (c) 1999 Oliver Montenbruck, Thomas Pfleger
//------------------------------------------------------------------------------

#include <cmath>
#include <fstream>
#include <iomanip>
#include <iostream>
#include "APC_Cheb.h"
#include "APC_Const.h"
#include "APC_IO.h"
#include "APC_Math.h"
#include "APC_Moon.h"
#include "APC_Phys.h"
#include "APC_PrecNut.h"
#include "APC_Spheric.h"
#include "APC_Sun.h"
#include "APC_Time.h"
#include "APC_VecMat3D.h"
using namespace std;

const double Interval   = 1.0/36525.0;    // 1d [cy]

Cheb3D ChebMoonEqu(MoonEqu, 10, Interval);
double RA_Star;

// Class Star
class Star { ... };

// Class Event
enum enEvent {dummy=-1, in=0, out=1};  // Identifies dis- or reappearance
class Event {
  public:
    Event();
    Event( const Star& aStar, double MjdUT, enEvent InOut, bool visible,
           double PosAng, double a, double b );
```

```
    inline bool IsVisible() { return m_visible; };
    friend ostream& operator << (ostream& os, Event& anEvent);
  private:
    bool      m_visible;
    Star      m_Star;
    double    m_MjdUT,m_PosAng,m_a,m_b;
    enEvent   m_InOut;
};

// Constructors
Event::Event() : m_visible(false), m_InOut(dummy) { }
Event::Event( const Star& aStar, double MjdUT, enEvent InOut, bool visible,
              double PosAng, double a, double b )
 : m_visible(visible),  m_Star(aStar),  m_MjdUT(MjdUT),  m_InOut(InOut),
   m_PosAng(PosAng),  m_a(a),  m_b(b) { }
// Output operator
ostream& operator << (ostream& os, Event& anEvent)
{
  if ( anEvent.m_InOut != dummy ) {
    // Output true events only
    os << " " << DateTime(anEvent.m_MjdUT)
       << "  " << anEvent.m_Star.Name()
       << fixed << setprecision(1)
       << setw(5) << anEvent.m_Star.Mag()
       << "    " << ( (anEvent.m_InOut==in)? "D" : "R" )
       << "    " << Time(24.0*Modulo(anEvent.m_MjdUT, 1.0),HHMMSS)
       << fixed << setprecision(1)
       << setw(7) << int(Deg*anEvent.m_PosAng + 0.5)
       << setw(8) << anEvent.m_a/Deg
       << setw(7) << anEvent.m_b/Deg
       << endl;
  ]
  return os;
}
...

//----------------------------------------------------------------------------
// GetInput: Prompts user for input
//   MjdStart  Start time for search for occultations
//   MjdEnd    End time for search for occultations
//   ET_UT     Difference ephemeris time - universal time in [s]
//   R_Obs     Geocentric position of the observer in [km]
//----------------------------------------------------------------------------
void GetInput ( double& MjdStart, double& MjdEnd,  double& ET_UT,
                Vec3D& R_Obs )
{
  int     year,month,day;
  bool    valid;
  double  Lambda, Phi;
  // Query prediction interval
  cout << " Period of time for prediction of occultations" << endl;
  cout << "  first date (yyyy mm dd)                    ... ";
  cin >> year >> month >> day; cin.ignore(81,'\n');
```

```
  MjdStart = Mjd(year,month,day);
  cout << "  last  date (yyyy mm dd)                    ... ";
  cin >> year >> month >> day; cin.ignore(81,'\n');
  MjdEnd = Mjd(year,month,day);
  // Query difference between ephemeris time and universal time
  ETminUT ( (0.5*(MjdStart+MjdEnd)-MJD_J2000)/36525.0, ET_UT, valid );
  if ( valid ) {
    cout << " Difference ET-UT (proposal:" << fixed << setw(6)
         << setprecision (1) << ET_UT << " sec)       ... ";
    cin >> ET_UT; cin.ignore(81,'\n');
  }
  else {
    cout << " Difference ET-UT (sec)" << setw(27) << right << "... ";
    cin >> ET_UT; cin.ignore(81,'\n');
  }
  // Query geographic coordinates
  cout << " Observer's coordinates: East longitude [deg] ... ";
  cin >> Lambda; cin.ignore(81,'\n');
  cout << "                         latitude       [deg] ... ";
  cin >> Phi; cin.ignore(81,'\n');
  Lambda *= Rad; Phi *= Rad;
  R_Obs = Site (Lambda, Phi);
}
...

//------------------------------------------------------------------------------
// FG: Compute relative fundamental plane coordinates
//   T_ET     Time ET in Julian centuries since J2000
//   ET_UT    Difference betweeen ephemeris time and universal time in [s]
//   e        Direction of star to check (equatorial coordinates)
//   R_Obs    Geocentric position of the observer in [km]
//   f, g     Relative fundamental plane coordinates  in [km]
//   s        Square of distance between shadow axis and observer in [km]
//------------------------------------------------------------------------------
void FG ( double T_ET, double ET_UT, const Vec3D& e, const Vec3D& R_Obs,
          double& f, double& g, double& s )
{
  double  MjdUT;
  Vec3D   e_x, e_y, r_Moon, r_Obs;
  // Fundamental plane unit vectors
  e_x = Cross ( Vec3D(0,0,1), e );
  e_x = e_x / Norm(e_x);
  e_y = Cross ( e, e_x );
  // Lunar position
  r_Moon = ChebMoonEqu.Value(T_ET);
  // Observer's position
  MjdUT = (36525.0*T_ET+MJD_J2000) - ET_UT/86400;
  r_Obs = R_z(-GMST(MjdUT)) * R_Obs;
  // Relative fundamental plane coordinates
  f = Dot ( e_x, r_Moon - r_Obs );
  g = Dot ( e_y, r_Moon - r_Obs );
  s = f*f + g*g - R_Moon*R_Moon;
}
```

```
//------------------------------------------------------------------------------
// Contact: Compute position angle, longitude and latitude coefficients and
//          check for visibility (concerning altitude and sky brightness)
//   T_ET       Time ET in Julian centuries since J2000
//   ET_UT      Difference betweeen ephemeris time and universal time [s]
//   e          Direction of star to check (equatorial coordinates)
//   R_Obs      Geocentric position of the observer in [km]
//   InOut      Indicates disappearance or reappearance
//   mag        Magnitude of the star in [mag]
//   PosAngle   Position angle w.r.t. North in [rad]
//   a          Longitude coefficient in [min/']
//   b          Latitude coefficient in [min/']
//   Visible    Flag indicating visibility of the occultation
//------------------------------------------------------------------------------
void Contact(double T_ET, double ET_UT, const Vec3D& e, const Vec3D& R_Obs,
             enEvent InOut, double mag,
             double& PosAngle, double& a, double& b, bool& Visible )
{
  // Constants
  const double cy = 36525.0*1440.0;   // Minutes per century
  const double dt = 1.0;              // Time step [min]
  //Variables
  double  MjdUT;
  Vec3D   e_x, e_y, r_Moon, r_Obs, e_Obs, r_Sun;
  Vec3D   dt_dr, dr_dlambda, dr_dphi;
  double  s_0, f, g, ff, gg, fdot, gdot;
  double  Elev, ElevSun, PosAngSun, DPosAng, i;
  bool    BrightLimb;

  // Fundamental plane unit vectors
  e_x = Cross ( Vec3D(0,0,1), e );
  e_x = e_x / Norm(e_x);
  e_y = Cross ( e, e_x );
  // Lunar position
  r_Moon = ChebMoonEqu.Value(T_ET);
  // Observer's geocentric position
  MjdUT = (36525.0*T_ET+MJD_J2000) - ET_UT/86400;
  r_Obs = R_z(-GMST(MjdUT)) * R_Obs;
  e_Obs = r_Obs / Norm(r_Obs);
  // Altitude of star above the horizon
  Elev = asin ( Dot(e_Obs,e) );
  // Relative fundamental plane coordinates
  f = Dot ( e_x, r_Moon - r_Obs );
  g = Dot ( e_y, r_Moon - r_Obs );
  // Position angle
  PosAngle = Modulo(atan2(-f, -g), pi2);
  // Time derivative of fundamental plane coordinates
  FG ( T_ET+dt/cy, ET_UT, e, R_Obs, ff, gg, s_0 );
  fdot = (ff-f)/dt;  // [km/min]
  gdot = (gg-g)/dt;
  // Derivatives of observer position w.r.t. geocentric coordinates
  dr_dlambda = Cross ( Vec3D(0,0,1), r_Obs );
  dr_dphi    = Cross ( r_Obs, dr_dlambda / Norm(dr_dlambda) );
```

```
  // Compute longitude and latitude coefficients
  dt_dr = (f*e_x+g*e_y) / (f*fdot+g*gdot);
  a = Dot ( dt_dr, dr_dlambda );
  b = Dot ( dt_dr, dr_dphi    );
  // Compute solar coordinates, altitude and position angle
  r_Sun = AU*SunEqu(T_ET);
  ElevSun = asin ( Dot ( e_Obs, r_Sun/Norm(r_Sun) ) );
  PosAngSun = PosAng(r_Moon-r_Obs, r_Sun-r_Moon);
  DPosAng = Modulo(PosAngle-PosAngSun+pi, pi2) - pi;
  BrightLimb = ( fabs(DPosAng) < pi/2.0 );
  // Phase angle of Moon
  i = acos(Dot(-r_Moon,r_Sun-r_Moon)/(Norm(r_Sun-r_Moon)*Norm(r_Moon)));
  // Visibility conditions
  Visible = ( // Minimum altitude of star
              ( Elev > +5.0*Rad )                        ||
              ( Elev > +2.0*Rad ) && ( mag <= 1.9 )
            ) &&
            ( // Minimum depression of the Sun
              ( ElevSun < -6.0*Rad ) && ( mag <= 7.5 ) ||
              ( ElevSun < -3.0*Rad ) && ( mag <= 5.5 ) ||
              ( ElevSun < -0.5*Rad ) && ( mag <= 4.5 ) ||
                                        ( mag <= 1.9 )
            ) &&
            ( // Lunar phase
              ( i > 36.0*Rad )                       ||
              ( i > 24.0*Rad ) && ( mag <= 6.5 ) ||
              ( i > 12.0*Rad ) && ( mag <= 5.5 ) ||
                                  ( mag <= 3.5 )
            ) &&
            ( // Limb illumination
              ( !BrightLimb ) && (InOut==in )                 ||
              ( !BrightLimb ) && (InOut==out) && (mag<=6.5) ||
              (  BrightLimb ) && (InOut==in ) && (mag<=4.5) ||
              (  BrightLimb ) && (InOut==out) && (mag<=3.5)
            );
}
//------------------------------------------------------------------------------
// Examine: Checks for possibility of a lunar occultation
//   T1       Begin of search interval
//   T2       End of search interval
//   RA1      Right ascension of the Moon at T1 in [rad]
//   RA2      Right ascension of the Moon at T2 in [rad]
//   ET_UT    Difference ephemeris time - universal time in [s]
//   R_Obs    Geocentric position of the observer in [km]
//   aStar    Star to examine
//   In       Disappearance event
//   Out      Reappearance event
//------------------------------------------------------------------------------
void Examine ( double T1, double T2, double RA1, double RA2, double ET_UT,
               const Vec3D& R_Obs, Star& aStar, Event& In, Event& Out )
{
  const double cy = 876600.0;     // Hours per century
  const double dT = 0.25/cy;      // Step size
```

```
  const double DT = 2.25/cy;      // Search interval [-DT-dT,+DT+dT]
  Vec3D   e;
  bool    Conj, Visible;
  int     n_found, n_root, i;
  double  T_Conj, T, T_Cont[2], MjdUT;
  double  s_minus, s_0, s_plus, xe, ye, root[2];
  double  f, g, PosAng, a, b;
  // Default values
  In = Out = Event();
  // Apparent coordinates of star
  e = aStar.Apparent();
  // Search for time of conjunction
  Conjunct ( T1, T2, RA1, RA2, e, Conj, T_Conj );
  if (!Conj) return;
  // Search for times of contact
  n_found = 0;
  T = T_Conj - DT;
  FG (T-dT, ET_UT, e, R_Obs, f, g, s_minus);
  while (true) {
    // Interpolate fundamental plane distance and find times of contact
    FG (T   , ET_UT, e, R_Obs, f, g, s_0   );
    FG (T+dT, ET_UT, e, R_Obs, f, g, s_plus);
    Quad (s_minus, s_0, s_plus, xe, ye, root[0], root[1], n_root);
    for (i=0;i<n_root;i++)
      T_Cont[n_found+i] = T + root[i]*dT;
    n_found += n_root;
    Conj = (n_found==2);
    if ( (Conj) || (T_Conj+DT<T) ) break; // Exit loop
    T+=2.0*dT; // Increment time
    s_minus = s_plus;
  };
  // Contacts (disappearance, reappearance)
  if (Conj) {
    for (i=in;i<=out;i++)  {
      MjdUT = 36525.0*T_Cont[i] + MJD_J2000 - ET_UT/86400.0;
      Contact ( T_Cont[i], ET_UT, e, R_Obs, enEvent(i), aStar.Mag(),
                PosAng, a, b, Visible );
      if (i==in)  In  = Event(aStar, MjdUT, in , Visible, PosAng, a, b);
      if (i==out) Out = Event(aStar, MjdUT, out, Visible, PosAng, a, b);
    }
  }
}
//------------------------------------------------------------------------
// Main program
//------------------------------------------------------------------------
void main(int argc, char* argv[])
{
  const double Margin    = 3.0 / (36525.0*24.0);  // 3h [cy]
  Star*     Stars = NULL;      // Star catalogue
  int       NStars, iStar;
  double    MjdStart, MjdEnd, ET_UT, T, TEnd, RA_min, RA_max;
  Vec3D     R_Obs;
  Event     In, Out;
```

```
  char      InputFile[APC_MaxFilename] = "";
  char      OutputFile[APC_MaxFilename] = "";
  bool      FoundInputfile  = false, FoundOutputfile = false;
  ofstream  OutFile;
  // Title
  cout << "          OCCULT: occultations of stars by the Moon  " << endl
       << "         (c) 1999 Oliver Montenbruck, Thomas Pfleger " << endl
       << endl;
  // Find catalogue file and (optional) name of output file
  GetFilenames( argc, argv, "Occult.dat", InputFile, FoundInputfile,
                OutputFile, FoundOutputfile );
  if (!FoundInputfile) {
    cerr << "Terminating program." << endl;  exit(-1);
  }
  ReadCatalogue (InputFile, Stars, NStars);
  if (NStars == 0) cerr << "Terminating program." << endl; exit(-1);
  // Get user input (search interval, observer location, etc.)
  GetInput (MjdStart, MjdEnd, ET_UT, R_Obs);
  // Redirect output if output file shall be created
  if (FoundOutputfile) {
    OutFile.open(OutputFile); if (OutFile.is_open()) cout = OutFile;
  }
  // Print header
  cout
   << endl
   << "  Date        Name       m_v   D/R       UT        Pos      a      b  "
   << endl
   << "                         mag                       deg    m/deg  m/deg"
   << endl;
  // Search occultations in subsequent time intervals (ephemeris time)
  T    = (MjdStart-MJD_J2000)/36525.0;
  TEnd = (MjdEnd  -MJD_J2000)/36525.0;
  do {
    // Chebyshev approximation of lunar coordinates
    ChebMoonEqu.Fit(T-Margin, T+Interval+Margin);
    RA_min = ChebMoonEqu.Value(T)[phi];
    RA_max = ChebMoonEqu.Value(T+Interval)[phi];
    // Set epoch for computation of apparent coordinates
    Star::SetEpoch(T+Interval/2.0);
    // Star catalogue loop
    for (iStar=0; iStar<NStars; iStar++) {
      // Check for occultation
      Examine ( T, T+Interval, RA_min, RA_max,
                ET_UT, R_Obs, Stars[iStar], In, Out );
      // Print disappearance and reappearance, if visible
      if (In.IsVisible() ) cout<<In;  if (Out.IsVisible()) cout<<Out;
    }
    T += Interval;  // Next interval
  }
  while (T<TEnd);
  if (OutFile.is_open()) OutFile.close();
  if (Stars!=NULL) delete[] Stars;
}
```

As an example, we will now use `Occult` to calculate two series of Pleiades occultations that occurred in 1989. As search period we will take the interval between 1989 September 15 and November 15. The predictions are for Munich, which is at longitude 11°.6 East, and latitude 48°.1 North. We will assume that the difference between Ephemeris Time and Universal Time amounts to 57 seconds as suggested by the polynomial approximation in Table 9.2. Data entered by the user are printed in italics.

```
        OCCULT: occultations of stars by the Moon
       (c) 1999 Oliver Montenbruck, Thomas Pfleger

9 stars read from catalogue

Period of time for prediction of occultations
 first date (yyyy mm dd)                     ...  1989 09 15
 last  date (yyyy mm dd)                     ...  1989 11 15
Difference ET-UT (proposal:  57.0 sec)       ...  57.0
Observer's coordinates: East longitude [deg] ...  11.60
                        latitude       [deg] ...  48.10
```

Date	Name	m_v mag	D/R	UT	Pos deg	a m/deg	b m/deg
1989/09/19	Celaeno	5.4	R	22:41:00	222	0.0	2.1
1989/09/19	Taygeta	4.4	D	22:01:57	67	0.1	1.7
1989/09/19	Taygeta	4.4	R	23:00:37	251	0.5	1.7
1989/09/19	Maia	4.0	D	22:15:34	98	0.5	1.3
1989/09/19	Maia	4.0	R	23:07:32	220	0.1	2.2
1989/09/19	Asterope	5.9	R	23:20:36	256	0.6	1.6
1989/11/13	Alcyone	3.0	D	19:00:10	122	1.1	0.6
1989/11/13	Alcyone	3.0	R	19:36:07	198	-0.3	2.9

We find that in the two months a total of five occultations of Pleiades stars take place, where the disappearance, reappearance, or both, fulfil the chosen criteria, and may be well observed.

To minimize the computation time and the output, the input file used in this example contains only a limited number of stars. The accompanying CD-ROM provides an additional file `ZC.dat` for this purpose, which contains 3546 stars from the Zodiacal Catalogue for general occultation predictions. The positions have been brought up to date using the PPM Catalogue, which has a greater positional accuracy than the original Zodiacal Catalogue. To use `ZC.dat` or any other specific stellar catalogue, on running `Occult` the desired input file is entered after the program name on the command line (see Sect. A.1.3).

10.6 Estimation of ΔT=ET-UT from Observations

After observing a stellar occultation, it requires little effort to estimate the difference between Universal Time and Ephemeris Time and thus keep track of the decrease in the Earth's rate of rotation.

To do this, we compare the time of the observed appearance or disappearance at the lunar limb with the value $t_{\mathrm{ET=UT}}$ predicted on the asssumption that $\Delta T = \mathrm{ET} - \mathrm{UT} = 0$. Provided the time of observation t_{UT} is referred to Universal Time, the following relation holds for the same instant expressed in Ephemeris Time:

$$t_{\mathrm{ET}} = t_{\mathrm{UT}} + \Delta T \quad .$$

The Ephemeris Time t_{ET} of the occulation, however, is not the same as the predicted value t_0. This is because the computed Greenwich sidereal time differs from the actual value by $\Delta\Theta_0 \approx \Delta T$, when assuming ET = UT in the prediction. Therefore the computed occultation refers to a point on Earth that is located $\Delta\lambda = 15°/\mathrm{h}\cdot\Delta T$ to the east of the real observing site. This difference may, however, easily be accounted for by the station coefficient a, yielding

$$t_{\mathrm{ET}} = t_0 - a\,(0\overset{\circ}{.}25/\mathrm{m}\cdot\Delta T) \quad .$$

After inserting and rearranging these equations, one obtains the relation

$$\Delta T = \frac{t_0 - t_{\mathrm{UT}}}{1 + a\cdot 0\overset{\circ}{.}25/\mathrm{m}} \quad , \tag{10.14}$$

from which ΔT may be estimated.

Table 10.2. Examples for the determination of the difference between Universal Time and Ephemeris Time from stellar occultations

Star	Observing site λ φ	Observing time Date UT	Computed ET=UT	a	ΔT
17 Tau	Roy. Obsy Berlin $+13\overset{\circ}{.}396$ $+52\overset{\circ}{.}505$	26.09.1896 20:32:10	20:32:02	$-0\overset{\mathrm{m}}{.}7/°$	-10 s
57B Sco	Univ. Warschau $+21\overset{\circ}{.}030$ $+52\overset{\circ}{.}218$	24.01.1930 05:30:45	05:31:08	$-0\overset{\mathrm{m}}{.}3/°$	25 s
μ Cet	Stuttgart Obsy $+9\overset{\circ}{.}197$ $+48\overset{\circ}{.}784$	10.02.1962 21:12:45	21:13:22	$+0\overset{\mathrm{m}}{.}6/°$	32 s
δ Psc	Stuttgart Obsy $+9\overset{\circ}{.}197$ $+48\overset{\circ}{.}784$	13.01.1970 19:35:12	19:36:03	$+1\overset{\mathrm{m}}{.}1/°$	40 s
ε Ari	St. Augustin $+7\overset{\circ}{.}177$ $+50\overset{\circ}{.}775$	30.11.1990 22:01:32	22:02:59	$+1\overset{\mathrm{m}}{.}9/°$	59 s

As an example, five stellar occultations made between 1896 and 1990 that have been observed at various European sites are listed in Table 10.2. It is easy to see that the difference between Ephemeris Time and Universal Time has increased by approximately one minute over the last century, which is a visible sign of the deceleration of the Earth's rotation. Compared to the actual values, the ΔT values estimated from the observations exhibit differences of 1–5 seconds. These errors correspond to the accuracy of predictions made by `Occult`, and are essentially caused by small errors in the stellar and lunar coordinates and the neglect of the lunar limb profile.

11. Orbit Determination

The classical aim of orbit determination is to obtain the orbital elements of a planet, comet, or minor planet from the smallest possible number of observed positions. This is therefore essentially the opposite to determining an ephemeris, where positions are obtained from known orbital elements. Any observation made from Earth at a specific time gives two spherical coordinates. We may choose whether these relate to the celestial equator (right ascension and declination) or to the ecliptic (ecliptic longitude and latitude). On the other hand, the distance cannot be measured, so knowledge of it cannot be used in orbit determination. To derive six orbital elements, the same number of independent observational values are required, so three observations must be available.

The method of determining orbits described here goes back to C. F. Gauss, but several points were revised and modernized by H. Bucerius. The core of the procedure consists, on the one hand, of establishing the orbital elements from two position vectors, and on the other, of iterating the so-called triangle-area ratios which represent the geometrical position of the three observed points in the orbital plane. The procedure is supplemented by introducing the solution of the Gauss-Lagrangian equation, which gives a reliable method of distinguishing between multiple solutions of the problem of determining orbits.

11.1 Determining an Orbit from Two Position Vectors

The orbital elements are generally employed to describe a planetary or cometary orbit, because they enable one to obtain a particularly clear interpretation of the individual values. For orbit determination, however, it is more convenient to use another method of description. The orbit is equally well defined if we know the position and velocity of a celestial body at a specific instant, or alternatively, two positions on the orbit and the times at which they were reached. The former description is used in Laplace's method of orbit determination, but Gauss's method employs two position vectors. This section will deal with calculating the elements of an orbit that is described by two known positions $\boldsymbol{r}_a$ and $\boldsymbol{r}_b$ at times t_a and t_b. The problem of determining an orbit then reduces to determining two heliocentric positions from three observed directions.

First, we will discuss the intermediate step of calculating the *sector-triangle ratio*, which is the most difficult part of determining an orbit. Yet this value plays an exceedingly important part in the later steps in determining an orbit, as will be seen from later sections.

11.1.1 The Sector-Triangle Ratio

The area Δ of the triangle defined by the vectors $\boldsymbol{r}_a$ and $\boldsymbol{r}_b$ (Fig. 11.1) depends on the length of the sides r_a and r_b, and the included angle $\nu_b - \nu_a$, which must, however, always be less than 180° in what follows:

$$\Delta = \frac{1}{2} r_a r_b \cdot \sin(\nu_b - \nu_a) \quad . \tag{11.1}$$

Here ν_a and ν_b are the values of the true anomaly at the end points of the portion of the orbit under consideration.

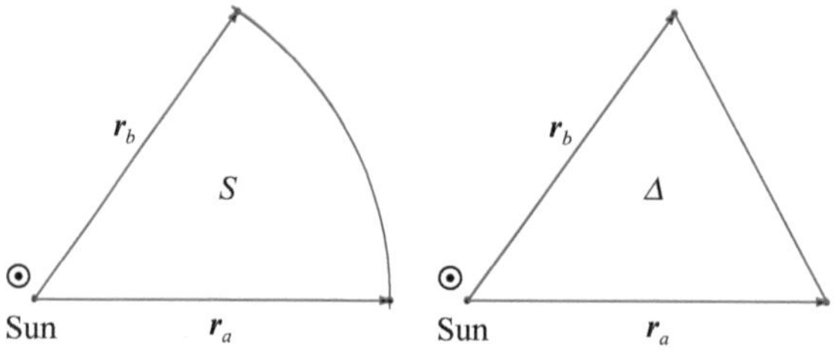

Fig. 11.1. Areas of sectors and triangles

The area S of the sector that is bounded by $\boldsymbol{r}_a$ and $\boldsymbol{r}_b$ and the arc of the orbit between them, is – because of Kepler's Second Law (the law of areas) – proportional to the difference between the times t_a and t_b:

$$S = \frac{1}{2}\sqrt{GM_\odot} \cdot \sqrt{a(1-e^2)} \cdot (t_b - t_a) \quad . \tag{11.2}$$

Here a and e denote the semi-major axis and the eccentricity of the orbit that joins the given points (see Chap. 4). If we substitute the semi-latus rectum $p = a(1-e^2)$, then we obtain the expression

$$\eta = \frac{S}{\Delta} = \frac{\sqrt{p} \cdot \tau}{r_a r_b \cdot \sin(\nu_b - \nu_a)} \quad , \tag{11.3}$$

for the ratio η between the two areas, where, for simplicity, the interval is defined by

$$\tau = \sqrt{GM_\odot} \cdot (t_b - t_a) \quad . \tag{11.4}$$

As will be seen, the equation for η contains the semi-latus rectum p, which has not previously been expressed in terms of $\boldsymbol{r}_a$ and $\boldsymbol{r}_b$. If we try to eliminate the semi-latus rectum by using the known equations for the two-body problem, then we find that it is no longer possible to express η as a solvable algebraic equation. Instead we obtain the transcendental equation[1]

$$\eta = 1 + \frac{m}{\eta^2} \cdot W\left(\frac{m}{\eta^2} - l\right) \quad , \tag{11.5}$$

[1] A derivation of this equation would occupy several pages and is therefore omitted here. The reader will find the details in one of the reference works included in the bibliography.

with the (positive) auxiliary variables

$$
\begin{aligned}
m &= \frac{\tau^2}{\sqrt{2(r_a r_b + \boldsymbol{r}_a \cdot \boldsymbol{r}_b)}^3} \\
l &= \frac{r_a + r_b}{2\sqrt{2(r_a r_b + \boldsymbol{r}_a \cdot \boldsymbol{r}_b)}} - \frac{1}{2} \quad ,
\end{aligned}
\tag{11.6}
$$

from which η has to be determined. Here the function W is defined by:

$$
W(w) = \begin{cases}
\dfrac{2g - \sin(2g)}{\sin^3(g)} \; , \quad g = 2 \arcsin \sqrt{w} & 0 < w < 1 \\
\dfrac{4}{3} + \dfrac{4 \cdot 6}{3 \cdot 5} w + \dfrac{4 \cdot 6 \cdot 8}{3 \cdot 5 \cdot 7} w^2 + \ldots & w \approx 0 \\
\dfrac{\sinh(2g) - 2g}{\sinh^3(g)} \; , \quad g = 2 \operatorname{arsinh} \sqrt{-w} & w < 0 \quad .
\end{cases}
\tag{11.7}
$$

If we write

$$
f(x) = 1 - x + \frac{m}{x^2} \cdot W\left(\frac{m}{x^2} - l\right) \quad ,
$$

then the desired value of η is that at which the function f becomes zero.

```
// F=1-eta+(m/eta**2)*W(m/eta**2-1); local function for use by FindEta()
double F (double eta, double m, double l) {
  const double eps=100.0*eps_mach;  double w,W,a,n,g;
  w = m/(eta*eta)-l;
  if (fabs(w)<0.1) {   // Series expansion
    W = a = 4.0/3.0; n = 0.0;
    do { n+=1.0; a*=w*(n+2.0)/(n+1.5); W+=a; } while (fabs(a)>=eps);
  }
  else {
    if (w > 0.0) {
      g = 2.0*asin(sqrt(w));  W = (2.0*g - sin(2.0*g)) / pow(sin(g), 3);
    }
    else {
      g = 2.0*log(sqrt(-w)+sqrt(1.0-w));      // =2.0*arsinh(sqrt(-w))
      W = (sinh(2.0*g) - 2.0*g) / pow(sinh(g), 3);
    }
  }
  return ( 1.0 - eta + (w+l)*W );
}
```

Because the C++ standard library does not support inverse hyperbolic functions, the function $\operatorname{arsinh} x = \ln(x + \sqrt{1 + x^2})$ has been expressed in terms of the natural logarithm.

To determine η iteratively, it is appropriate to use the secant procedure. Using two approximations η_{i-1} and η_i, we obtain an improved value η_{i+1} via

$$
\eta_{i+1} = \eta_i - f(\eta_i) \cdot \frac{\eta_i - \eta_{i-1}}{f(\eta_i) - f(\eta_{i-1})} \quad .
$$

Geometrically, this gives us the zero-point of the secant that passes through the points $(\eta_{i-1}, f(\eta_{i-1}))$ and $(\eta_i, f(\eta_i))$ on the curve given by f. If this step is repeatedly carried out, the iteration soon converges to the desired value of the sector-triangle ratio. Suitable starting values

$$\eta_1 = \eta_{\text{Hansen}} + 0.1 \quad \text{and} \quad \eta_2 = \eta_{\text{Hansen}}$$

are given by what is known as the Hansen Approximation

$$\eta_{\text{Hansen}} = \frac{12}{22} + \frac{10}{22}\sqrt{1 + \frac{44}{9}\frac{m}{l + 5/6}} \quad . \tag{11.8}$$

We then obtain the following function for calculating the sector-triangle ratio:

```
//------------------------------------------------------------------------------
// FindEta: computes the sector-triangle ratio for two given positions and
//          the time between them
//   r_a        Position at first instant in [AU]
//   r_a        Position at second instant in [AU]
//   tau        Time between the positions r_a and r_b (kGauss * dT in [d])
//   <return>   Sector-triangle ratio
//------------------------------------------------------------------------------
double FindEta (const Vec3D& r_a, const Vec3D& r_b, double tau)
{
  const int maxit = 30;
  const double delta = 100.0*eps_mach;
  int    i;
  double kappa, m, l, s_a, s_b, eta_min, eta1, eta2, F1, F2, d_eta;
  s_a = Norm(r_a);  s_b = Norm(r_b);
  kappa = sqrt ( 2.0*(s_a*s_b+Dot(r_a,r_b)) );
  m = tau*tau / pow(kappa,3);   l = (s_a+s_b) / (2.0*kappa) - 0.5;
  eta_min = sqrt(m/(l+1.0));
  // Start with Hansen's approximation
  eta2 = ( 12.0 + 10.0*sqrt(1.0+(44.0/9.0)*m /(l+5.0/6.0)) ) / 22.0;
  eta1 = eta2 + 0.1;
  // Secant method
  F1 = F(eta1, m, l);  F2 = F(eta2, m, l);  i = 0;
  while (fabs(F2-F1) > delta) {
    d_eta = -F2*(eta2-eta1)/(F2-F1);
    eta1 = eta2;  F1 = F2;  while (eta2+d_eta<=eta_min) d_eta *= 0.5;
    eta2 += d_eta;  F2 = F(eta2,m,l); ++i;
    if (i==maxit ) {cerr<<" Convergence problems in FindEta"<<endl; break;}
  }
  return eta2;
}
```

11.1.2 Orbital Elements

The orbit of a celestial body that passes through the points $\boldsymbol{r}_a$ and $\boldsymbol{r}_b$ is always restricted to the plane determined by these two points and the Sun. To determine the inclination i of this plane to the ecliptic as well as the longitude of the ascending

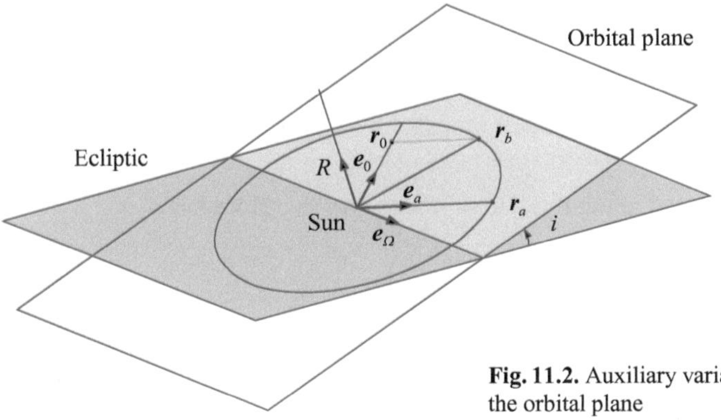

Fig. 11.2. Auxiliary variables used to determine the orbital plane

node, we first obtain the unit vectors $\boldsymbol{e}_a$ and $\boldsymbol{e}_0$, which both lie in the orbital plane:

$$\boldsymbol{e}_a = \frac{\boldsymbol{r}_a}{|\boldsymbol{r}_a|} \tag{11.9}$$

$$\boldsymbol{e}_0 = \frac{\boldsymbol{r}_0}{|\boldsymbol{r}_0|} \quad \text{where} \quad \boldsymbol{r}_0 = \boldsymbol{r}_b - (\boldsymbol{r}_b \cdot \boldsymbol{e}_a)\boldsymbol{e}_a \quad . \tag{11.10}$$

The nature of these vectors is shown in Fig. 11.2. $\boldsymbol{e}_a$ is aligned with $\boldsymbol{r}_a$, $\boldsymbol{r}_0$ and $\boldsymbol{e}_0$ are perpendicular to it. If we now form the cross-product of $\boldsymbol{e}_a$ and $\boldsymbol{e}_0$, the result obtained is the Gaussian vector $\boldsymbol{R}$, which is perpendicular to the orbital plane and is likewise normalized to unit length ($|\boldsymbol{R}| = 1$):

$$\boldsymbol{R} = \boldsymbol{e}_a \times \boldsymbol{e}_0 \quad , \quad \begin{pmatrix} R_x \\ R_y \\ R_z \end{pmatrix} = \begin{pmatrix} y_a z_0 - z_a y_0 \\ z_a x_0 - x_a z_0 \\ x_a y_0 - y_a x_0 \end{pmatrix} \quad . \tag{11.11}$$

$\boldsymbol{R}$ is directed to ecliptic longitude $l = \Omega - 90°$ and ecliptic latitude $b = 90° - i$, and may therefore be expressed in terms of the elements Ω and i:

$$\boldsymbol{R} = \begin{pmatrix} R_x \\ R_y \\ R_z \end{pmatrix} = \begin{pmatrix} +\cos(90°-i)\cos(\Omega-90°) \\ +\cos(90°-i)\sin(\Omega-90°) \\ +\sin(90°-i) \end{pmatrix} = \begin{pmatrix} +\sin i \sin\Omega \\ -\sin i \cos\Omega \\ +\cos i \end{pmatrix} \quad . \tag{11.12}$$

We thus obtain three equations, which may be solved unambiguously for the longitude of the node and the orbital inclination:

$$\Omega = 90° + \arctan(R_y/R_x) = \arctan(-R_x/R_y) \tag{11.13}$$

$$i = 90° - \arcsin(R_z) \quad . \tag{11.14}$$

Both angles refer to the same equinox as the vectors $\boldsymbol{r}_a$ and $\boldsymbol{r}_b$. From the position of the line of nodes the argument of latitude u_a may now be determined, it being given by the angle between the position vector $\boldsymbol{r}_a$ and the direction of the ascending node of the orbit. For this angle we have

$$\cos u_a = \boldsymbol{e}_a \cdot \boldsymbol{e}_\Omega = x_a \cdot \cos\Omega + y_a \cdot \sin\Omega$$

$$\cos(u_a + 90°) = \boldsymbol{e}_0 \cdot \boldsymbol{e}_\Omega = x_0 \cdot \cos\Omega + y_0 \cdot \sin\Omega \quad ,$$

where

$$e_\Omega = \begin{pmatrix} \cos\Omega \\ \sin\Omega \\ 0 \end{pmatrix}$$

is the unit vector directed along the line of nodes. We therefore have

$$\begin{aligned} u_a &= \arctan\left(\frac{-x_0 \cdot \cos\Omega - y_0 \cdot \sin\Omega}{+x_a \cdot \cos\Omega + y_a \cdot \sin\Omega}\right) \\ &= \arctan\left(\frac{+x_0 \cdot R_y - y_0 \cdot R_x}{-x_a \cdot R_y + y_a \cdot R_x}\right) \quad . \end{aligned} \tag{11.15}$$

To determine the remaining orbital elements we require the sector-triangle ratio, which is calculated in the manner described in the previous section. Using this, we are next able to express the semi-latus rectum

$$p = \left(\frac{2 \cdot \Delta \cdot \eta}{\tau}\right)^2$$

in terms of the area of the triangle defined by the vectors $\boldsymbol{r}_a$ and $\boldsymbol{r}_b$

$$\Delta = \frac{1}{2} r_a r_b \cdot \sin(\nu_b - \nu_a) = \frac{1}{2} r_a r_0 \quad ,$$

and the interval τ.

The shape of the orbit is defined by the eccentricity e, which may be determined from the equation for the conic section

$$r = \frac{p}{1 + e \cdot \cos\nu} \quad .$$

Solving for $e\cos\nu$, we have

$$\begin{aligned} e \cdot \cos\nu_a &= p/r_a - 1 \\ e \cdot \cos\nu_b &= p/r_b - 1 \quad . \end{aligned}$$

If we take into account the fact that

$$\begin{aligned} \cos\nu_b &= \cos\nu_a \cos(\nu_b - \nu_a) - \sin\nu_a \sin(\nu_b - \nu_a) \\ &= \cos\nu_a \cdot \left(\frac{\boldsymbol{r}_b \cdot \boldsymbol{e}_a}{r_b}\right) - \sin\nu_a \cdot \left(\frac{r_0}{r_b}\right) \quad , \end{aligned}$$

then, by substitution, we obtain the two equations

$$\begin{aligned} e \cdot \cos\nu_a &= p/r_a - 1 \\ e \cdot \sin\nu_a &= \left\{(p/r_a - 1)\left(\frac{\boldsymbol{r}_b \cdot \boldsymbol{e}_a}{r_b}\right) - (p/r_b - 1)\right\} / \left(\frac{r_0}{r_b}\right) \quad , \end{aligned}$$

which may themselves be solved for the eccentricity and the true anomaly at time t_a:

$$
\begin{aligned}
e &= \sqrt{(e \cdot \cos(\nu_a))^2 + (e \cdot \sin(\nu_a))^2} \\
\nu_a &= \arctan\left(\frac{e \cdot \sin(\nu_a)}{e \cdot \cos(\nu_a)}\right) \quad .
\end{aligned}
$$

The argument and longitude of perihelion are obtained from the difference between the argument of latitude and the true anomaly, being

$$\omega = u_a - \nu_a \tag{11.16}$$

$$\varpi = u_a - \nu_a + \Omega \quad . \tag{11.17}$$

The eccentricity enables us to determine whether the orbit between $\boldsymbol{r}_a$ and $\boldsymbol{r}_b$ is an ellipse ($e < 1$) or a hyperbola ($e > 1$). Parabolic orbits are not considered further here, because in determining an orbit in practice it is highly unlikely that e will turn out to be exactly equal to one. From the semi-latus rectum and the eccentricity we can now also obtain the semi-major axis and the perihelion distance:

$$a = \frac{p}{1 - e^2} \tag{11.18}$$

$$q = \frac{p}{1 + e} \quad . \tag{11.19}$$

Here it should be noted that, by definition, the semi-major axis of a hyperbolic orbit is a negative value.

We now know all the orbital elements that determine the orbit's shape (e), size (a), and orientation in space (i, Ω, ω). The sixth and last element to be determined is therefore the time of perihelion passage, which defines when the body will pass (or passed) the point closest to the Sun. For elliptical orbits, we first need to determine the eccentric anomaly E_a from the equations

$$
\begin{aligned}
\cos E_a &= \frac{\cos \nu_a + e}{1 + e \cdot \cos \nu_a} \\
\sin E_a &= \frac{\sqrt{1 - e^2} \sin \nu_a}{1 + e \cdot \cos \nu_a} \quad .
\end{aligned}
$$

We obtain these equations by eliminating the radius from (4.5) by using the equation for the conic section, and then solving for the eccentric anomaly. The value for the mean anomaly M corresponding to E is obtained from the Kepler equation

$$M_a = E_a - e \cdot \sin E_a \qquad \text{(radians)} \quad .$$

For an orbital period of

$$T = 2\pi \cdot \sqrt{\frac{a^3}{GM_\odot}} \quad ,$$

as obtained from Kepler's Third Law, the mean anomaly varies daily by

$$n = 2\pi \cdot \frac{1^{\mathrm{d}}}{T} = \sqrt{\frac{GM_\odot}{a^3}} \cdot 1^{\mathrm{d}} \quad ,$$

so the time of perihelion passage is given by

$$t_0 = t_a - M_a / \sqrt{GM_\odot / a^3} \quad . \tag{11.20}$$

For hyperbolic orbits the corresponding equations are

$$\sinh H_a = \frac{\sqrt{e^2-1}\sin\nu_a}{1+e\cdot\cos\nu_a} \tag{11.21}$$

$$M_a = e \cdot \sinh H_a - H_a \tag{11.22}$$

$$t_0 = t_a - M_a / \sqrt{GM_\odot / |a|^3} \quad . \tag{11.23}$$

The complete method of determining orbital elements from two given points on the orbit is given in function `Elements`.

```
//------------------------------------------------------------------------------
// Elements: computes orbital elements from two given positions
//   GM        Product of gravitat. constant and centre of mass [AU^3*d^-2]
//   Mjd_a     Time of the passage of position A [MJD]
//   Mjd_b     Time of the passage of position B [MJD]
//   r_a       Heliocentric ecliptical position A in [AU]
//   r_b       Heliocentric ecliptical position B in [AU]
//   Mjd_p     Time of perihelion passage
//   q         Perihelion distance in [AU]
//   e         Eccentricity of the orbit
//   i         Inclination of the orbit to the ecliptic in [rad]
//   Omega     Longitude of the ascending node of the orbit in [rad]
//   omega     Argument of perihelion in [rad]
//------------------------------------------------------------------------------
void Elements ( double GM, double Mjd_a, double Mjd_b,
                const Vec3D& r_a, const Vec3D& r_b,
                double& Mjd_p, double& q, double& e,
                double& i, double& Omega, double& omega )
{
  double  tau, eta, p;
  double  a, n, nu, E, M, u;
  double  s_a, s_b, s_0, fac, sinhH;
  double  cos_dnu, sin_dnu, ecos_nu, esin_nu;
  Vec3D   e_a, r_0, e_0, R;
  // Calculate vector r_0 (fraction of r_b perpendicular to r_a)
  // and the magnitudes of r_a,r_b and r_0
  s_a = Norm(r_a);  e_a = r_a/s_a;
  s_b = Norm(r_b);
  fac = Dot(r_b,e_a); r_0 = r_b-fac*e_a;
  s_0 = Norm(r_0);  e_0 = r_0/s_0;
  // Inclination and ascending node
  R     = Cross(e_a,e_0);
  i     = pi/2.0 - R[theta];
```

```
  Omega = Modulo ( pi/2.0 + R[phi], 2.0*pi );
  if (i==0.0)
    u = atan2 ( r_a[y], r_a[x] );
  else
    u = atan2 ( (+e_O[x]*R[y]-e_O[y]*R[x]) , (-e_a[x]*R[y]+e_a[y]*R[x]) );
  // Semilatus rectum
  tau = sqrt(GM) * fabs(Mjd_b-Mjd_a);
  eta = FindEta ( r_a, r_b, tau );
  p   = pow ( s_a*s_0*eta/tau, 2 );
  // Eccentricity, true anomaly and argument of perihelion
  cos_dnu = fac / s_b;
  sin_dnu = s_0 / s_b;
  ecos_nu = p / s_a - 1.0;
  esin_nu = ( ecos_nu * cos_dnu - (p/s_b-1.0) ) / sin_dnu;
  e  = sqrt ( ecos_nu*ecos_nu + esin_nu*esin_nu );
  nu = atan2(esin_nu,ecos_nu);
  omega = Modulo(u-nu,2.0*pi);
  // Perihelion distance, semi-major axis and mean motion
  q = p/(1.0+e);
  a = q/(1.0-e);
  n = sqrt ( GM / fabs(a*a*a) );
  // Mean anomaly and time of perihelion passage
  if (e<1.0) {
    E = atan2 ( sqrt((1.0-e)*(1.0+e)) * esin_nu,  ecos_nu + e*e );
    M = E - e*sin(E);
  }
  else {
    sinhH = sqrt((e-1.0)*(e+1.0)) * esin_nu / ( e + e * ecos_nu );
    M = e * sinhH - log ( sinhH + sqrt(1.0+sinhH*sinhH) );
  }
  Mjd_p = Mjd_a - M / n;
}
```

11.2 The Shortened Gauss Method

11.2.1 The Geometry of Geocentric Observations

If the vector $\boldsymbol{r}$ is the position of a planet relative to the Sun, and $\boldsymbol{R}$ the geocentric position of the Sun, then the planet's geocentric position is given by

$$\rho \boldsymbol{e} = \boldsymbol{R} + \boldsymbol{r} \quad . \tag{11.24}$$

Here ρ is the distance of the planet from the Earth, and $\boldsymbol{e}$ is a vector of unit length, directed from the Earth towards the planet. In ecliptic or equatorial coordinates, $\boldsymbol{e}$ has the components

$$\begin{pmatrix} \cos\lambda\cos\beta \\ \sin\lambda\cos\beta \\ \sin\beta \end{pmatrix} \quad \text{or} \quad \begin{pmatrix} \cos\alpha\cos\delta \\ \sin\alpha\cos\delta \\ \sin\delta \end{pmatrix} \quad ,$$

where λ and β are the ecliptic longitude and latitude and α and δ are the right ascension and declination of the planet as seen from Earth.

If the coordinates of the planet on the apparent celestial sphere (i.e., α and δ) are determined, then the observing direction – and thus $\boldsymbol{e}$ – is defined. The distance ρ is unknown, and must be calculated during the process of determining the orbit. For a planetary orbit to be determined unambiguously at least three observations $\boldsymbol{e}_1$, $\boldsymbol{e}_2$ and $\boldsymbol{e}_3$ must be available. We may also assume that the coordinates of the Sun at the times of observation, $\boldsymbol{R}_1$, $\boldsymbol{R}_2$ and $\boldsymbol{R}_3$, are known. We must try to calculate the distances ρ_1, ρ_2 and ρ_3 from these data. Only when the distances are known are we able to determine the heliocentric positions that define the orbit and derive the orbital elements. The geometrical relationships that are obtained from a set of three planetary positions, now need to be expressed in a form suitable for orbit determination.

At the times $t_1 < t_2 < t_3$ the planet is at positions $\boldsymbol{r}_1, \boldsymbol{r}_2, \boldsymbol{r}_3$ with respect to the Sun. Because, for unperturbed Keplerian motion, all the points and the Sun lie in a single plane, it is always possible to express one position vector using an appropriate combination of the other two. For this purpose we choose $\boldsymbol{r}_2$, and may then write

$$\boldsymbol{r}_2 = n_1\boldsymbol{r}_1 + n_3\boldsymbol{r}_3 \quad \text{(equation of the orbital plane)} \quad . \tag{11.25}$$

The factors n_1 and n_3 depend on the relative position of $\boldsymbol{r}_1$, $\boldsymbol{r}_2$, and $\boldsymbol{r}_3$. In what follows, if we assume that the entire arc of the orbit is less than 180°, then both factors are positive. We now combine (11.24) and (11.25), obtaining

$$(\rho_2\boldsymbol{e}_2 - \boldsymbol{R}_2) = n_1 \cdot (\rho_1\boldsymbol{e}_1 - \boldsymbol{R}_1) + n_3 \cdot (\rho_3\boldsymbol{e}_3 - \boldsymbol{R}_3)$$

or, rearranging,

$$n_1\rho_1\boldsymbol{e}_1 - \rho_2\boldsymbol{e}_2 + n_3\rho_3\boldsymbol{e}_3 = n_1\boldsymbol{R}_1 - \boldsymbol{R}_2 + n_3\boldsymbol{R}_3 \quad . \tag{11.26}$$

If we now define the vectors

$$\boldsymbol{d}_1 = \boldsymbol{e}_2 \times \boldsymbol{e}_3 \quad \boldsymbol{d}_2 = \boldsymbol{e}_3 \times \boldsymbol{e}_1 \quad \boldsymbol{d}_3 = \boldsymbol{e}_1 \times \boldsymbol{e}_2 \quad ,$$

then, because of the properties of the cross-products, $\boldsymbol{d}_1$ is perpendicular to $\boldsymbol{e}_2$ and $\boldsymbol{e}_3$, $\boldsymbol{d}_2$ is perpendicular to $\boldsymbol{e}_3$ and $\boldsymbol{e}_1$, and $\boldsymbol{d}_3$ perpendicular to $\boldsymbol{e}_1$ and $\boldsymbol{e}_2$. Consequently, the dot product $\boldsymbol{e}_i \cdot \boldsymbol{d}_j$ only differs from zero for $i = j$. If we multiply (11.26) by $\boldsymbol{d}_1$, $\boldsymbol{d}_2$ and $\boldsymbol{d}_3$ individually, then we obtain the equations

$$\begin{aligned} n_1\rho_1 \cdot (\boldsymbol{e}_1 \cdot \boldsymbol{d}_1) &= (n_1\boldsymbol{R}_1 - \boldsymbol{R}_2 + n_3\boldsymbol{R}_3) \cdot \boldsymbol{d}_1 \\ -\rho_2 \cdot (\boldsymbol{e}_2 \cdot \boldsymbol{d}_2) &= (n_1\boldsymbol{R}_1 - \boldsymbol{R}_2 + n_3\boldsymbol{R}_3) \cdot \boldsymbol{d}_2 \\ n_3\rho_3 \cdot (\boldsymbol{e}_3 \cdot \boldsymbol{d}_3) &= (n_1\boldsymbol{R}_1 - \boldsymbol{R}_2 + n_3\boldsymbol{R}_3) \cdot \boldsymbol{d}_3 \quad . \end{aligned}$$

If we introduce the abbreviations

$$\begin{aligned} D &= \boldsymbol{e}_1 \cdot (\boldsymbol{e}_2 \times \boldsymbol{e}_3) &&= \boldsymbol{e}_2 \cdot (\boldsymbol{e}_3 \times \boldsymbol{e}_1) &&= \boldsymbol{e}_3 \cdot (\boldsymbol{e}_1 \times \boldsymbol{e}_2) \\ &= \boldsymbol{e}_1 \cdot \boldsymbol{d}_1 &&= \boldsymbol{e}_2 \cdot \boldsymbol{d}_2 &&= \boldsymbol{e}_3 \cdot \boldsymbol{d}_3 \end{aligned}$$

and

$$D_{ij} = \boldsymbol{d}_i \cdot \boldsymbol{R}_j \quad ,$$

then we obtain the three equations

$$\begin{aligned}\rho_1 &= \frac{1}{n_1 D}(n_1 D_{11} - D_{12} + n_3 D_{13}) \\ \rho_2 &= \frac{1}{-D}(n_1 D_{21} - D_{22} + n_3 D_{23}) \\ \rho_3 &= \frac{1}{n_3 D}(n_1 D_{31} - D_{32} + n_3 D_{33}) \quad .\end{aligned} \tag{11.27}$$

The distances ρ_1, ρ_2 and ρ_3 may therefore be expressed in terms of n_1 and n_3 as well as the vectors $\boldsymbol{e}_{1...3}$ and $\boldsymbol{R}_{1...3}$. At first it would seem that we have not gained much, because n_1 and n_3 are unknown. By using the equation of the orbital plane, we have at least reduced the number of unknowns from three ($\rho_{1...3}$) to two ($n_{1,3}$). The newly introduced coefficients are particularly important, however, because – as will now be shown – they may be closely approximated by expressions involving the known intervals between the observations.

Consider the equation of the orbital plane (11.25). If we form the cross-product of both sides with $\boldsymbol{r}_3$ or $\boldsymbol{r}_1$, and, remind ourselves that the cross-product of a vector with itself cancels out, we obtain the expressions

$$(\boldsymbol{r}_2 \times \boldsymbol{r}_3) = n_1 \cdot (\boldsymbol{r}_1 \times \boldsymbol{r}_3) \qquad (\boldsymbol{r}_1 \times \boldsymbol{r}_2) = n_3 \cdot (\boldsymbol{r}_1 \times \boldsymbol{r}_3)$$

and

$$n_1 = \frac{|\boldsymbol{r}_2 \times \boldsymbol{r}_3|}{|\boldsymbol{r}_1 \times \boldsymbol{r}_3|} \qquad n_3 = \frac{|\boldsymbol{r}_1 \times \boldsymbol{r}_2|}{|\boldsymbol{r}_1 \times \boldsymbol{r}_3|} \quad .$$

Generalizing, as the area Δ of the triangle bounded by two vectors $\boldsymbol{r}_a$ and $\boldsymbol{r}_b$, is equal to

$$\Delta = \frac{1}{2}|\boldsymbol{r}_a \times \boldsymbol{r}_b| \quad ,$$

n_1 and n_3 may be interpreted as expressions of the areas of the triangles bounded by $\boldsymbol{r}_1$, $\boldsymbol{r}_2$, and $\boldsymbol{r}_3$ (see Fig. 11.3):

$$n_1 = \frac{\Delta_1}{\Delta_2} \qquad n_3 = \frac{\Delta_3}{\Delta_2} \quad .$$

For small arcs of the orbit in particular, the areas of the triangles differ only slightly from the corresponding sector areas $S_i = \eta_i \Delta_i$, which are themselves proportional to the intervals τ_i:

$$n_1 = \frac{\eta_2}{\eta_1} \cdot \frac{\tau_1}{\tau_2} \approx \frac{\tau_1}{\tau_2} \qquad n_3 = \frac{\eta_2}{\eta_3} \cdot \frac{\tau_3}{\tau_2} \approx \frac{\tau_3}{\tau_2} \quad . \tag{11.28}$$

We do, therefore, at least know approximate values for n_1 and n_3, which enable us to determine first approximations for the geocentric distances (ρ_1, ρ_2, ρ_3).

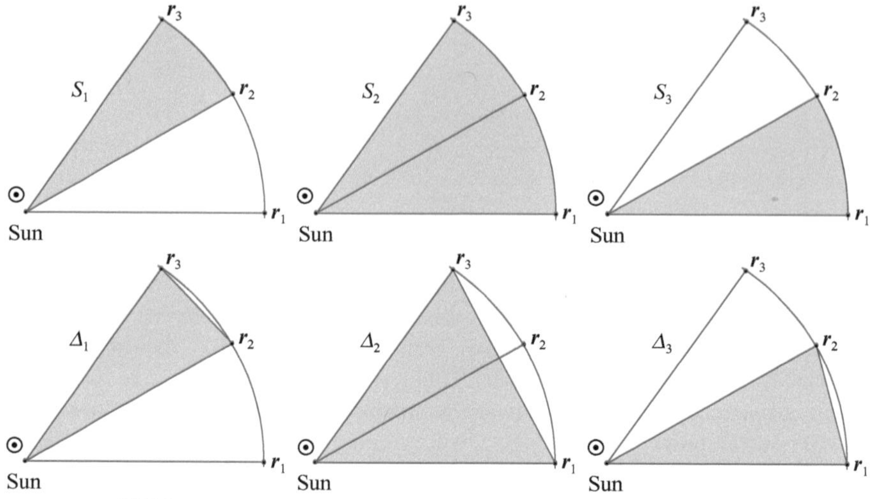

Fig. 11.3. Sector areas, triangle areas, and intervals for three heliocentric positions

11.2.2 Successive Improvement of the Sector-Triangle Ratios

Two statements essentially summarize the various considerations that we have covered so far: If we know the heliocentric position of a celestial body at two given times, then the whole orbit is unambiguously determined. The same applies to the orbital elements and the sector-triangle ratio. If, on the other hand, we know the value of the sector-triangle ratio for a set of three observed positions, then the geocentric and heliocentric position vectors may be calculated. The shortened Gaussian method of orbit determination to be described now is based upon this.

Let three geocentric observed directions $(\boldsymbol{e}_1, \boldsymbol{e}_2, \boldsymbol{e}_3)$, the corresponding geocentric coordinates of the Sun $(\boldsymbol{R}_1, \boldsymbol{R}_2, \boldsymbol{R}_3)$ and the intervals (τ_1, τ_2, τ_3) be given. The steps in the calculation, using these initial data, are:

1. Set $n_1 = \tau_1/\tau_2$ and $n_3 = \tau_3/\tau_2$ as initial approximations for the ratios of the areas of the triangles.
2. Repeat steps (a)...(d), until n_1, n_3 and the other values no longer vary significantly.
   (a) Calculate the geocentric distances (ρ_1, ρ_2, ρ_3) with (11.27).
   (b) Then calculate the heliocentric position vectors $(\boldsymbol{r}_1, \boldsymbol{r}_2, \boldsymbol{r}_3)$ from (11.24).
   (c) Calculate the sector-triangle ratios (η_1, η_2, η_3) from each pair of heliocentric position vectors and the corresponding interval with (11.5).
   (d) Calculate improved values $n_1 = (\eta_2/\eta_1) \cdot (\tau_1/\tau_2)$ and $n_3 = (\eta_2/\eta_3) \cdot (\tau_3/\tau_2)$ for the triangle area ratios.
3. Calculate the orbital elements from the last values of $\boldsymbol{r}_1, \boldsymbol{r}_3$ and τ_2.

The orbit calculated in this way has the properties desired, in that any body moving along it is located at the times t_1, t_2 and t_3 at the points $\boldsymbol{r}_1$, $\boldsymbol{r}_2$ and $\boldsymbol{r}_3$, which themselves lie in the directions $\boldsymbol{e}_1$, $\boldsymbol{e}_2$ and $\boldsymbol{e}_3$ relative to the Earth. In principle, the orbit determination problem solved by now. For practical applications, however, the successive improvement of the sector-triangle ratios is subject to serious restrictions.

11.2.3 Multiple Solutions

Apart from exceptional cases when the iteration does not converge, the simplified Gaussian method always provides a result that correctly satisfies all the equations involved in determining an orbit. In doubtful cases it may always be confirmed by appropriate subsequent calculation of the three observations from the orbital elements that have been derived. However, we cannot rule out the fact that at times one obtains solutions that appear, at first sight, false or absurd. This is particularly obvious when we encounter hyperbolic orbits with high eccentricities, which contradict actual experience[2]. The reason for this is that determining orbits from three observations is not always unambiguous. This is particularly easily seen in the *Earth-orbit solution*. The observer is moving in an almost unperturbed Keplerian orbit around the Sun, and is simultaneously always situated in the direction of observation. The quickest way of recognizing the Earth-orbit solution is from the semi-major axis, ($a \approx 1$), the eccentricity ($e \approx 0$), and the inclination to the ecliptic ($i \approx 0°$).

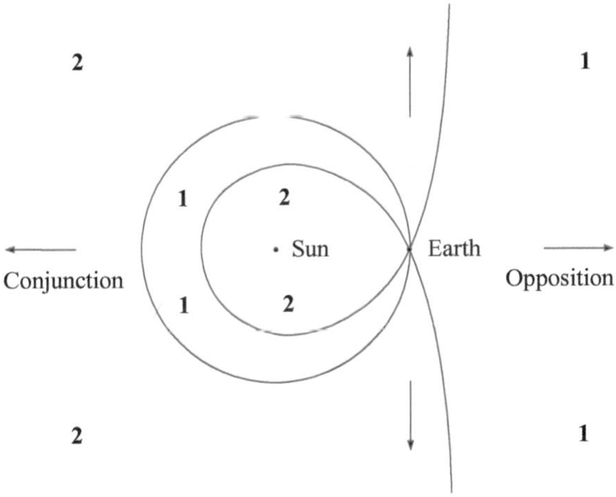

Fig. 11.4. Charlier's boundary line (—) separates regions in which orbit determination is unambiguous (1), from regions in which it is ambiguous (2)

A general discussion of possible ambiguities in orbit determination is normally restricted to small arcs of an orbit. With this assumption, if we consider the relative positions of the Sun, Earth and the observed object, then a total of four separate regions are defined by the Earth's orbit and what is known as *Charlier's boundary*

[2]All known comets, without exception, have eccentricities that are smaller than $e = 1.1$.

line (Fig. 11.4). For a minor planet that is observed at around the time of opposition, for example, the result is unambiguous, with the exception of the Earth-orbit solution that has been mentioned. If, on the other hand, at the time of observation a comet is within the teardrop-shaped zone that immediately surrounds the Sun, or lies in the region of conjunction outside the Earth's orbit, in principle two solutions to the orbit determination are possible.

11.3 The Comprehensive Gaussian Method

The method of determining an orbit described in Sect. 11.2.2 is based upon continuing to refine approximations for the triangle-area ratios n_1 and n_3 by repeated substitution in the orbit determination equations, until all the equations are equally satisfied. In doing so, however, there is no point at which one obtains any information about whether the solution that has been found is the sole solution, nor, in the event of there being multiple solutions, how one should compute the others.

An overall view of the possible solutions of the problem of orbit determination may be obtained if we make a few simplifications to the basic equations. The solutions of this approximate problem may be obtained from the Gauss-Lagrangian equations and essentially agree with the solutions to the rigorous equations. Finally, the Gauss-Lagrangian equations may be built into the old calculation procedure for determining orbits (see. Sect. 11.2.2) and then permit a deliberate choice and improvement of individual solutions at each iteration step.

11.3.1 The Gauss-Lagrangian Equation

To derive the Gauss-Lagrangian equations, we first look at an improved approximation for the triangle-area ratios that goes back to Encke:

$$\begin{aligned} n_1 &= \frac{\tau_1}{\tau_2} + \frac{1}{6}\tau_1\tau_3\left(1+\frac{\tau_1}{\tau_2}\right)\cdot\frac{1}{r_2^3} \\ n_3 &= \frac{\tau_3}{\tau_2} + \frac{1}{6}\tau_1\tau_3\left(1+\frac{\tau_3}{\tau_2}\right)\cdot\frac{1}{r_2^3} \quad . \end{aligned} \tag{11.29}$$

When compared with (11.28), the Encke approximation gives good results even with large intervals, but does depend on the still unknown distance r_2.

If we now insert the Encke approximation in the relationship (11.27) for the geocentric distance, then, using the expressions

$$\begin{aligned} n_{10} &= \frac{\tau_1}{\tau_2} & \mu_1 &= \frac{1}{6}\tau_1\tau_3\left(1+\frac{\tau_1}{\tau_2}\right) \\ n_{30} &= \frac{\tau_3}{\tau_2} & \mu_3 &= \frac{1}{6}\tau_1\tau_3\left(1+\frac{\tau_3}{\tau_2}\right) \end{aligned}$$

as well as

$$\rho_0 = -\frac{1}{D}(n_{10}D_{21} - D_{22} + n_{30}D_{23}) \qquad \sigma = +\frac{1}{D}(\mu_1 D_{21} + \mu_3 D_{23}) \quad ,$$

we obtain the simple relationship

$$\rho_2 = \rho_0 - \frac{\sigma}{r_2^3} \tag{11.30}$$

between the geocentric distance ρ_2 and the heliocentric distance $\boldsymbol{r}_2$ at the time of the second observation. Here ρ_0 and σ are known values, that are given directly from the observations.

On the other hand, we obtain a second expression for the two values from the equation $\boldsymbol{r}_2 = \rho_2 \boldsymbol{e}_2 - \boldsymbol{R}_2$, which implies that ρ_2, $\boldsymbol{r}_2$ and $\boldsymbol{R}_2$ are the sides of the Sun–Earth–Planet triangle. By squaring, we obtain from this the relationship

$$r_2 = \sqrt{(\rho_2 - \gamma R_2)^2 + R_2^2(1-\gamma^2)} \quad , \tag{11.31}$$

in which

$$\gamma = \boldsymbol{e}_2 \cdot \boldsymbol{R}_2 / R_2$$

indicates the cosine of the angle between the observed direction $\boldsymbol{e}_2$ and the direction of the Sun $\boldsymbol{R}_2$. At the time of opposition we have $\gamma = -1$, and at the time of conjunction $\gamma = +1$.

By substituting the values for r_2 from (11.30) and (11.31), we obtain the Gauss-Lagrangian equation

$$\sqrt[3]{\frac{\sigma}{\rho_0 - \rho_2}} = \sqrt{(\rho_2 - \gamma R_2)^2 + R_2^2(1-\gamma^2)} \quad , \tag{11.32}$$

which is the key to the determination of the unknowns ρ_2 and r_2. Here there may be, in total, up to three positive pairs of values that fulfill both equations. For each r_2, the triangle-area ratios n_1 and n_3 are calculated, together with ρ_1 and ρ_3. With the accompanying ρ_2, we thence obtain the heliocentric positions $\boldsymbol{r}_1$, $\boldsymbol{r}_2$ and $\boldsymbol{r}_3$, and have thus solved the orbit determination within the framework of the Encke approximation. Because each (ρ_2, r_2) pair leads to other values of the heliocentric positions $\boldsymbol{r}_{1,2,3}$, we may obtain up to three different orbits that correctly reflect the given observations.

Instead of the (11.30, 11.31) system of equations, most of the well-known procedures for orbital determination employ other, but essentially equivalent, forms of the Gauss-Lagrangian equation. To determine the orbit of Ceres, Gauss himself developed a trigonometrical formulation, which was particularly suitable for the calculation procedure that he had developed. An eighth-order equation

$$r_2^8 - \{\rho_0^2 - 2\gamma\rho_0 R_2 + R_2^2\} r_2^6 + \{2\sigma(\rho_0 - \gamma R_2)\} r_2^2 - \{\sigma^2\} = 0 \tag{11.33}$$

for the heliocentric distance r_2 goes back to Lagrange. This is obtained when one substitutes the value of ρ_2 from equation (11.30) into (11.31). With the simplification $\sigma \approx \rho_0 R_2^3$ – although this is valid only for very small sections of an orbit – this eighth-order equation also forms the centre of Laplace's methods of determining orbits.

To solve the Gauss-Lagrangian equation it is helpful first to look briefly at the graph of the two functions

$$\hat{r}(\rho) = \sqrt[3]{\frac{\sigma}{\rho_0 - \rho}} \qquad \text{and} \qquad \tilde{r}(\rho) = \sqrt{(\rho - \gamma R)^2 + R^2(1-\gamma^2)} \quad ,$$

whose intersections indicate the required solutions. While the function $\hat{r}$ increases or decreases strictly monotonically and has a pole at $\rho = \rho_0$, the function $\tilde{r}$ is in the form of an upwardly concave hyperbola, with the minimum at $\rho = \gamma R$. From this it follows that, in principle, there are one or three intersections of the two graphs, the positions of which (because $R \approx 1$ AU), are solely dependent on σ, ρ_0 and γ.

With the simplifying assumption $\sigma \approx \rho_0 R^3$ – which, as already mentioned, is valid only for small arcs of an orbit – we always find the Earth-orbit solution, with $\rho = 0$ AU and $r = 1$ AU, independently of the remaining values. There then remain at most two positive solutions ρ, of which one represents the actual orbit of the object observed. Interestingly, however, the above assumption is not valid for the Ceres problem investigated by Gauss, where the Gauss-Lagrangian equation has only one solution.

In what follows, we will employ the function `SolveGL` for the computation of the Gauss-Lagrangian equation. Through a combination of bisection and Newton's method it determines all (up to three) solutions of (11.32), eliminates negative solutions (i.e., those with no physical reality), and sorts the remaining solutions according to their value. Because of the multiplicity of steps necessary to discriminate between cases and the resulting length of code, we reproduce here just the specification of the function's header, from which the individual parameter calls are apparent.

```
//------------------------------------------------------------------------------
// SolveGL: Finds the (up to three) solutions of the Gauss-Lagrangian
//          equation. Only positive solutions are returned.
//   gamma    Cosine of elongation from the Sun at time of 2nd observation
//   R        Earth-Sun distance at time of 2nd observation
//   rho_0    0th-order approximation of geocentric distance
//   sigma    Auxiliary quantity
//   rho_a    Geocentric distance at time of 2nd observation (1st solution)
//   rho_b    Geocentric distance at time of 2nd observation (2nd solution)
//   rho_c    Geocentric distance at time of 2nd observation (3rd solution)
//   n        Number of solutions actually found
//------------------------------------------------------------------------------
void SolveGL ( double gamma, double R, double rho_0, double sigma,
               double& rho_a, double& rho_b, double& rho_c, int& n );
```

11.3.2 Improved Iteration of the Triangle-Area Ratios

Because of the Encke approximation, the Gauss-Lagrangian equation does not yield any exact solutions of the basic orbit determination problem. It may, however, be so combined with the iteration of the triangle-area ratios in the shortened Gauss method, that the respective advantages are sensibly expanded. We thus obtain the process for the comprehensive Gaussian orbit determination method that is about to be described.

For this, we are again given three geocentric observations ($\boldsymbol{e}_1, \boldsymbol{e}_2, \boldsymbol{e}_3$), the respective geocentric coordinates of the Sun ($\boldsymbol{R}_1, \boldsymbol{R}_2, \boldsymbol{R}_3$) and the intervals ($\tau_1, \tau_2, \tau_3$). Using these starting data, the following computational steps are carried out:

1. Set $\mu_1 = (\tau_1\tau_3/6)(1 + \tau_1/\tau_2)$ and $\mu_3 = (\tau_1\tau_3/6)(1 + \tau_3/\tau_2)$ as initial approximation for the correction terms for the triangle-area ratios.
2. Repeat steps (a)…(e), until μ_1, μ_3 and the remaining values no longer change significantly.
   (a) Calculate ρ_0 and σ and solve the respective Gauss-Lagrangian equation. Choose the first of the three possible solutions for ρ_2, and calculate the corresponding heliocentric distance r_2.
   (b) Calculate the triangle-area ratios $n_1 = n_{10} + \mu_1/r_2^3$ and $n_3 = n_{30} + \mu_3/r_2^3$ and with them the geocentric distances (ρ_1, ρ_2, ρ_3) using (11.27).
   (c) Calculate from this the heliocentric position vectors ($\boldsymbol{r}_1, \boldsymbol{r}_2, \boldsymbol{r}_3$) from (11.24).
   (d) Calculate the sector to triangle ratios (η_1, η_2, η_3) for every two heliocentric position vectors and the corresponding interval using (11.5).
   (e) Calculate improved values $\mu_1 = (\eta_2/\eta_1 - 1) \cdot (\tau_1/\tau_2) \cdot r_2^3$ and $\mu_3 = (\eta_2/\eta_3 - 1) \cdot (\tau_3/\tau_2) \cdot r_2^3$ for the correction terms for the triangle-area ratios.
3. Calculate the orbital elements from the last values of $\boldsymbol{r}_1, \boldsymbol{r}_3$ and τ_2.

If the Gauss-Lagrangian equation exhibits more than one solution, then the whole process is repeated, with the second or the third solution being chosen in step (a). If necessary, considerations of plausibility may then be used to eliminate the Earth orbit solution or strongly hyperbolic orbits as a sensible solution. Otherwise, in doubtful cases, a fourth observation will help to identify the correct orbit of the comet or planet out of a set of multiple solutions.

11.3.3 Light-Time

A minor change is still required for this method to be employed in practice. The way in which the problem of determining an orbit has been handled until now has assumed that the observed position of a planet or comet is always identical with the instantaneous, geometrical position. In reality, however, there is a certain lapse of time between the light being emitted and its being observed, and this interval is determined by the finite speed of light

$$
\begin{aligned}
c &= 173.1\,\mathrm{AU/d} \\
1/c &= 0\overset{\mathrm{d}}{.}05776/\mathrm{AU} \quad .
\end{aligned}
$$

The vectors $\boldsymbol{e}_i$, which represent the directions of observation, therefore are directed from the heliocentric positions $-\boldsymbol{R}_i$, where the Earth is located at the times of

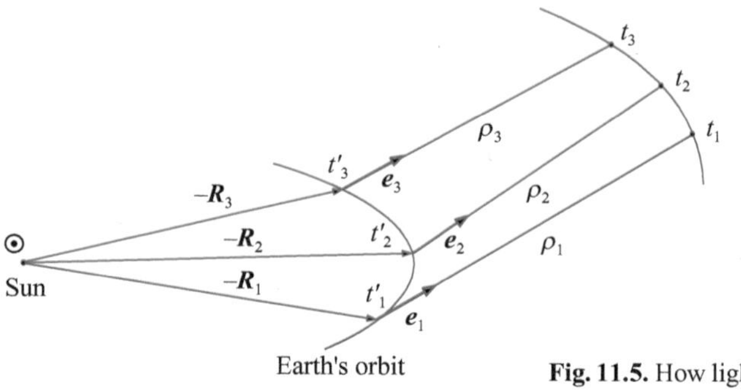

Fig. 11.5. How light-time is taken into account

observation, to the three points $\boldsymbol{r}_i = \rho_i \boldsymbol{e}_i - \boldsymbol{R}_i$, where the planet was at the slightly earlier times $t_i < t_i'$ (see Fig. 11.5). The time $t_i' - t_i$ that elapses between when the light is emitted and when it is observed, is thus equal to the time light requires to cover the distance ρ_i:

$$\rho_i = c \cdot (t_i' - t_i) \quad .$$

We immediately have the problem that the times of emission of the light t_i that are required for calculating the intervals τ_i are unknown. In the course of determining the orbit, however, we obtain increasingly accurate values for the geocentric distances (ρ_1, ρ_2, ρ_3), so the effect of light-time may be regarded as not posing any problems. We employ the observational times t_i' solely in determining the initial approximations for the ratios of the areas of the triangles. Afterwards the intervals

$$\tau_1 = \sqrt{GM_\odot}(t_3 - t_2) \quad \tau_2 = \sqrt{GM_\odot}(t_3 - t_1) \quad \tau_3 = \sqrt{GM_\odot}(t_2 - t_1)$$

are recalculated at each iteration using

$$t_i = t_i' - \rho_i \cdot 0\overset{\mathrm{d}}{.}05776/\mathrm{AU} \quad (i = 1 \ldots 3) \quad ,$$

as soon as the actual geocentric distances are known. The auxiliary geometrical variables $\boldsymbol{d}_i$, D_{ij} and D depend only on the direction vectors $\boldsymbol{e}_i$ and the solar coordinates $\boldsymbol{R}_i$, and therefore do not require correction.

11.4 The GAUSS Program

Using the `Gauss` program described here, the orbital elements of a cometary or minor-planet orbit may be determined from three observations. First the unit vectors of the directions of observation and the coordinates of the Sun are determined from the given values of right ascension and declination (`Start`). Because the required orbital elements must, as usual, be relative to the ecliptic, all the values are calculated in a system of ecliptic coordinates, whose equinox may be chosen at liberty. The orbit determination itself is performed in the `GaussMethod` function using the comprehensive Gaussian procedure. For solving the Gauss-Lagrangian equation the `SolveGL` function is used, which is not, however, printed in full because of its size.

Because of the amount of data that have to be entered, these are assembled in an input file `Gauss.dat`. The data file begins with a single comment line, which enables the data to be easily identified. The three following lines each contain the date (Year, Month, Day, and Hour with decimal fraction), the observed equatorial coordinates (right ascension being given as: Hours, Minutes, and Second with decimal fraction; declination as: Degrees, Minutes, and Seconds with decimal fraction), as well as an optional comment. The next two lines contain the equinox to which the observed positions relate, and the equinox in which the desired orbital elements should be specified. The following data file, for example, contains some observations of Ceres, which Gauss used as an example of his method of determining orbits:

```
Ceres (Gauss's example)
 1805 09 05 24.165   6 23 57.54  22 21 27.08  ! Three observations
 1806 01 17 22.095   6 45 14.69  30 21 24.20  ! Format: Date (ymdh),
 1806 05 23 20.399   8 07 44.60  28 02 47.04  !         RA (hms), Dec (dms)
 1806.0                                       ! Equinox
 1806.0                                       ! Required Equinox
```

In determining orbits all observations are expected to be in the form of astrometric coordinates, i.e., free from the effects of stellar aberration. Amongst others, this is always the case when the coordinates are derived by comparison with surrounding stars from a star catalogue. One example is the measurement of the photograph of a star field, which is the method most frequently employed to follow the movement of a comet or a minor planet.

```
//------------------------------------------------------------------------
// Module:  Gauss program (Gauss.cpp)
// Purpose: Orbit determination from three observations
// (c) 1999 Oliver Montenbruck, Thomas Pfleger
//------------------------------------------------------------------------

#include <cmath>
#include <fstream>
#include <iomanip>
#include <iostream>

#include "APC_Const.h"
#include "APC_IO.h"
#include "APC_Kepler.h"
#include "APC_Math.h"
#include "APC_PrecNut.h"
#include "APC_Spheric.h"
#include "APC_Sun.h"
#include "APC_Time.h"
#include "APC_VecMat3D.h"

using namespace std;
```

```
//------------------------------------------------------------------------
// Start: reads the input file and preprocesses the observational data
//   InputFile  Name of input file
//   Header     Comment
//   R_Sun[]    Three Sun position vectors (ecliptic coordinates)
//   e[]        Three observation direction unit vectors
//   Mjd0[]     Three observation times (Modified Julian Date)
//   T_eqx      Equinox of R_Sun and e (in Julian centuries since J2000)
//------------------------------------------------------------------------
void Start ( char* InputFile,
             char* Header,
             Vec3D R_Sun[], Vec3D e[], double Mjd0[], double& T_eqx )
{
  int       i, Year, Month, Day, deg, min;
  double    Hour, secs, year, T_eqx0, T, RA, Dec;
  ifstream  inp;
  inp.open(InputFile);  // Open observation file
  inp.get(Header,81);   // Read header
  // Read observations
  for (i=0;i<=2;i++) {
    inp >> Year >> Month >> Day >> Hour;
    inp >> deg >> min >> secs;
    RA = Rad*15.0*Ddd(deg,min,secs);
    inp >> deg >> min >> secs;
    Dec = Rad*Ddd(deg,min,secs);
    inp.ignore(81,'\n');
    Mjd0[i] = Mjd(Year,Month,Day) + Hour/24.0;
    e[i]    = Vec3D(Polar(RA,Dec));
  }
  // Equinox of observations and target equinox
  inp >> year; inp.ignore(81,'\n');  T_eqx0 = (year-2000.0)/100.0;
  inp >> year; inp.ignore(81,'\n');  T_eqx = (year-2000.0)/100.0;
  // Initial data for orbit determination (observations and Sun position
  // referred to ecliptic and equinox of T_eqx)
  for (i=0;i<=2;i++) {
    T        = (Mjd0[i]-MJD_J2000) / 36525.0;
    e[i]     = PrecMatrix_Ecl(T_eqx0,T_eqx) * Equ2EclMatrix(T_eqx0) * e[i];
    R_Sun[i] = PrecMatrix_Ecl(T,T_eqx) * SunPos(T);
  }
  // Print summary of initial data
  cout << " Summary of orbit determination" << endl << endl
       << "  " << Header << endl  << endl
       << " Initial data (geocentric ecliptic coordinates; Equinox J"
       << fixed << setprecision(2) << 100.0*T_eqx+2000.0 << ")" << endl
       << endl;
  cout << "  Observation"
       << setw(25) << "1" << setw(12) << "2" << setw(12) << "3" << endl;
  cout << "  Julian Date                 ";
  for (i=0;i<=2;i++) { cout << setw(12) << Mjd0[i]+2400000.5; };
  cout << endl;
  cout << "  Solar longitude [deg]       " << setprecision(4);
  for (i=0;i<=2;i++) { cout << setw(12) << Deg*(R_Sun[i][phi]); };
  cout << endl;
```

```
  cout << "  Planet/Comet longitude [deg]";
  for (i=0;i<=2;i++) { cout << setw(12) << Deg*((e[i])[phi]); };
  cout << endl;
  cout << "  Planet/Comet latitude [deg] ";
  for (i=0;i<=2;i++) { cout << setw(12) << Deg*((e[i])[theta]); };
  cout << endl;
}

//------------------------------------------------------------------------------
// DumpElem: Formatted output of orbital elements
//   Mjd_p      Modified Julian Date of perihelion passage
//   q          Perihelion distance in [AU]
//   e          Eccentricity
//   i          Inclination in [rad]
//   Omega      Longitude of ascending node in [rad]
//   omega      Argument of perihelion in [rad]
//   T_eqx      Reference equinox (Julian centuries since J2000)
//------------------------------------------------------------------------------
void DumpElem ( double Mjd_p, double q, double e, double i,
                double Omega, double omega, double T_eqx )
{ ... }

//------------------------------------------------------------------------------
// Retard: Light time correction and computation of time differences
//   Mjd0[]   Times of observation (t1',t2',t3') (Modified Julian Date)
//   rho[]    Three geocentric distances in [AU]
//   Mjd[]    Times of light emittance (t1,t2,t3) (Modified Julian Date)
//   tau[]    Scaled time differences
//------------------------------------------------------------------------------
void Retard ( const double Mjd0[], const double rho[],
              double Mjd[], double tau[] )
{
  for (int i=0;i<=2;i++)  Mjd[i] = Mjd0[i] - rho[i]/c_light;
  tau[0] = kGauss * (Mjd[2]-Mjd[1]);
  tau[1] = kGauss * (Mjd[2]-Mjd[0]);
  tau[2] = kGauss * (Mjd[1]-Mjd[0]);
}

//------------------------------------------------------------------------------
// SolveGL: Finds the (up to three) solutions of the Gauss-Lagrangian
//          equation. Only positive solutions are returned.
//   gamma    Cosine of elongation from the Sun at time of 2nd observation
//   R        Earth-Sun distance at time of 2nd observation
//   rho_0    0th-order approximation of geocentric distance
//   sigma    Auxiliary quantity
//   rho_a    Geocentric distance at time of 2nd observation (1st solution)
//   rho_b    Geocentric distance at time of 2nd observation (2nd solution)
//   rho_c    Geocentric distance at time of 2nd observation (3rd solution)
//   n        Number of solutions actually found
//------------------------------------------------------------------------------
void SolveGL ( double gamma, double R, double rho_0, double sigma,
               double& rho_a, double& rho_b, double& rho_c, int& n )
{ ... }
```

```
//------------------------------------------------------------------------
// Select: Selection of a single solution rho of the Gauss-Lagrangian
//         equation
//   rho_a      First solution
//   rho_b      Second solution
//   rho_c      Third solution
//   n_sol      Total number of solutions
//   n          Number of selected solution
//   <return>   Selected solution rho
//------------------------------------------------------------------------
double Select ( double rho_a, double rho_b, double rho_c,
                int n_sol, int n )
{
  if ( (n<1) || (n_sol<n) ) {
    cerr << " Error in Select: n = " << setw(2) << n << " n_sol = "
         << setw(2) << n_sol << endl;
    exit(1);
  }
  else
    switch(n) {
      case 1: return rho_a;
      case 2: return rho_b;
      case 3: return rho_c;
    };
  return 0.0;
}

//------------------------------------------------------------------------
// GaussMethod: Orbit determination using Gauss's method
//   R_Sun[]    Three Sun position vectors (ecliptic coordinates)
//   e[]        Three observation direction unit vectors
//   Mjd0[]     Three observation times (Modified Julian Date)
//   N_run      Number of current run (=selected number of solution of
//              the Gauss-Lagrangian equation)
//   N_sol      Total number of solutions of the Gauss-Lagrangian equation
//   Mjd_p      Modified Julian Date of perihelion passage
//   q          Perihelion distance [AU]
//   ecc        Eccentricity
//   inc        Inclination [rad]
//   Omega      Longitude of ascending node [rad]
//   omega      Argument of perihelion [rad]
//------------------------------------------------------------------------
void GaussMethod(const Vec3D R_Sun[], const Vec3D e[], const double Mjd0[],
                 int& N_run, int& N_sol,
                 double& Mjd_p, double& q, double& ecc,
                 double& inc, double& Omega, double& omega )
{
  const double eps_rho = 1.0e-8;
  const int    maxit   = 20;
  int     i,j,iterat;
  double  rho_old, Det,l,L,gamma;
  double  rho_max,rho_mean,rho_min;
  double  Mjd[3],rho[3],n[3],n0[3],tau[3],eta[3];
```

```
double  mu[3];
Vec3D   d[3],r[3];
double  D[3][3];

// Matrix D and determinant Det
d[0] = Cross(e[1],e[2]);
d[1] = Cross(e[2],e[0]);
d[2] = Cross(e[0],e[1]);
for (i=0;i<=2;i++) for (j=0;j<=2;j++)  D[i][j] = Dot(d[i],R_Sun[j]);
Det = Dot(e[2],d[2]);
// Direction cosine of observation unit vector with respect to the Sun
// direction at time of second observation
gamma  = Dot(e[1],R_Sun[1])/Norm(R_Sun[1]);
// Time differences tau[i] and initial approximations of mu[0] and mu[2]
tau[0] = kGauss*(Mjd0[2]-Mjd0[1]);
tau[1] = kGauss*(Mjd0[2]-Mjd0[0]);
tau[2] = kGauss*(Mjd0[1]-Mjd0[0]);
mu[0] = (1.0/6.0) * tau[0]*tau[2] * (1.0+tau[0]/tau[1]);
mu[2] = (1.0/6.0) * tau[0]*tau[2] * (1.0+tau[2]/tau[1]);

// Header
cout << endl << endl
  << " Solution No." << setw(1) << N_run << endl << endl
  << " Iteration of the geocentric distances " << endl
  << " Solutions (rho_2 in [AU]) of the Gauss-Lagrangian equation"
  << endl << endl
  << "  Iteration             Number   1st Sol.    2nd Sol.    3rd Sol."
  << endl;

// Iterative improvement of tau[i] and mu[i]
rho[1] = 0.0;  iterat = 0;
do {
  // Save previous value and increment iteration count
  rho_old = rho[1];
  iterat++;
  // Determine geocentric distance rho at time of second observation
  // from the Gauss-Lagrangian equation
  n0[0] = tau[0]/tau[1];   n0[2] = tau[2]/tau[1];
  L = - ( n0[0]*D[1][0]-D[1][1]+n0[2]*D[1][2] ) / Det;
  l =    ( mu[0]*D[1][0]    +    mu[2]*D[1][2] ) / Det;
  SolveGL (gamma,Norm(R_Sun[1]),L,l,rho_min,rho_mean,rho_max,N_sol);
  rho[1] = Select ( rho_min,rho_mean,rho_max, N_sol,N_run );
  r[1]   = rho[1]*e[1] - R_Sun[1];
  cout << setw(4) << iterat << setw(23) << N_sol
       << fixed << setprecision(8)
       << setw(15) << rho_min << setw(12) << rho_mean
       << setw(12) << rho_max << endl;
  // Compute n1 and n3
  n[0] = n0[0] + mu[0]/pow(Norm(r[1]),3);
  n[2] = n0[2] + mu[2]/pow(Norm(r[1]),3);
  // Geocentric distances rho_1 and rho_3 from n_1 and n_3
  rho[0] = ( n[0]*D[0][0] - D[0][1] + n[2]*D[0][2] ) / (n[0]*Det);
  rho[2] = ( n[0]*D[2][0] - D[2][1] + n[2]*D[2][2] ) / (n[2]*Det);
```

```
    // Apply light time correction and compute scaled time differences
    Retard ( Mjd0, rho, Mjd, tau );
    // Heliocentric coordinate vector
    for (i=0;i<=2;i++)  r[i] = rho[i]*e[i] - R_Sun[i];
    // Sector/triangle ratios eta_i
    eta[0] = FindEta ( r[1],r[2], tau[0] );
    eta[1] = FindEta ( r[0],r[2], tau[1] );
    eta[2] = FindEta ( r[0],r[1], tau[2] );
    // Improved values of mu_1, mu_3
    mu[0] = ( eta[1]/eta[0] - 1.0 ) * (tau[0]/tau[1]) * pow(Norm(r[1]),3);
    mu[2] = ( eta[1]/eta[2] - 1.0 ) * (tau[2]/tau[1]) * pow(Norm(r[1]),3);
    if (maxit<=iterat) {
      cerr << "    (Convergence problems; iteration stopped after "
           << setw(2) << maxit << " steps)" << endl;
      break;
    }
  }
  while (fabs(rho[1]-rho_old) > eps_rho);
  cout << endl << " Geocentric and heliocentric distances" << endl << endl;
  cout << "  Observation"
       << setw(25) << "1" << setw(12) << "2" << setw(12) << "3" << endl;
  cout << "  rho [AU]" << setw(20) << " " << fixed << setprecision(8);
  for (i=0;i<=2;i++) cout << setw(12) << rho[i]; cout << endl;
  cout << "  r [AU]" << setw(22) << " " << fixed << setprecision(8);
  for (i=0;i<=2;i++) cout << setw(12) << Norm(r[i]); cout << endl;
  // Derive orbital elements from first and third observation
  Elements ( GM_Sun, Mjd[0], Mjd[2], r[0], r[2],
             Mjd_p, q, ecc, inc, Omega, omega );
}

//------------------------------------------------------------------------------
// Main program
//------------------------------------------------------------------------------
void main(int argc, char* argv[])
{
  // Variables
  char      Header[81];
  int       N_run, N_sol;
  double    T_eqx;
  double    Mjd0[3];
  Vec3D     R_Sun[3],e[3];
  double    Mjd_p, q, ecc, inc, Omega, omega;
  char      InputFile[APC_MaxFilename] = "";
  char      OutputFile[APC_MaxFilename] = "";
  bool      FoundInputfile = false;
  bool      FoundOutputfile = false;
  ofstream  OutFile;

  // Title
  cout << endl
   << "      GAUSS: orbit determination from three observations " << endl
   << "             (c) 1999 Oliver Montenbruck, Thomas Pfleger "
   << endl << endl;
```

```
  // Find input and optional output files
  GetFilenames( argc, argv, "Gauss.dat", InputFile, FoundInputfile,
                OutputFile, FoundOutputfile );
  // Terminate program if input file could not be found
  if (!FoundInputfile) {
    cerr << " Terminating program." << endl;  exit(-1);
  }
  // Redirect output if output file shall be created
  if (FoundOutputfile) {
    OutFile.open(OutputFile);  if (OutFile.is_open()) cout = OutFile;
  }

  // Initialization
  Start ( InputFile, Header, R_Sun, e, Mjd0, T_eqx );

  // Handle all solutions
  N_run = 0;  N_sol = 1;
  do {
    N_run++;
    // Determine the orbit using Gauss's method for solution N_run of the
    // Gauss-Lagrangian equation
    GaussMethod( R_Sun,e,Mjd0, N_run, N_sol, Mjd_p,q,ecc,inc,Omega,omega );
    // Print orbital elements
    DumpElem ( Mjd_p,q,ecc,inc,Omega,omega, T_eqx );
  }
  while (N_run<N_sol);

  if (OutFile.is_open()) OutFile.close();

}
```

Once Gauss is called, no input is required from the user, because all the necessary information is held in the file Gauss.dat. For the values given above we obtain the following report:

```
      GAUSS: orbit determination from three observations
             (c) 1999 Oliver Montenbruck, Thomas Pfleger

 Using default input file Gauss.dat
 Summary of orbit determination

  Ceres (Gauss's example)

 Initial data (geocentric ecliptic coordinates; Equinox J1806.00)

  Observation                            1            2            3
  Julian Date                    2380570.51   2380704.42   2380830.35
  Solar longitude [deg]            162.9166     297.2126      61.9405
  Planet/Comet longitude [deg]      95.5389      99.8239     118.1018
  Planet/Comet latitude [deg]       -0.9918       7.2775       7.6473

 Solution No.1
```

```
Iteration of the geocentric distances
Solutions (rho_2 in [AU]) of the Gauss-Lagrangian equation

 Iteration                Number   1st Sol.    2nd Sol.    3rd Sol.
  1                         1     1.63338760  0.00000000  0.00000000
  2                         1     1.63476170  0.00000000  0.00000000
  3                         1     1.63599887  0.00000000  0.00000000
  4                         1     1.63655182  0.00000000  0.00000000
  5                         1     1.63676833  0.00000000  0.00000000
  6                         1     1.63684968  0.00000000  0.00000000
  7                         1     1.63687981  0.00000000  0.00000000
  8                         1     1.63689091  0.00000000  0.00000000
  9                         1     1.63689499  0.00000000  0.00000000
 10                         1     1.63689649  0.00000000  0.00000000
 11                         1     1.63689704  0.00000000  0.00000000
 12                         1     1.63689724  0.00000000  0.00000000
 13                         1     1.63689731  0.00000000  0.00000000
 14                         1     1.63689734  0.00000000  0.00000000
 15                         1     1.63689735  0.00000000  0.00000000

Geocentric and heliocentric distances

 Observation                           1           2           3
 rho [AU]                         2.90181811  1.63689735  2.95876464
 r [AU]                           2.68083076  2.58786985  2.54398416

Orbital elements (Equinox J1806.00)

 Perihelion date       tp         1806/06/28.039  (JD 2380865.539)
 Perihelion distance   q [AU]       2.541676
 Semi-major axis       a [AU]       2.767165
 Eccentricity          e            0.081487
 Inclination           i           10.6178 deg
 Ascending node        Omega       80.9788 deg
 Long. of perihelion   pi         147.0173 deg
 Arg. of perihelion    omega       66.0385 deg
```

As may be seen from the ecliptic coordinates of the Sun and Ceres, the minor planet was close to opposition at the time of observation. In the example chosen, there is no danger of a double solution to the orbit determination.

As an example of an orbit determination with two, non-trivial, positive solutions of the Gauss-Lagrangian equation, let us consider three observations of Comet Orkisz (C/1925 G1) from April 1925:

```
Comet Orkisz (Example from Stracke: Bahnbestimmung, p.182)
 1925 04 05  2.786  22 26 43.51  16 37 16.00  ! Three observations
 1925 04 08  2.731  22 29 42.90  19 46 25.10  ! Format: Date (ymdh),
 1925 04 11  2.614  22 32 55.00  23 04 52.30  !         RA (hms), Dec (dms)
 1925.0                                       ! Equinox
 1925.0                                       ! Required equinox
```

If we run the Gauss program with additional specification of the file Orkisz.dat (see Sect. A.1.3), then we obtain the following result:

```
     GAUSS: orbit determination from three observations
           (c) 1999 Oliver Montenbruck, Thomas Pfleger

Summary of orbit determination

 Comet Orkisz (Example from Stracke: Bahnbestimmung, p.182)

Initial data (geocentric ecliptic coordinates; Equinox J1925.00)

 Observation                            1           2           3
 Julian Date                   2424245.62  2424248.61  2424251.61
 Solar longitude [deg]            14.8175     17.7642     20.7031
 Planet/Comet longitude [deg]    345.0988    347.2423    349.5855
 Planet/Comet latitude [deg]      24.4159     27.0070     29.6985


Solution No.1

Iteration of the geocentric distances
Solutions (rho_2 in [AU]) of the Gauss-Lagrangian equation

 Iteration            Number   1st Sol.    2nd Sol.    3rd Sol.
  1                     2    1.67288439  6.37592623 -0.00141974
  2                     2    1.67314456  6.37552347 -0.00162268
  3                     2    1.67314511  6.37552331 -0.00162312
  4                     2    1.67314510  6.37552331 -0.00162312
  5                     2    1.67314511  6.37552331 -0.00162313
  6                     2    1.67314510  6.37552331 -0.00162312
  7                     2    1.67314509  6.37552331 -0.00162311

Geocentric and heliocentric distances

 Observation                            1           2           3
 rho [AU]                      1.71490811  1.67314509  1.63322160
 r [AU]                        1.10821587  1.10918577  1.11233811

Orbital elements (Equinox J1925.00)

 Perihelion date      tp        1925/04/05.280  (JD 2424245.780)
 Perihelion distance  q [AU]      1.108212
 Semi-major axis      a [AU]    -80.901263
 Eccentricity         e           1.013698
 Inclination          i         101.2244 deg
 Ascending node       Omega     318.9892 deg
 Long. of perihelion  pi        359.8990 deg
 Arg. of perihelion   omega      40.9098 deg


Solution No.2

Iteration of the geocentric distances
Solutions (rho_2 in [AU]) of the Gauss-Lagrangian equation
```

```
 Iteration               Number    1st Sol.     2nd Sol.     3rd Sol.
  1                        2     1.67288439  6.37592623 -0.00141974
  2                        2     1.67368234  6.37178371 -0.00197746
  3                        2     1.67368197  6.37178566 -0.00197720
  4                        2     1.67368197  6.37178566 -0.00197720

Geocentric and heliocentric distances

 Observation                          1           2           3
 rho [AU]                        6.53717259  6.37178566  6.22572525
 r [AU]                          5.77840936  5.63979853  5.52113670

Orbital elements (Equinox J1925.00)

 Perihelion date       tp        1925/04/25.025  (JD 2424265.525)
 Perihelion distance   q [AU]      5.262900
 Semi-major axis       a [AU]     -0.020764
 Eccentricity          e         254.460587
 Inclination           i           67.3802 deg
 Ascending node        Omega      326.7908 deg
 Long. of perihelion   pi          21.6630 deg
 Arg. of perihelion    omega       54.8721 deg
```

Here, the Gauss-Lagrangian equation has a total of three solutions, of which one ($\rho_2 \approx 0.0$ AU) represents the Earth orbit solution. The remaining solutions ($\rho_2 \approx 1.7$ AU) and ($\rho_2 \approx 6.4$ AU) both describe physically possible cometary orbits. On the grounds of plausibility ($e \gg 1$), however, the second solution may be excluded as unlikely.

12. Astrometry

Photography provides a relatively simple method of determining the coordinates of a comet or planet with respect to the known positions of stars. All that is required apart from the plate itself is a stellar catalogue and possibly an atlas to make it easier to identify the neighbouring stars.

An actual photograph of Comet Bradfield (C/1972 E1) in 1982 is shown diagrammatically in Fig. 12.1 to help in the following discussion. Comparison with a star atlas shows that the picture covers an area $\pm 4^{\mathrm{m}}$ in right ascension and $\pm 1^\circ$ in declination, and that it is centred on the point ($\alpha_{2000} = 12^{\mathrm{h}}28^{\mathrm{m}}45^{\mathrm{s}}, \delta_{2000} = 44^\circ 00'$). Apart from the comet, a number of stars may be identified, whose coordinates may be obtained from the *Position and Proper Motion Catalogue* (PPM). This covers the whole of the northern and southern skies and contains about 470 000 stars. The average distance between two neighbouring stars is about half a degree. At least three reference stars with known coordinates are required to determine the position of a celestial object from a photograph. With a view to obtaining the most accurate results, we nevertheless usually try to determine the required coordinates using the largest possible number of neighbouring stars. First, we will discuss how a section of the sky containing stars of different right ascensions and declinations will be reproduced on the plate.

12.1 Photographic Imaging

An ideal camera or telescope objective brings the rays of light from a star to a point on the film-plane. This lies at a distance F, known as the focal length, behind the objective. The point P, at which the star is imaged, may be found by the projection of the ray of light through the centre of the objective O (Fig. 12.2).

In a coordinate system defined by $\boldsymbol{u}$, $\boldsymbol{v}$, and $\boldsymbol{w}$,

$$\boldsymbol{e} = \begin{pmatrix} \cos(\delta)\cos(\alpha - \alpha_0) \\ \cos(\delta)\sin(\alpha - \alpha_0) \\ \sin(\delta) \end{pmatrix}$$

denotes the direction of a star located at right ascension α and declination δ. Correspondingly,

$$\boldsymbol{e}_0 = \begin{pmatrix} \cos(\delta_0) \\ 0 \\ \sin(\delta_0) \end{pmatrix}$$

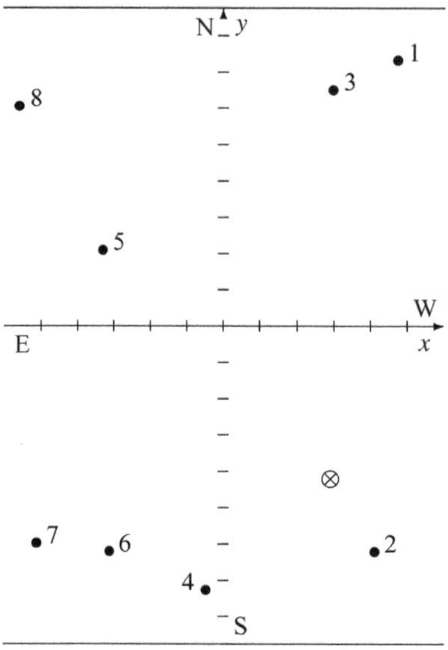

Comet Bradfield on 1982 Sept. 4

Reference stars:

1	PPM 052974
2	PPM 052987
3	PPM 052990
4	PPM 053019
5	PPM 053028
6	PPM 053029
7	PPM 053050
8	PPM 053051
⊗	Comet

Plate centre:

$\alpha \approx 12^{\mathrm{h}}28^{\mathrm{m}}45^{\mathrm{s}}$

$\delta \approx 44°00'$

Fig. 12.1. An example of a star field

defines the point (α_0,δ_0) on the sky at which the axis of the camera is pointing. The vectors $\boldsymbol{F} = -F \cdot \boldsymbol{e}_0$ and $\boldsymbol{p} = -p \cdot \boldsymbol{e}$ describe the paths followed by a ray of light from the objective to the centre of the plate, and to the image P of the star. They include the angle φ, where

$$\begin{aligned} \cos(\varphi) &= \boldsymbol{e}_0 \cdot \boldsymbol{e} \\ &= \cos(\delta_0)\cos(\delta)\cos(\alpha-\alpha_0) + \sin(\delta_0)\sin(\delta) \quad . \end{aligned}$$

Within the film-plane, the vectors

$$\boldsymbol{e}_X = \begin{pmatrix} 0 \\ 1 \\ 0 \end{pmatrix} \quad \text{and} \quad \boldsymbol{e}_Y = \begin{pmatrix} +\sin(\delta_0) \\ 0 \\ -\cos(\delta_0) \end{pmatrix}$$

define a system of coordinates that is used to measure the plate, and which is oriented North-South and East-West. If the coordinates X and Y of point P are measured in units of the focal length F, $\boldsymbol{p}$ may be expressed as:

$$\boldsymbol{p} = \boldsymbol{F} + (F \cdot X) \cdot \boldsymbol{e}_X + (F \cdot Y) \cdot \boldsymbol{e}_Y \quad .$$

If we write this equation in component form, the relationship between (α, δ) on one side and (X, Y) on the other may be expressed by the three equations

$$\begin{aligned} p\cos(\delta)\cos(\alpha-\alpha_0) &= F\cos(\delta_0) - FY\sin(\delta_0) \\ p\cos(\delta)\sin(\alpha-\alpha_0) &= - FX \\ p\sin(\delta) &= F\sin(\delta_0) + FY\cos(\delta_0) \end{aligned} \tag{12.1}$$

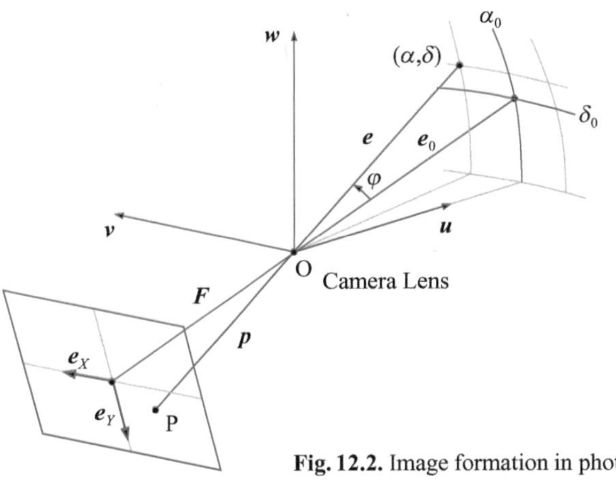

Fig. 12.2. Image formation in photographing a star field

where

$$p = |\boldsymbol{p}| = F\sqrt{1+X^2+Y^2}$$

or

$$\begin{aligned} p &= F/\cos(\varphi) \\ &= F/(\cos(\delta_0)\cos(\delta)\cos(\alpha-\alpha_0)+\sin(\delta_0)\sin(\delta)) \quad . \end{aligned}$$

Solving for the spherical coordinates gives us the equations

$$\begin{aligned} \alpha &= \alpha_0 + \arctan\left\{\frac{-X}{\cos(\delta_0)-Y\sin(\delta_0)}\right\} \\ \delta &= \arcsin\left\{\frac{\sin(\delta_0)+Y\cos(\delta_0)}{\sqrt{1+X^2+Y^2}}\right\} \quad . \end{aligned} \tag{12.2}$$

The inverse of (12.2) can similarly be obtained from (12.1) as

$$\begin{aligned} X &= -\frac{\cos(\delta)\sin(\alpha-\alpha_0)}{\cos(\delta_0)\cos(\delta)\cos(\alpha-\alpha_0)+\sin(\delta_0)\sin(\delta)} \\ Y &= -\frac{\sin(\delta_0)\cos(\delta)\cos(\alpha-\alpha_0)-\cos(\delta_0)\sin(\delta)}{\cos(\delta_0)\cos(\delta)\cos(\alpha-\alpha_0)+\sin(\delta_0)\sin(\delta)} \quad . \end{aligned} \tag{12.3}$$

Both of these transforms may be simply expressed in functions:

```
//---------------------------------------------------------------------------
// StdEqu: Calculation of equatorial coordinates from standard coordinates
//         used in astrographic plate measurement
//   RA0       Right ascension of optical axis [rad]
//   Dec0      Declination of optical axis [rad]
```

```
//   X            Standard coordinate X
//   Y            Standard coordinate Y
//   RA           Right ascension [rad]
//   Dec          Declination [rad]
//---------------------------------------------------------------------------
void StdEqu ( double RA0, double Dec0, double X, double Y,
              double& RA, double& Dec )
{
  RA  = RA0 + atan ( -X / (cos(Dec0)-Y*sin(Dec0)) );
  Dec = asin ( (sin(Dec0)+Y*cos(Dec0))/sqrt(1.0+X*X+Y*Y) );
}

//---------------------------------------------------------------------------
// EquStd: Calculation of standard coordinates from equatorial coordinates
//   RA0          Right ascension of optical axis [rad]
//   Dec0         Declination of optical axis [rad]
//   RA           Right ascension [rad]
//   Dec          Declination [rad]
//   X            Standard coordinate X
//   Y            Standard coordinate Y
//---------------------------------------------------------------------------
void EquStd ( double RA0, double Dec0, double RA, double Dec,
              double& X, double& Y )
{
  const double c =
    cos(Dec0)*cos(Dec)*cos(RA-RA0)+sin(Dec0)*sin(Dec);
  X = - ( cos(Dec)*sin(RA-RA0) ) / c;
  Y = - ( sin(Dec0)*cos(Dec)*cos(RA-RA0)-cos(Dec0)*sin(Dec) ) / c;
}
```

12.2 Plate Constants

The dimensionless coordinates X and Y are described as standard coordinates, because, by definition, they are not dependent on the focal length of the optics employed, and refer to a system of coordinates that is oriented exactly parallel to the meridian that passes through the centre $((\alpha_0, \delta_0)$ of the plate. If this coordinate system is also to be used as the basis for the plate reduction, then the measured coordinates x and y have only to be divided by the focal length to obtain the standard coordinates:

$$X = x/F \qquad Y = y/F \quad .$$

In general, the origin of the coordinate system used will, nevertheless, not precisely coincide with the intersection of the optical axis with the film-plane. An offset of this nature, amounting to $(\Delta x, \Delta y)$ may, however, be accommodated by a small alteration to the equation just given:

$$X = x/F - (\Delta x)/F \qquad Y = y/F - (\Delta y)/F \quad .$$

If, in addition, the coordinate axes are rotated through an angle γ with respect to the North-South line, then the transformation equations become:

$$\begin{aligned} X &= (x \cdot \cos(\gamma) - y \cdot \sin(\gamma))/F - (\Delta x)/F \\ Y &= (x \cdot \sin(\gamma) + y \cdot \cos(\gamma))/F - (\Delta y)/F \quad . \end{aligned}$$

However, offset and rotation of the coordinate system are not the only factors that affect the relationship between the measured coordinates and the standard coordinates. Optical errors, or possible tilt or distortion of the film generally mean that further corrections have to be made. In general, therefore, the equations are expressed in the form:

$$\begin{aligned} X &= a \cdot x + b \cdot y + c \\ Y &= d \cdot x + e \cdot y + f \end{aligned} \tag{12.4}$$

with six *plate constants* a, b, c, d, e, and f, allowing conversion between the (x,y) and (X,Y) coordinate pairs. The plate constants cannot be known in advance, but must be determined by the use of reference stars. If we know the equatorial coordinates $(\alpha_i, \delta_i)_{i=1,2,3}$ of three reference stars, then from (12.3), we obtain the three equations

$$\begin{aligned} X_1 &= x_1 \cdot a + y_1 \cdot b + c \\ X_2 &= x_2 \cdot a + y_2 \cdot b + c \\ X_3 &= x_3 \cdot a + y_3 \cdot b + c \quad , \end{aligned} \tag{12.5}$$

which may be solved for a, b, and c by appropriate manipulation. The remaining plate constants d, e, and f are correspondingly obtained from the equations for Y_i. The focal length of the camera optics is not required. This is of particular advantage when one is working with an enlargement whose scale is unknown, rather than with an original plate.

As we have just seen, the plate constants may be determined by solving two equations, each with three unknowns, if at least three reference stars with known right ascension and declination are available. As the measurement of stellar coordinates can never be completely free from errors, it is desirable to use as many reference stars as possible in determining the plate constants. Normally, however, a set of equations of the form

$$\begin{aligned} X_1 &= x_1 \cdot a + y_1 \cdot b + c \\ &\dots \\ X_n &= x_n \cdot a + y_n \cdot b + c \quad , \end{aligned} \tag{12.6}$$

with $n > 3$ equations for the three desired plate constants is overdetermined and may therefore no longer be solved. To avoid this problem, the reduction procedure about to be described uses the *method of least squares*.

12.3 Least Squares Adjustment

In general[1], if we want to determine m unknowns $\boldsymbol{x} = (x_1, \ldots, x_m)$ from n equations

$$\boldsymbol{a}_i \cdot \boldsymbol{x} = a_{i1} \cdot x_1 + \ldots + a_{im} \cdot x_m \;=\; b_i \quad (i = 1, \ldots, n) \tag{12.7}$$

with given coefficients x_i and a_{ij}, then when $n > m$, we cannot expect all the equations to be satisfied simultaneously. We therefore first introduce terms known as *residuals*, the values of which are

$$\nu_i \;=\; a_{i1} \cdot x_1 + \ldots + a_{im} \cdot x_m \;-\; b_i \quad (i = 1, \ldots, n) \quad . \tag{12.8}$$

Instead of trying to obtain $x_1, \ldots, x_m$ where all the values of ν_i are zero (which would mean that all the equations were satisfied), we can content ourselves with minimizing the value of the sum

$$\chi \;=\; \sum_{i=1}^{n} \nu_i^2$$

of the squares of the residuals.

In order to do this, the equations (12.8) are first subjected to a series of transformations that give an equivalent system of the form

$$\begin{array}{lclcccccccl}
\mu_1 & = & r_{11} \cdot x_1 & + & r_{12} \cdot x_2 & + & \ldots & + & r_{1m} \cdot x_m & - & d_1 \\
\mu_2 & = & & & r_{22} \cdot x_2 & + & \ldots & + & r_{2m} \cdot x_m & - & d_2 \\
\vdots & = & & & & & \ddots & & \vdots & & \vdots \\
\mu_m & = & & & & & & & r_{mm} \cdot x_m & - & d_m \\
\mu_{m+1} & = & & & & & & & & - & d_{m+1} \\
\vdots & = & & & & & & & & & \vdots \\
\mu_n & = & & & & & & & & - & d_n \quad ,
\end{array} \tag{12.9}$$

where all coefficients r_{ij} are equal to zero whenever $(i > j)$, which means that the system is significantly easier to handle.

If c and s are two real number where $c^2 + s^2 = 1$, then the sum of the squares of the residuals does not alter, for example, when ν_1 is replaced by $\nu_1' = p \cdot \nu_1 - q \cdot \nu_2$ and ν_2 by $\nu_2' = q \cdot \nu_1 + p \cdot \nu_2$, because, given the assumption that we have made

$$\begin{aligned}
\nu_1'^2 + \nu_2'^2 \;&=\; c^2\nu_1^2 - 2cs\,\nu_1\nu_2 + s^2\nu_2^2 + s^2\nu_1^2 + 2cs\,\nu_1\nu_2 + c^2\nu_2^2 \\
&=\; (c^2 + s^2) \cdot (\nu_1^2 + \nu_2^2)
\end{aligned}$$

[1] To conform to a common notation, some of the symbols introduced earlier are used with different meanings for the following discussion of least squares adjustment. This is particularly true for the quantities a, b and x, which denote the coefficients and unknowns of the system of equations, instead of plate constants and stellar coordinates.

is equal to $\nu_1^2 + \nu_2^2$. If we specifically choose

$$\begin{aligned} c &= +a_{11}/h \quad \text{and} \\ s &= +a_{21}/h \quad \text{with} \quad h = \pm\sqrt{a_{11}^2 + a_{21}^2} \quad , \end{aligned}$$

then, as we require, a_{21} is replaced by

$$a'_{21} \;=\; -sa_{11} + ca_{21} \;=\; (-a_{21}a_{11} + a_{11}a_{21})/h \;=\; 0 \quad .$$

Using similar rearrangements (which are also known as Givens rotations), one after the other $a_{31} \ldots a_{n1}, a_{32} \ldots a_{n2}, \ldots$, and $a_{m+1,m} \ldots a_{nm}$ are also eliminated, until finally we obtain a system of equations in the form of (12.9).

Whereas the coefficient matrix $\boldsymbol{A} = (a_{ij})_{i=1\ldots n, j=1\ldots m}$ in the original system $\boldsymbol{Ax} = \boldsymbol{b}$ is generally completely filled, the transformed system

$$\begin{pmatrix} \boldsymbol{R} \\ \boldsymbol{0} \end{pmatrix} \boldsymbol{x} = \boldsymbol{d} \tag{12.10}$$

stands out because of the triangular nature of the quadratic matrix $\boldsymbol{R} = (r_{ij})$, where all elements beneath the diagonal disappear.

The sum of the residuals squared may now be written as two terms

$$\chi \;=\; |\boldsymbol{v}|^2 = |\boldsymbol{\mu}|^2 \;=\; \sum_{i=1}^{m} {\mu_i}^2 + \sum_{i=m+1}^{n} {d_i}^2$$

of which only the first is still dependent on the unknown x_k. The least sum will obviously be obtained when all values of μ_i disappear for $i = 1 \ldots m$. The corresponding values of x_k are obtained by back substitution:

$$\begin{aligned} x_m &= d_m/r_{mm} \\ x_k &= (d_k - \sum_{l=k+1}^{m} r_{kl}x_l)/r_{kk} \quad (k = m-1, \ldots, 1) \quad . \end{aligned} \tag{12.11}$$

The rearrangement of the set of equations to upper triangular form and the subsequent solution are combined in the `solverLSQ` class, which provides an easy method of solving any linear estimation problem of this sort.

```
// Least squares estimation class SolverLSQ
class SolverLSQ {
  public:
    // Constructor, destructor
    SolverLSQ(int n); // n: Number of estimation parameters
    ~SolverLSQ();
    // Reset to solve a new problem
    void Init();
    // Insert a data equation of form Ax=b to a least squares estimation
    // (performs a row-wise QR transformation using Givens rotations)
    void Accumulate (const double A[], double b);
    // Solve the LSQ problem for vector x[] by back substitution
```

```
    void Solve (double x[]);
  private:
    int      N;          // Number of estimation parameters
    double   *d;         // Right hand side of transformed equations
    double   **R;        // Upper right triangular matrix
};
```

To handle a system of equations $\boldsymbol{A}\boldsymbol{x} = \boldsymbol{b}$ with n unknowns, an object of the class `SolverLSQ` must be declared. Making use of the known dimensions the constructor will then reserve and initialize dynamic storage for the upper right triangular matrix $\boldsymbol{R}$ and the vector $\boldsymbol{d}$ of the transformed system.

```
// Constructor; n: Number of estimation parameters
SolverLSQ::SolverLSQ (int n)
  : N(n)
{
  d = new double[N];     // Allocate storage for R and d
  R = new double*[N];
  for (int i=0; i<N; i++) R[i] = new double[N];
  Init();                // Reset d and R
}
// Initialize new problem (Reset R and d)
void SolverLSQ::Init()
{
  for (int i=0; i<N; i++) { d[i]=0.0; for (int j=0;j<N;j++) R[i][j]=0.0; }
}
```

For each line $\boldsymbol{a}_i \cdot \boldsymbol{x} = b_i$ $(i = 1, \ldots, n)$ of the set of equations, the coefficients $\boldsymbol{a}_i$ and b_i will then be processed sequentially using the `Accumulate` method and transformed into the right upper triangular matrix $\boldsymbol{R}$ and the vector $\boldsymbol{d}$.

```
//  Add a data equation of form Ax=b to a least squares problem
//  using row-wise Givens rotations
void SolverLSQ::Accumulate (const double A[], double b)
{
  // Variables
  int      i,j;
  double   c,s,h;
  double*  a = new double[N];
  // Copy A
  for (i=0;i<N;i++) a[i]=A[i];
  // Construct and apply Givens plane rotation.
  for (i=0; i<N; i++)
  {
    // Construct the rotation and apply it to eliminate i-th element of a.
    if ( R[i][i]==0.0 &&  a[i]==0.0 ) {
      c = 1.0; s = 0.0; R[i][i] = 0.0;
    }
    else {
      h = sqrt ( R[i][i]*R[i][i] + a[i]*a[i] );
      if (R[i][i]<0.0) h=-h;
      c = R[i][i]/h;  s = a[i]/h;  R[i][i] = h;
    };
```

```
    a[i] = 0.0;
    // Apply the rotation to the remaining elements of a
    for (j=i+1; j<N; j++) {
      h = +c*R[i][j]+s*a[j];  a[j] = -s*R[i][j]+c*a[j];  R[i][j] = h;
    }
    // Apply the rotation to the i-th element of d
    h = +c*d[i]+s*b;  b = -s*d[i]+c*b;  d[i] = h;
  }
  delete[] a;
}
```

Using the `Solve` method, we finally obtain, through back substitution, the desired solution of the system of equations.

```
// Solve the LSQ problem for vector x[] by back substitution
void SolverLSQ::Solve (double x[])
{
  int i,j; i=j=0;
  double Sum=0.0;
  // Check for singular matrix
  for (i=0;i<N;i++)
    if ( R[i][i] == 0.0 ) {
      cerr << " ERROR: Singular matrix R in SolverLSQ::Solve()" << endl;
      exit(1);
    };
  //  Solve Rx=d for x_n,...,x_1 by backsubstitution
  x[N-1] = d[N-1] / R[N-1][N-1];
  for (i=N-2;i>=0;i--) {
    Sum = 0.0;
    for (j=i+1;j<N;j++) Sum += R[i][j]*x[j];
    x[i] = ( d[i] - Sum ) / R[i][i];
  };
}
```

Here array indexes, as normal in C++, run from 0 to $n-1$.

12.4 The FOTO Program

The `Foto` program enables accurate positions of stars, comets, or minor planets to be obtained from photographic plates of the sky. It will relieve the user of the considerable amount of computational work that is required in determining standard coordinates, and solving for the residuals.

Before calling `Foto`, however, a certain amount of preparation is required to obtain the necessary input data. First, using a star chart, a few of the brighter stars must be identified, thus establishing the approximate coordinates of the field covered by the plate. Then suitable reference stars are chosen, whose equatorial coordinates may be found in a star catalogue, such as the *PPM Catalogue*. If necessary, the proper motions of the stars between the epoch of the catalogue and the date of the exposure must be taken into account. The reference stars should be evenly distributed across the plate, and should be as point-like as possible, to make it easier to obtain accurate measurements. The positions of the reference stars and the object being examined may be determined without difficulty, by laying a transparent millimetre grid, oriented approximately North-South, over an enlargement. Finally, we determine the right ascension and declination of the plate centre, which are required in calculating the standard coordinates. As errors in the coordinates of the plate centre have little effect on position determination, it suffices to take these values from an atlas.

The various pieces of data are all entered in a data file `Foto.dat`. Its first line contains the right ascension (in $^{h}, ^{m}, ^{s}$) and declination (in $^\circ, ', ''$) of the plate centre. This is followed by details of the individual objects on the plate. Reference stars that are to be used to determine the plate constants are indicated by an asterisk ($*$) in the first column. After the name, which may contain up to 11 letters, the measured coordinates (x, y in mm) are given, and additionally for reference stars, their equatorial coordinates (α in $^{h}, ^{m}, ^{s}$, δ in $^\circ, ', ''$). The data given in the following example refer to the plate shown diagrammatically in Fig. 12.1.

```
CENTRE                               12 28 45.000  +44 00 00.00
*PPM 052974    +47.5  +73.3   12 25 06.952  +44 47 55.79
*PPM 052987    +41.2  -62.1   12 25 59.995  +43 07 05.93
*PPM 052990    +29.9  +65.2   12 26 22.451  +44 42 39.02
*PPM 053019     -4.8  -72.7   12 29 09.766  +43 00 54.07
*PPM 053028    -33.0  +21.0   12 30 52.966  +44 11 19.19
*PPM 053029    -31.2  -62.0   12 30 53.525  +43 09 30.08
*PPM 053050    -51.2  -59.7   12 32 17.476  +43 11 57.76
*PPM 053051    -55.7  +60.6   12 32 21.122  +44 40 54.24
BRADFIELD      +29.2  -42.1
```

The `Getinp` function reads and stores these values. Then the standard coordinates of the reference stars, which are required to evaluate the plate constants, are determined. From the plate constants, the procedure may be reversed: the standard coordinates and equatorial coordinates of all the objects on the plate being determined from the measured coordinates. With the reference stars, this enables one to obtain a good estimate of the error that one can expect in reducing the plate. In addition, any gross errors may be recognized, such as individual reference stars having been incorrectly identified or measured.

```
//------------------------------------------------------------------------
// Module:  Foto program (Foto.cpp)
// Purpose: Astrometric analysis of photographic plates
// (c) 1999 Oliver Montenbruck, Thomas Pfleger
//------------------------------------------------------------------------
#include <cmath>
#include <fstream>
#include <iomanip>
#include <iostream>
#include "APC_Const.h"
#include "APC_IO.h"
#include "APC_Math.h"
#include "APC_Spheric.h"
```

```
using namespace std;

const int MaxObj = 30;    // Maximum number of objects on the photo
const int StrLen = 13;

//------------------------------------------------------------------------------
// GetInput: Reads input from file
//   Filename  Name of input file
//   RA0       Right ascension of plate centre [rad]
//   Dec0      Declination of plate centre [rad]
//   NObj      Number of objects read
//   Name[]    Names of the objects
//   RA[]      Right ascensions of reference objects [rad]
//   Dec[]     Declinations of reference objects [rad]
//   x[]       Measured coordinates of reference objects
//   y[]         with respect to the plate centre [mm]
//------------------------------------------------------------------------------
void GetInput ( char* Filename,
                double& RA0, double& Dec0, int& NObj, char Name[][StrLen],
                double RA[], double Dec[], double x[], double y[] )
{
  int       i,k,h,m,deg,min;
  double    s,sec;
  char      c,sign;
  ifstream  inp;
  // Open file for reading
  inp.open(Filename);
  cout << " Input data file: " << Filename << endl << endl;
  // Read coordinates of the plate centre
  for (k=0;k<StrLen;k++) inp.get(c);
  inp >> h >> m >> s;       inp.get(c).get(c).get(sign);
  inp >> deg >> min >> sec; inp.ignore(81,'\n');
  RA0  = Rad*15.0*Ddd(h,m,s);
  Dec0 = Rad*Ddd(deg,min,sec); if (sign=='-') Dec0 = -Dec0;
  // Read name, plate coordinates (and equatorial coordinates)
  i = -1;
  do {
    ++i;
    inp.get(Name[i],StrLen);  if (inp.fail()) break;
    if ( Name[i][0] == '*' ) {
      // Reference star
      inp >> x[i] >> y[i];
      inp >> h >> m >> s;       inp.get(c).get(c).get(sign);
      inp >> deg >> min >> sec; inp.ignore(81,'\n');
      RA[i]  = Rad*15.0*Ddd(h,m,s);
      Dec[i] = Rad*Ddd(deg,min,sec); if (sign=='-') Dec[i] = -Dec[i];
    }
    else {
      // Unknown object
      inp >> x[i] >> y[i]; inp.ignore(81,'\n');
      RA[i]=0.0; Dec[i]=0.0;
    };
    NObj = i+1;
```

```
    c=inp.peek(); // Required to enable EOF check
  }
  while (!inp.eof() && i<MaxObj);
  inp.close();  // Close input file
}

//------------------------------------------------------------------------------
//
// Main program
//
//------------------------------------------------------------------------------
void main(int argc, char* argv[])
{
  // Constants
  const int NEst = 3;    // Number of estimation parameters
  // Variables
  int        i, NObj;
  char       Name[MaxObj][StrLen];
  double     RA0, Dec0, RA_Obs, Dec_Obs, D_RA, D_Dec;
  double     RA[MaxObj], Dec[MaxObj], Delta[MaxObj];
  double     x[MaxObj], y[MaxObj], X[MaxObj], Y[MaxObj];
  double     A[NEst],C[NEst];
  double     a,b,c,d,e,f, Det,FocLen,Scale;
  SolverLSQ LSQx(NEst), LSQy(NEst);
  char       InputFile[APC_MaxFilename] = "";
  char       OutputFile[APC_MaxFilename] = "";
  bool       FoundInputfile = false;
  bool       FoundOutputfile = false;
  ofstream   OutFile;
  // Title
  cout << endl
      << "     FOTO: astrometric analysis of photographic plates  " << endl
      << "            (c) 1999 Oliver Montenbruck, Thomas Pfleger  " << endl
      << endl;
  // Find input file
  GetFilenames( argc, argv, "Foto.dat", InputFile, FoundInputfile,
                OutputFile, FoundOutputfile );
  // Terminate program if input file could not be found
  if (!FoundInputfile) {
    cerr << " Terminating program." << endl;  exit(-1);
  }
  // Read input file
  GetInput (InputFile, RA0,Dec0, NObj, Name, RA,Dec, x,y);

  // Calculate standard coordinates of reference stars and
  // fill least squares systems
  for (i=0;i<NObj;i++)
    if (Name[i][0]=='*') {
      EquStd (RA0, Dec0, RA[i], Dec[i], X[i], Y[i]);
      A[0]=x[i]; A[1]=y[i]; A[2]=1.0;
      LSQx.Accumulate (A, X[i]);
      LSQy.Accumulate (A, Y[i]);
    };
```

```
  // Calculate plate constants
  LSQx.Solve(C); a=C[0]; b=C[1]; c=C[2];
  LSQy.Solve(C); d=C[0]; e=C[1]; f=C[2];
  // Calculate equatorial coordinates (and errors for reference stars(*) )
  for (i=0;i<NObj;i++) {
    X[i] = a*x[i]+b*y[i]+c;
    Y[i] = d*x[i]+e*y[i]+f;
    StdEqu (RA0, Dec0, X[i], Y[i], RA_Obs, Dec_Obs);
    if (Name[i][0]=='*') {
      D_RA  = (RA_Obs-RA[i]) * cos(Dec[i]);
      D_Dec = (Dec_Obs-Dec[i]);
      Delta[i] = Arcs * sqrt(D_RA*D_RA+D_Dec*D_Dec);  // in [arcsec]
    };
    RA[i] = RA_Obs;  Dec[i] = Dec_Obs;
  };
  // Calculate focal length and scale
  Det    = a*e-d*b;
  FocLen = 1.0/sqrt(fabs(Det)); // in [mm]
  Scale  = Arcs / FocLen;       // in ["/mm]
  // Redirect output if output file shall be created
  if (FoundOutputfile) {
    OutFile.open(OutputFile); if (OutFile.is_open())  cout = OutFile;
  }
  // Output
  cout << " Plate constants:" << endl << endl
       << fixed << setprecision(8)
       << "    a =" << setw(12) << a << "    b =" << setw(12) << b
       << "    c =" << setw(12) << c << endl
       << "    d =" << setw(12) << d << "    e =" << setw(12) << e
       << "    f =" << setw(12) << f << endl << endl;
  cout << " Effective focal length and image scale:" << endl << endl
       << fixed << setprecision(2)
       << "    F =" << setw(9) << FocLen << " mm" << endl
       << "    m =" << setw(7) << Scale  << " \"/mm" << endl << endl;
  cout << " Coordinates:" << endl << endl
       << "          Name         x      y        X        Y "
       << "           RA             Dec     Error" << endl
       << "                       mm     mm                  "
       << "       h  m  s        o  '  \"      \" " << endl;
  for (i=0;i<NObj;i++) {
    cout << "   " << Name[i] << fixed
         << setprecision(1) << setw(7) << x[i] << setw(7) << y[i]
         << setprecision(4) << setw(9) << X[i] << setw(8) << Y[i]
         << setprecision(2) << setw(14) << Angle(Deg*RA[i]/15,DMMSSs)
         << showpos << "  "
         << setprecision(1) << setw(11) << Angle(Deg*Dec[i],DMMSSs)
         << noshowpos;
    if (Name[i][0]=='*') cout << setprecision(1) << setw(6) << Delta[i];
    cout << endl;
  }
  if (OutFile.is_open()) OutFile.close();
}
```

Using the data given earlier, Foto produces the following output:

```
     FOTO: astrometric analysis of photographic plates
           (c) 1999 Oliver Montenbruck, Thomas Pfleger

Using default input file Foto.dat
Input data file: Foto.dat

Plate constants:

  a =  0.00021651   b =  0.00000827   c =  0.00035472
  d = -0.00000799   e =  0.00021639   f = -0.00149942

Effective focal length and image scale:

  F =  4616.79 mm
  m =  44.68 "/mm

Coordinates:

     Name          x      y        X       Y          RA           Dec       Error
                  mm     mm                         h  m  s       o  '  "       "
*PPM 052974     47.5   73.3   0.0112  0.0140  12 25 07.11  +44 47 50.9    5.2
*PPM 052987     41.2  -62.1   0.0088 -0.0153  12 25 59.96  +43 07 23.9   18.0
*PPM 052990     29.9   65.2   0.0074  0.0124  12 26 22.46  +44 42 25.9   13.2
*PPM 053019     -4.8  -72.7  -0.0013 -0.0172  12 29 09.18  +43 00 54.0    6.4
*PPM 053028    -33.0   21.0  -0.0066  0.0033  12 30 51.88  +44 11 18.0   11.7
*PPM 053029    -31.2  -62.0  -0.0069 -0.0147  12 30 55.30  +43 09 30.5   19.5
*PPM 053050    -51.2  -59.7  -0.0112 -0.0140  12 32 16.68  +43 11 38.5   21.1
*PPM 053051    -55.7   60.6  -0.0112  0.0121  12 32 21.67  +44 41 14.4   21.0
BRADFIELD       29.2  -42.1   0.0063 -0.0108  12 26 45.28  +43 22 39.7
```

After the plate constants, the effective focal length (=Camera focal length × Enlargement) and the image scale are displayed. These are followed by the measured rectangular coordinates (x, y) for each object, together with the standard coordinates (X, Y) calculated from them, and the equatorial coordinates right ascension and declination. The reference stars show typical departures from the catalogued places of between $10''$ to $20''$. This corresponds to measurement errors of about $1/4$–$1/2$ mm, which are to be expected in determining errors with millimetre grid overlays. More accurate results can generally be obtained by using a special measuring machine.

The proper motion during the interval between the epoch of the PPM Catalogue (2000.0) and the time of the exposure (1982.8) was not taken into account. It amounts to $2''$ at most for the chosen reference stars, and is therefore smaller than the scatter found in determining the positions.

12.5 The Position and Proper Motion Catalogue

Among stellar catalogues that are currently available, the *Position and Proper Motion Catalogue* by S. Röser and U. Bastian provides a good basis for astrometric reduction of photographs of star fields, because of its accuracy and the number of stars included. It makes sense to use it wherever a limiting magnitude brighter than 11^{m} or an accuracy of around $1''$ will suffice for the aim in mind. Only when more stringent requirements need to be met do we need to turn to specialized catalogues like the Hipparcos Star Catalogue or the Hubble Guide Star Catalogue. By using secondary reference stars, it is then possible to measure even CCD-images, for example, with small fields of view.

With the agreement of the authors and the Astronomisches Rechen-Institut Heidelberg, the full PPM Catalogue, including the corresponding descriptive information is available on the accompanying CD-ROM. The Catalogue originally consisted of four sections, which have here been combined into a single database `PPM.dat` for easier operation:

- PPM North (No. 1 – 181731)
- PPM South (No. 181732 – 378910)
- PPM Supplement of 275 bright stars (No. 400001 – 400321)
- PPM Supplement of 90000 southern stars (No. 700001 - 789676)

The file includes the data for a total of 468 861 stars, and consists of a corresponding number of lines, each containing 131 characters. The meaning of the most important data is summarized in Table 12.1.

Table 12.1. Structure of the Position and Proper Motion star catalogue (excerpt)

Cols	Description
2– 7	PPM catalogue number
10–18	Bonner/Cordoba Durchmusterung number
20–23	Brightness [mag]
25–26	Spectral type
28–29	Right ascension J2000 (hours)
31–32	Right ascension J2000 (minutes)
34–39	Right ascension J2000 (seconds)
42–	Declination J2000 (sign)
43–44	Declination J2000 (degrees)
46–47	Declination J2000 (minutes)
49–53	Declination J2000 (seconds)
56–62	Proper motion in RA [s/year]
64–69	Proper motion in Dec [$''$/year]

Although the file contains printable ASCII characters only, and could, in principle, be handled with a text editor, reading it would be extremely costly in time, because of its size of about 60 Mb. For this reason, the two auxiliary programs `PPMbin` and `PPMcat` are also provided on the CD-ROM.

Using `PPMbin`, the most important information (Number, Position, Proper Motion, and Magnitude) are read from the Catalogue and stored in a binary file `PPM.bin`:

```
C> PPMbin D:/APCe/Data/PPM/PPM.dat PPM.bin
```

```
     PPMbin: Conversion of the PPM Catalogue (ASCII->binary)
          (c) 1999 Oliver Montenbruck, Thomas Pfleger

 **********************************************
 468861 stars processed
```

This requires less space and may be processed correspondingly faster. While the program is running, an asterisk (∗) is output to the screen for every 10 000 lines of data processed, so that even on slow computers the progress of the task may be followed. The compression normally takes several minutes to complete. To make things easier, previously prepared versions of `PPM.bin` for use with both the Windows and Linux operating systems are also available on the CD-ROM.

Finally, the program `PPMcat` allow all PPM stars within the predetermined field of view to be selected from the catalogue.

```
C> PPMcat D:/APCe/Data/PPM/Win32/PPM.bin

       PPMcat: Selection of stars from the PPM Catalogue
          (c) 1999 Oliver Montenbruck, Thomas Pfleger

 Image centre
   Right ascension [hh mm.m] ... 12 28.75
   Declination [deg]         ... 44.0
 Radius [deg]                ... 1.0
 Epoch [yyyy.y]              ... 1982.7

 PPM Number        RA           Dec        PM(RA)  PM(Dec)    m
              h  m  s        o  '  "       s/cy    "/cy     mag
 PPM  52981  12 25 45.107  +44 44 03.74  -0.730    -1.90    8.2
 PPM  52983  12 25 53.345  +44 37 34.07  +0.120    -3.60   11.6
...
 PPM  53039  12 31 39.100  +44 04 05.16  -0.140    -0.30   10.9
 PPM  53044  12 31 41.746  +43 13 06.26  -0.440    -1.90   10.6
 PPM  53046  12 31 59.476  +43 28 58.62  -0.260    +0.10   10.6
 PPM  53051  12 32 21.186  +44 40 54.45  -0.370    -1.20    9.7
 PPM  53062  12 33 39.312  +44 05 47.71  -1.300    -6.90    7.9
 PPM  53065  12 34 00.512  +44 09 06.19  -0.220    -1.50   10.2
 PPM 400176  12 28 04.711  +44 47 39.35  -1.800    -2.00    7.5
```

The output format is chosen so that the coordinates may be further processed by the `Foto` program in a simple manner. For this reason, when `PPMcat` is run, the name of an output file may also be entered (see Sect. A.1.3). The source code for `PPMcat` is contained, together with all the other programs, on the CD-ROM accompanying this book.

Appendix

A.1 The Accompanying CD-ROM

A.1.1 Contents

The accompanying CD-ROM contains the complete source code for all the programs described in this book, together with the corresponding input data files. For IBM-compatible PCs there are also executable versions of the programs, which may be run directly under the Windows 98/2000/XP and Linux operating systems. The following table gives a summary of the individual directories (folders) and files. The letter designating the CD-ROM drive (here D:), may, of course vary depending on the PC's configuration.

Folder/Directory	Description
`D:\APCe\Win32\`	Executable programs (Windows 32-bit Version) `AOEcat.exe`, `Coco.exe`, `Comet.exe`, `Eclipse.exe`, `EclTimer.exe`, `Foto.exe`, `Gauss.exe`, `Luna.exe`, `Numint.exe`, `Occult.exe`, `Phases.exe`, `Phys.exe`, `Planpos.exe`, `Planrise.exe`, `PPMbin.exe`, `PPMcat.exe`, `Sunset.exe`
`\Linux\`	Executable programs (Linux Version) `AOEcat`, `Coco`, `Comet`, `Eclipse`, `EclTimer`, `Foto`, `Gauss`, `Luna`, `Numint`, `Occult`, `Phases`, `Phys`, `Planpos`, `Planrise`, `PPMbin`, `PPMcat`, `Sunset`
	Unix archives of data files, executable programs (Linux only) and source code `apcedat.tar`, `apceexe.tar`, `apcesrc.tar` (incl. makefile)
`\Source\`	Source code for the library modules (header and implementation files) `APC_Cheb.h`, `APC_Cheb.cpp`, `APC_Const.h`, `APC_DE.h`, `APC_DE.cpp`, `APC_IO.h`, `APC_IO.cpp`, `APC_Kepler.h`, `APC_Kepler.cpp`, `APC_Math.h`, `APC_Math.cpp`, `APC_Moon.h`, `APC_Moon.cpp`, `APC_Phys.h`, `APC_Phys.cpp`, `APC_Planets.h`, `APC_Planets.cpp`, `APC_PrecNut.h`, `APC_PrecNut.cpp`, `APC_Spheric.h`, `APC_Spheric.cpp`, `APC_Sun.h`, `APC_Sun.cpp`, `APC_Time.h`, `APC_Time.cpp`, `APC_VecMat3D.h`, `APC_VecMat3D.cpp` `GNU_iomanip.h`

Source code of the application programs

`AOEcat.cpp`, `Coco.cpp`, `Comet.cpp`, `Eclipse.cpp`, `EclTimer.cpp`, `Foto.cpp`, `Gauss.cpp`, `Luna.cpp`, `Numint.cpp`, `Occult.cpp`, `Phases.cpp`, `Phys.cpp`, `Planpos.cpp`, `Planrise.cpp`, `PPMbin.cpp`, `PPMcat.cpp`, `Sunset.cpp`

`D:\APCe\Data\` Input data files

`Comet.dat`, `Foto.dat`, `Gauss.dat`, `Numint.dat`, `Occult.dat`, `Orkisz.dat`, `ZC.dat`

Asteroid Orbital Elements database and documentation

`\Aoe\astorb.dat`, `\Aoe\astorb.html`

Position and Proper Motion Catalogue and documentation

`\Ppm\PPM.dat`, `\Ppm\PPMbss.txt`, `\Ppm\PPMnorth.txt`, `\Ppm\PPMsouth.txt`, `\Ppm\PPMsupp.txt`, `\Ppm\ReadMe.txt`, `\Ppm\Win32\PPM.bin`, `\Ppm\Linux\PPM.bin`

The `\Win32\*.exe` programs and the `\Data\Ppm\Win32\Ppm.bin` binary files are solely for use under Microsoft Windows (32-bit version for Intel processors). Likewise the binary file `\Data\Ppm\Linux\PPM.bin` and the programs in the `\Linux\` directory may be used only on Intel x86 PCs running Linux. All files having extensions `*.h`, `*.cpp`, and `*.dat`, however, are ASCII text files, which may also be read and processed under other computer architectures and operating systems (such as Unix, Macintosh, and VMS). Here, however, the file and directory names may be translated in accord with the specific operating-system conventions.

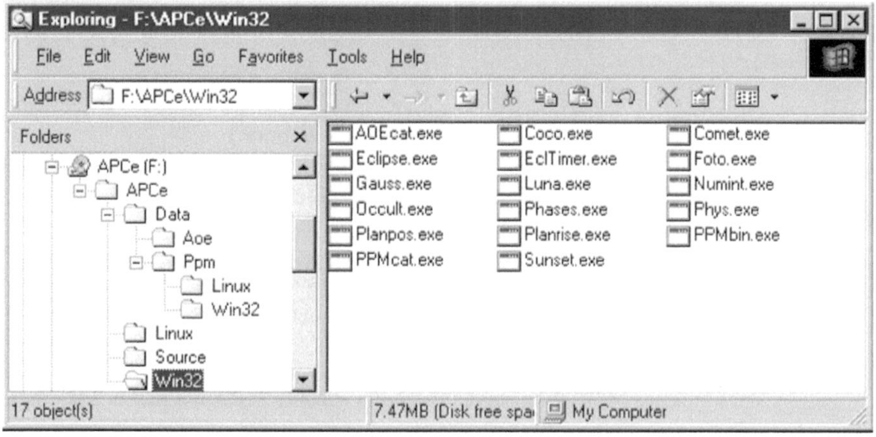

Fig. A.1. Directory tree of the book's CD-ROM as displayed by Windows Explorer

A.1.2 System Requirements

For optimum performance of the programs in this book we would recommend installation on a PC with the following specification:

- Intel Pentium Processor 500 MHz or similar
- Memory 128 Mb
- Free hard-disk space 10–150 Mb
- CD-ROM drive
- Windows 98/2000/XP or SuSE Linux Version 8.1 operating system
- Optional: Microsoft Visual C++ Version 6.0 or GNU C++ 3.2

The higher requirements are particularly recommended to users who want to compile the source code of the programs themselves, or want to use them as the basis for building new programs. The larger memory and hard-disk storage recommendations arise here because of the requirements of the C++ compiler's development environment and also the demands of temporary data files (such as for classes and browser information).

A smaller PC system (Intel 486, 50 MHz, 8Mb) suffices for simply running the programs provided. The installation of a 32-bit operating system (Windows or Linux) is, however, essential. Under DOS (16-bit) or Windows 3.1, the programs provided (`*.exe`) will not run. If hard-disk space is limited, the programs may also be run directly from the CD-ROM.

In principle, various common compilers and development environments may be used (see Sect. A.2.2 and A.2.3), to modify the programs and to develop individual applications. Because of the various differences in implementation that exist, small alterations may, however, be required in the source code. Under Visual C++ 6.0, which was used for the development of our programs, all of them may be compiled without any modifications.

Apart from Windows and Linux machines, the source code and input data files may also be read and processed on many other computers, provided these support the ISO 9660 format (with extensions) of the CD ROM. Because of the multiplicity of possible systems, we are unable to give here general and definite information about suitable compilers and development environments. Specific alterations to the source code may also be required.

It should be noted that the file names on the CD-ROM may not be reproduced correctly on other computers, but are translated differently depending on the operating system (e.g., upper-case letters only). For Unix systems, in particular, it is therefore recommended that the source code should be extracted from the tar archives in the `Linux` directory, because in this way the correct transcription will be retained.

A.1.3 Executing the Programs

To run the programs on a PC with the Windows operating system, the files in the `D:\APCe\Win32` folder on the CD-ROM should be copied to an appropriate folder (e.g., `C:\APCe\Exe`) on the hard disk. The input data files from the folder `D:\APCe\Data` should then be copied to the same folder. After the command interpreter has been started (by switching to DOS mode or running `command.exe` under Windows 98, `cmd.exe` under Windows 2000/XP) and switching to the chosen program folder, the individual programs may be run as in the following example:

```
C:\>cd C:\APCe\Exe

C:\APCe\Exe>Sunset

      SUNSET: solar and lunar rising and setting times
        (c) 1999 Oliver Montenbruck, Thomas Pfleger

 First date (yyyy mm dd)              ...
```

In the case of many programs, whenever it is sensible or essential, the input or output files may be included in the command line (see Table A.1).

Table A.1. Calling programs with optional arguments

Name	Argument(s)	Normal input	Normal output
AOEcat	[<in-file> [<out-file>]]	astorb.dat	Screen
Comet	[<in-file> [<out-file>]]	Comet.dat	Screen
Eclipse	[<out-file>]		Screen
Foto	[<in-file> [<out-file>]]	Foto.dat	Screen
Gauss	[<in-file> [<out-file>]]	Gauss.dat	Screen
Numint	[<in-file> [<out-file>]]	Numint.dat	Screen
Occult	[<in-file> [<out-file>]]	Occult.dat	Screen
PPMbin	[<in-file> [<out-file>]]	PPM.dat	PPM.bin
PPMcat	[<in-file> [<out-file>]]	PPM.bin	Screen

The advice just given applies in a similar manner to running the programs under Linux. After creating a suitable directory and copying the files from the CD-ROM, the programs may be started by entering the corresponding program names, provided the working directory is contained in the path. As under Windows, input and output files may be specified in the command line under Linux.

If the names of the programs and input data files are not correctly reproduced under Linux upon reading the CD, there are two Unix tar archives available. These may be extracted with the commands

```
tar -xvf /cdrom/APCe/Linux/apcedat.tar
tar -xvf /cdrom/APCe/Linux/apceexe.tar
```

and copied into the current working directory.

A.2 Compiling and Linking the Programs

The following sections describe the steps required to compile and link the individual programs, and also give some advice about any possible necessary modifications to suit various operating systems and compilers. Only the most important problems that may occur in compiling the programs are, however, discussed here. Further information may normally be found in the manuals that accompany the respective compilers.

A.2.1 General Advice on Computer-Specific Modifications

Although programs written in C++ can generally be employed on different computers, there are, however, some small differences between the various implementations of the language. Blame for this lies, among other things, with the introduction of new language elements (name space and exception handling, for example) and the extension of the standard C++ libraries within the framework of current efforts at standardization. Particularly affected here are <iostream> and related library modules, the differing versions of which under many compilers require small adaptations to the programs.

As an example, previous versions of the GNU C++ compiler (prior to gcc 3.0) do not support the entire `iomanip` module of the standard C++ library, which provides the output manipulators `right`, `fixed`, `showpos` and `noshowpos`. As a remedy, the corresponding functions

```
ostream& left (ostream& o){o.setf(ios::left ,ios::adjustfield); return o;};
ostream& right(ostream& o){o.setf(ios::right,ios::adjustfield); return o;};
ostream& fixed(ostream& o){o.setf(ios::fixed,ios::floatfield);  return o;};
ostream& showpos   (ostream& o){o.setf(ios::showpos); return o;};
ostream& noshowpos(ostream& o){o.unsetf(ios::showpos); return o;};
```

are defined in the file `GNU_iomanip.h`, which is included by the command

```
#include "GNU_iomanip.h"
```

into all source code.

On the other hand, new versions of the GNU C++ compiler do no longer support the `ostream_withassign` class, following an update of the language standard. This class earlier allowed a redirection of screen output to a file by assigning the `cout` stream at run time. The commands

```
  if (OutFile.is_open()) cout=OutFile;
  cout << ...;
```

must now be replaced by individual output commands

```
  if (OutFile.is_open()) {
    OutFile << ...;
  }
  else {
    cout << ...;
  };
```

for screen and file output. All programs provided on the CD have been reworked accordingly for the present edition of this book, but the previous, shorter notation is still employed in the printed listings.

Problems may also be arising from the machine and operating system architectures, and thus in the related system-specific ranges of values for the basic integer and floating-point data types. We assume a 32-bit operating system, which allows the use of 4-byte `int` numbers with a value range of −2 147 483 648

...+2 147 483 647, and 8-byte `double` numbers with 15 to 16 digits accuracy. Most of the functions and programs also run correctly, however, if the compiler supports 2-byte `int` numbers as standard. Only in the `CalDat` function is it absolutely essential to use the `long` data type instead of `int` to ensure that there is a sufficiently large range of values.

In general, adaptation of the tolerance limits for iterations is not required, because in the programs sensible values are derived from the relative machine accuracies. These are available in the `<limits>` module in the C++ standard library through the `numeric_limits<double>::epsilon` function. Alternatively, with older C++ implementations, the C-macro `DBL_EPSILON` from `<float.h>` may be employed, or through the use of the function

```
// GetAccuracy: Determines machine accuracy
double GetAccuracy() {
  double eps = 1.0e-6;
  while ( 1.0 + eps > 1.0 )  eps *= 0.5;
  return 2.0*eps;
}
```

one can determine for oneself the smallest `double` number that when added to 1.0 returns a value differing from 1.0. In the case of the normal IEEE data types with a 53-bit mantissa normally employed today, the value of the relative machine accuracy $u \approx 2.2\cdot 10^{-16}$.

A.2.2 Microsoft Visual C++ for Windows 98/2000/XP

In the following, we will describe the basic steps that are required to produce the executable programs with Microsoft Visual C++ 6.0, with reference to the English version of the development environment.

First, create the folder `C:\APCe\Source\` (with Windows Explorer, for example), and copy into it all the files from the similarly named folder on the book's CD-ROM. After calling the development environment (MS Visual Studio), the first step is to set up (via the menu item File.New.Workspaces) a new workspace to serve for the management of the individual programs and projects. Here, its designation (`Prj` is recommended) and its working directory (again `C:\APCe\Prj` is recommended) should be entered.

The first project to be created in the workspace just established will be a library of the general purpose modules that may be linked to the applications. This considerably simplifies development of the applications. With File.New.Projects create a project of type *Win32 Static Library*, giving it the name `APC_Lib`. Choose the option *Add to current workspace* and confirm with *OK* the creation of a workspace for this project.

Finally, the source files of the library modules are added to the `APC_Lib` project. For this, select Project.Add to Project.Files from the menu, and choose all the files from the folder `C:\APCe\Source\` that appear on the list of names for `APC_*.cpp`. Optionally, the header files (`*.h`) may also be selected. This last step is not absolutely essential, because the compiler itself will search the source folder for the header files. The library may now be generated with Build.Build. Thereupon, the development environment first compiles all the source code and finally links the resulting object files into a static library `APC_Lib.lib`.

The applications `AOEcat` to `Sunset` may each be created in an identical manner as independent projects within the `APCe\Prj` workspace. As an example we take `Coco`. Via File.New.Projects a project of type *Win32 Console Application* is selected. The project name `Coco` is entered in the appropriate field. As previously, in creating the `APC_Lib` project, choose the option *Add to current workspace*, and confirm with *OK*, to create the project. Now, via Project.Add To Project.Files..., add the file `Coco.cpp` from the `C:\APCe\Source` directory to the project, and select the menu item Project.Dependencies.... For the current project `Coco` select the option `APC_Lib`, to indicate to the development environment that it is necessary to link `Coco` to the chosen library. The executable program `Coco.exe` will be created by selecting Build.Build, and may, finally, be run using Build.Execute.

For further information we would refer you to the comprehensive documentation on the Microsoft C++ compiler and the Visual Studio environment.

A.2.3 GNU C++ for Linux

To compile and link the programs under the Linux operating system it is advisable to first combine all modules `APC_*` into a library `libAPC.a`:

```
> g++ -c APC*.cpp              # Compile all library modules
> ar rc  libAPC.a APC*.o       # Create the library from the object files
```

Assuming that all files are in the same directory, then the main program may be compiled and linked with the commands

```
> g++ AOEcat.cpp -o AOEcat -lAPC -L. # Compile and link AOEcat
> g++ Coco.cpp   -o Coco   -lAPC -L. # Compile and link Coco
> ...
> g++ Sunset.cpp -o Sunset -lAPC -L. # Compile and link Sunset
```

To simplify this process, the `apcesrc.tar` archive on the CD-ROM contains a makefile, which carries out the required steps upon invoking the `make` command.

In case the names indicated are not correctly reproduced when reading the CD-ROM under Linux or other Unix systems (if, for example, the CD-ROM driver does not support extensions to the ISO 9660 format), a Unix tar archive with all the source code is available as an alternative. This may be extracted with the command

```
tar -xvf /cdrom/APCe/Linux/apcesrc.tar
```

and copied into the current working directory.

A.3 List of the Library Functions

The APC_Lib library contains basic operators, functions and classes, which are not linked to any specific application program and which are therefore particularly suitable for the development of separate programs. The library contains a total of 14 modules, which each cover a specific range of tasks:

APC_Cheb.h	Chebyshev approximation of functions
APC_Const.h	Mathematical and astronomical constants
APC_DE.h	Numerical integration of differential equations
APC_IO.h	Auxiliary functions for input and output
APC_Kepler.h	Keplerian orbits
APC_Math.h	Technical and scientific functions
APC_Moon.h	Lunar coordinates
APC_Phys.h	Physical ephemerides
APC_Planets.h	Planetary coordinates
APC_PrecNut.h	Precession and nutation
APC_Spheric.h	Spherical astronomy
APC_Sun.h	Solar coordinates
APC_Time.h	Time and calendar calculation
APC_VecMat3D.h	Vector and matrix operations

Using the header files mentioned, the required modules may be incorporated into application programs in a simple manner. The respective implementations are contained in the corresponding APC_*.cpp files, which are separately compiled and may be linked as object or library files. The sole exception is the APC_Const module, which does not require any implementation and which simply consists of the header file.

The following listing gives a summary of all the names of public types, constants, functions, and classes that are available in the various modules.

<<	output operator for date, time, angles, vectors and matrices
[]	access to vector components (Cartesian, polar)
+	addition of vectors and matrices
+=	vector addition
-	subtraction of vectors and matrices
-	unary minus (vector, matrix)
-=	vector subtraction
*	multiplication (scalar, vector, matrix)
/	division by a scalar
Accumulate	method of the SolverLSQ class for the addition of a data equation
AddThe	addition theorem for angular functions
Angle	auxiliary class for the output of angles
AngleFormat	enumeration type for indicating format of angle output

APC_MaxFilename	string length for filenames in the command line
Arcs	arc-seconds per radian
AU	astronomical unit [km]
Bright	apparent magnitude of a planet
c_light	velocity of light [AU/d]
CalDat	calendar date and time
Cheb3D	Chebyshev approximation in three dimensions
Col	column vectors of a matrix
Cross	cross product
DateTime	auxiliary class for the output of date and time
Dd	enumerator of type AngleFormat for decimal representation
DDd	enumerator of type TimeFormat for time as a fraction of a day
Ddd	fractional degrees from minutes and seconds
DE_ACCURACY_NOT_ACHIEVED	enumerator of type DE_STATE (flag for accuracy requirement that is too high)
DE_DONE	enumerator of type DE_STATE (flag for successful integration step)
DE_INIT	enumerator of type DE_STATE (restart of the integrator)
DE_INVALID_PARAMS	enumerator of type DE_STATE (incorrect input)
DE_STATE	enumeration type for status code of the SolverDE integrator
DE_STIFF	enumerator of type DE_STATE (stiff differential equations suspected)
DE_TOO_MANY_STEPS	enumerator of type DE_STATE (too large a number of steps)
Deg	$180^\circ/\pi$
Direct	enumerator of type RotationType (direct rotation)
DMM	enumerator of type AngleFormat for output in degrees and whole arc-minutes
DMMm	enumerator of type AngleFormat for output in degrees and decimal arc-minutes
DMMSS	enumerator of type AngleFormat for output in degrees, minutes, and whole seconds of arc
DMMSSs	enumerator of type AngleFormat for output in degrees, minutes, and decimal seconds of arc
DMS	minutes and seconds from fractions of a degree
Dot	scalar product of two vectors
Earth	enumerator of type PlanetType (Earth)
EccAnom	eccentric anomaly for elliptical orbits
Ecl2EquMatrix	transformation matrix ecliptic → equatorial
Elements	orbital elements from position and velocity
Elements	orbital elements from two position vectors
Ellip	position and velocity for elliptical orbits

`Equ2EclMatrix`	transformation matrix equatorial → ecliptic
`Equ2Hor`	transformation equatorial → horizontal system
`EquStd`	standard coordinates from equatorial coordinates
`ETminUT`	approximation of the difference ET-UT
`FileExists`	check whether a file exists
`FindEta`	sector-triangle ratio
`Fit`	method of class `Cheb3D` for Chebyshev approximation of a function
`Frac`	decimal part of a number
`GaussVec`	transformation matrix orbital plane → ecliptic
`GetFilenames`	extraction of filenames from the command line
`GM_Sun`	product of gravitational constant × solar mass [AU^3/d^2]
`GMST`	Greenwich Mean Sidereal Time
`HHh`	enumerator of type `TimeFormat` for output of times in decimal hours
`HHMM`	enumerator of type `TimeFormat` for output of time in hours and minutes
`HHMMSS`	enumerator of type `TimeFormat` for output of time in hours, minutes and seconds
`Hor2Equ`	transformation horizontal → equatorial system
`HypAnom`	eccentric anomaly for hyperbolic orbits
`Hyperb`	position and velocity for hyperbolic orbits
`Id3D`	identity matrix (unit matrix)
`Illum`	illumination parameter for planets
`Init`	method of the class `SolverLSQ` to initialize an approximation problem
`Integ`	method of the class `SolverDE` to carry out an integration step
`Jupiter`	enumerator of type `PlanetType`
`JupiterPos`	heliocentric Jupiter coordinates
`Kepler`	position and velocity for Keplerian orbits
`KepPosition`	planetary position from Keplerian elements
`KepVelocity`	planetary velocity from Keplerian elements
`kGauss`	Gaussian gravitational constant [AU^3/d^2]
`Mars`	enumerator of type `PlanetType` (Mars)
`MarsPos`	heliocentric Mars coordinates
`Mat3D`	class of 3×3-matrices
`Mercury`	enumerator of type `PlanetType` (Mercury)
`MercuryPos`	heliocentric Mercury coordinates
`MiniMoon`	low-accuracy lunar coordinates
`MiniSun`	low-accuracy solar coordinates
`Mjd`	Modified Julian Date
`MJD_J2000`	Modified Julian Date for Epoch 2000
`Modulo`	modulo function

`MoonEqu`	high-accuracy equatorial lunar coordinates
`MoonPos`	high-accuracy ecliptic lunar coordinates
`Neptune`	enumerator of type `PlanetType` (Neptune)
`NeptunePos`	heliocentric Neptune coordinates
`None`	enumerator of type `TimeFormat` for output of the date without time
`Norm`	length of a vector
`NutMatrix`	nutation matrix
`Orient`	orientation of the planetocentric reference frame
`Parab`	position and velocity for parabolic orbits
`Pegasus`	root-finder using the Pegasus method
`PertPosition`	high-accuracy planetary positions
`phi`	enumerator of type `pol_index` for the azimuth of a vector
`pi`	π
`pi2`	2π
`PlanetType`	enumeration type for the Sun and planets
`Pluto`	enumerator of type `PlanetType` (Pluto)
`PlutoPos`	heliocentric Pluto coordinates
`pol_index`	enumeration type for polar components of `Vec3D` objects
`Polar`	auxiliary class for defining vectors from polar coordinates
`PosAng`	position angle
`PrecMatrix_Ecl`	precession matrix (ecliptic coordinates)
`PrecMatrix_Equ`	precession matrix (equatorial coordinates)
`Quad`	quadratic interpolation
`r`	enumerator of type `pol_index` for length of a vector
`R_Earth`	radius of the Earth [km]
`R_Moon`	radius of the Moon [km]
`R_Sun`	Solar radius [km]
`R_x`	matrix for simple rotation around the x-axis
`R_y`	matrix for simple rotation around the y-axis
`R_z`	matrix for simple rotation around the z-axis
`Rad`	$\pi/180°$
`Retrograde`	enumerator of type `RotationType` (retrograde rotation)
`Rotation`	planetographic coordinates
`RotationType`	enumerated type to indicate direction of rotation
`Row`	row vectors in a matrix
`Saturn`	enumerator of type `PlanetType` (Saturn)
`SaturnPos`	heliocentric Saturn coordinates
`Set`	method of the classes `Angle`, `Time` and `DateTime` to select output format
`Shape`	form and size of the planets
`Site`	coordinates of a point on the Earth's surface

`Solve`	method of the class `SolverLSQ` for back substitution
`SolverDE`	class for the numerical solution of differential equations
`SolverLSQ`	class for the solution of approximation problems
`StdEqu`	equatorial coordinates from standard coordinates
`Stumpff`	Stumpff functions
`Sun`	enumerator of type `PlanetType` (Sun)
`SunEqu`	equatorial solar coordinates
`SunPos`	ecliptic solar coordinates
`Sys_I`	enumerator of type `SystemType` (System I)
`Sys_II`	enumerator of type `SystemType` (System II)
`Sys_III`	enumerator of type `SystemType` (System III)
`SystemType`	enumeration type for the designation of the rotation system
`T_B1950`	Epoch B1950 (in Julian centuries since J2000)
`T_J2000`	Epoch J2000 (in Julian centuries since J2000)
`theta`	enumerator of type `pol_index` for the altitude of a vector
`Time`	auxiliary class for the output of times
`TimeFormat`	enumeration type for format entry of time and date output
`Transp`	transposed matrix
`Uranus`	enumerator of type `PlanetType` (Uranus)
`UranusPos`	heliocentric Uranus coordinates
`Value`	method of the class `Cheb3D` to evaluate a Chebyshev approximation
`Vec3D`	class of three-dimensional vectors
`Venus`	enumerator of type `PlanetType` (Venus)
`VenusPos`	heliocentric Venus coordinates
`x`	enumerator of type `index` for the x-coordinate of a vector
`xyz_index`	enumerated type for indexing the Cartesian components of `Vec3D` objects
`y`	enumerator of type `index` for the y-coordinate of a vector
`z`	enumerator of type `index` for the z-coordinate of a vector

Symbols

a	semi-major axis
a	station coefficient in longitude
A	azimuth
$\boldsymbol{A}$	coefficient matrix of a linear system of equations
b	ecliptic latitude
b	station coefficient in latitude
$\dot{b}$	change in ecliptic latitude with time
$\boldsymbol{b}$	data vector of a linear system of equations
B	ecliptic latitude of the Sun
c	velocity of light ($c = 299\,792\,458$ m/s)
$c_i(E)$	Stumpff functions
C	parallactic angle
d	diameter of the umbra on the Earth
$\boldsymbol{d}$	data vector of a system of linear equations in triangular form
$\boldsymbol{d}$	rotation axis unit vector
db	perturbation of ecliptic latitude
dl	perturbation of ecliptic longitude
dr	perturbation of distance
D	mean elongation of the Moon from the Sun
D	diameter of the penumbra on the Earth
e	eccentricity
$\boldsymbol{e}$	vector of unit length
E	eccentric anomaly
E	elongation
$\boldsymbol{E}$	transformation matrix
ET	Ephemeris Time
f	flattening of the Earth or a planet
f	half apex angle of umbral and penumbral cone
f	coordinate difference in the fundamental plane
F	general: gravitational attraction between two bodies
F	mean distance of the Moon from the ascending node of its orbit
F	focal length
g	coordinate difference in the fundamental plane
G	gravitational constant ($G = 2.959122083 \cdot 10^{-4}\mathrm{AU}^3 M_\odot^{-1}\mathrm{d}^{-2}$)

GMST	Greenwich Mean Sidereal Time
h	altitude
h	step size
H	eccentric anomaly for a hyperbolic orbit
i	orbital inclination to the ecliptic
i	phase angle
$\boldsymbol{i}$	unit vector in the fundamental plane (x-axis)
$\boldsymbol{j}$	unit vector in the fundamental plane (y-axis)
JD	Julian Date
k	Gaussian gravitational constant ($k = 0.01720209895$)
k	phase
k	ratio of lunar radius to radius of the Earth ($k \approx 0.2725$)
$\boldsymbol{k}$	unit vector perpendicular to the fundamental plane (z-axis)
l	ecliptic longitude
l	shadow radius in the fundamental plane
l	mean anomaly of the Moon
l'	mean anomaly of the Sun
$\dot{l}$	change in ecliptic longitude with time
L	ecliptic longitude of the Sun
L	shadow radius at the observer's site
L_0	mean longitude of the Moon
m	apparent magnitude
m	observer's distance from the shadow axis
M	mean anomaly
M	magnitude of a solar eclipse
M_{h}	mean anomaly (hyperbola)
$M_{\odot}$	mass of the Sun
MJD	Modified Julian Date
n	triangle area ratio
n	daily motion
$\boldsymbol{N}$	nutation matrix
p	semi-latus rectum
P	position angle of the contact points of a solar eclipse
$\boldsymbol{P}$	Gaussian vector (in direction of perihelion)
$\boldsymbol{P}$	precession matrix
q	perihelion distance
$\boldsymbol{Q}$	Gaussian vector (perpendicular to perihelion)
r	general: distance
$\boldsymbol{r}$	general: position vector
$\dot{r}$	change in distance
R	Earth-Sun distance
R	radius of a celestial body
R	reduction to the ecliptic
R	refraction

$\boldsymbol{R}$	Gaussian vector perpendicular to the orbital plane
$\boldsymbol{R}$	vector of the geocentric position of the Sun
$\boldsymbol{R}$	upper triangular matrix
$\boldsymbol{R}_{x,y,z}$	elementary rotation matrices
S	area of sector
t	general: time
T	general: time in Julian centuries since J2000
T	orbital period
T_0	time of perihelion passage
$T_n(x)$	Chebyshev polynomial of the n-th order
ΔT	Ephemeris Time − Universal Time difference
TDB	Barycentric Dynamical Time
TDT	Terrestrial Dynamical Time
u	argument of latitude
U	auxiliary variable for near-parabolic orbits
$\boldsymbol{U}$	transformation matrix
UT	Universal Time
UTC	Coordinated Universal Time
$\boldsymbol{v}$	general: velocity vector
V	position angle of the contact points of a solar eclipse
$V(1, 0)$	visual magnitude at unit distance
W	orientation of the prime meridian
x, y, z	general: Cartesian coordinates
$\boldsymbol{y}$	state vector
X, Y	standard coordinates (astrometry)
z	auxiliary angle in description of precession

α	right ascension
α_0	right ascension of the rotation axis
β	ecliptic latitude
γ	cosine of the elongation from the Sun
δ	declination
δ_0	declination of the rotation axis
Δ	geocentric distance
Δ	area of triangle
ε	obliquity of the ecliptic
$\Delta\varepsilon$	nutation in ecliptic latitude
ζ	auxiliary angle in description of precession
η	sector-triangle ratio
ϑ	general: spherical coordinate
ϑ	auxiliary angle in description of precession
ϑ	position angle
Θ	local sidereal time

Θ_0	Greenwich sidereal time
λ	ecliptic longitude
λ	geographical longitude (positive eastwards)
λ	planetographic longitude (positive opposite to the rotation)
λ'	planetocentric longitude (positive towards east)
Λ	auxiliary angle in description of precession
μ	auxiliary quantity for expressing the triangle area ratio
ν	true anomaly
π	3.1415926…
π	equatorial horizontal parallax
π	auxiliary angle in description of precession
Π	auxiliary angle in description of precession
ϖ	longitude of perihelion
ρ	geocentric distance
σ	parameter of the Gauß-Lagrangian equation
τ	hour angle
τ	interval
φ	general: spherical coordinate
φ	geographical (planetographic) latitude
φ'	geocentric (planetocentric) latitude
ξ	residuals square sum
$\Delta\psi$	nutation in ecliptic longitude
ω	argument of perihelion
Ω	longitude of ascending node

Ø	diameter
♈	vernal equinox (First Point of Aries)
⊕	Earth
⊙	Sun
^{cy}	century
^{d}	days
^{h}	hours
^{m}	minutes
^{m}	magnitudes
^{r}	revolutions
^{s}	seconds
$^{\circ}$	degrees

Glossary

Aberration: The shift of the apparent position of a celestial body with respect to the geometric position, caused by the finite speed of light. (→stellar aberration; →light-time).

Altitude: In spherical astronomy, the angle between a celestial body and the →horizon, as measured by an observer at the surface of the Earth. The actual and the observed altitude of a body differ, because of →refraction, by up to 35′.

Apparent Coordinates: The coordinates of a celestial body that are required, for example, for use with setting circles on a telescope. Apparent coordinates refer to the actual orientation of Earth's axis, and include corrections for →precession and →nutation. In addition, →stellar aberration is taken into account, and, in the case of Solar-System bodies, →light-time.

Ascending Node: The point on an orbit at which the celestial body crosses the ecliptic from south to north.

Astrometric Coordinates: Coordinates to be compared with catalogued stellar positions. Astrometric coordinates generally relate to the mean →vernal equinox of a standard epoch (→J2000, →B1950), and take into account →light-time in the case of planets or comets.

Astrometry: The measurement of stellar positions using setting circles or photographic plates.

Astronomical Unit: A unit of length that is used to specify distances within the Solar System. One astronomical unit (AU) is approximately equal to the mean distance between the Earth and the Sun, and amounts to approximately 149.6 million km.

Azimuth: One of the coordinates in the →horizontal system. Azimuth is the angle, measured positive towards the west, between south and the direction in which an object is observed.

B1950: The beginning of the Besselian year 1950 (B1950 = JD 2433282.423 = Jan $0.^{d}923$, 1950). A reference epoch that has been used for a very long time, but which has now been superseded by the standard epoch of →J2000.

Central Meridian: The meridian passing through the centre of the apparent disk of a planet as seen from Earth..

Culmination: The moment at which a celestial body reaches its greatest (or least) altitude above the →horizon. Culmination occurs when a body crosses the →meridian.

Declination: The angle, at right-angles to the →equator, measured between the equator and a celestial body. Together with →right ascension, declination forms the equatorial system of coordinates. (→equinox)

Dynamical Time (TDB, TDT): A physical measure of time, such as that obtained by the use of atomic clocks, for example. Dynamical Time is, like →Ephemeris Time, a uniform measure of time, but which takes into account the relationship between time and space required by relativity theory. While Barycentric Dynamical Time (TDB) gives the time that would be measured by an observer at the Solar System's centre of gravity, Terrestrial Dynamical Time (TDT) gives the time that would be shown by a clock at the centre of the Earth. Both definitions of time differ between themselves and from →Ephemeris Time by just a few milliseconds. In 1991 it was agreed that Terrestrial Dynamical Time (TDT) should just be called →Terrestrial Time (TT).

Ecliptic: The imaginary great circle, representing the intersection of the plane of the Earth's orbit with the celestial sphere. As seen from the Earth, the Sun appears to travel along the ecliptic in the course of a year. The ecliptic serves as reference plane for the ecliptic system of coordinates, which uses the coordinates ecliptic longitude and ecliptic latitude (→ecliptic coordinates).

Ecliptic Coordinates: The position of a body established relative to the →ecliptic and the equinox of some reference epoch, using the coordinates longitude and latitude. Ecliptic longitude is measured from the →vernal equinox, positive towards the east. Ecliptic latitude is the angle, measured perpendicular to the ecliptic, between the body and the ecliptic.

Elongation: The angle between two bodies, as seen by an observer.

Ephemeris: A table giving the coordinates of a planet, comet or minor planet for a specific period of time.

Ephemeris Time (ET): A uniform measure of time, which is used to calculate the coordinates of a planet, comet, or minor planet. Ephemeris Time was introduced to be independent of the irregular, unpredictable variations in the rotation of the Earth, which forms the basis for measuring →Universal Time. The difference between Universal Time (UT) and Ephemeris Time (ET) was, by definition, zero at the beginning of the 20th century, and now amounts to about one minute. Within the framework of relativistic theories of motion, Ephemeris Time has been replaced by →Terrestrial Time (TT).

Equator: An imaginary great circle on the celestial sphere, which is perpendicular to the Earth's axis of rotation. The equator separates the northern and southern celestial hemispheres, and is simultaneously the reference plane for the equatorial system of coordinates, which uses the coordinates →right ascension and →declination.

Equatorial Coordinates: Coordinates referred to the equator (→right ascension, →declination).

Equinox: →vernal equinox.

Geocentric Coordinates: Coordinates referred to the centre of the Earth.

Geographic Coordinates: Two values (geographic longitude and geographic latitude), which specify the position of a point on the Earth's surface. The reference plane is the terrestrial equator, and longitude is, by international agreement, measured positively eastwards from the Greenwich →meridian.

Heliocentric Coordinates: Coordinates referred to the centre of the Sun.

Heliographic Coordinates: Coordinates for uniquely describing a point on the surface of the Sun, defined by analogy with →geographic and →planetographic coordinates.

Horizon: The imaginary line of intersection between a plane tangent to the surface of the Earth at the observer and the celestial sphere. The horizon is the reference plane for the →horizontal system with coordinates →azimuth and →altitude.

Horizontal System: A coordinate system related to the local horizon of an observer, and where the coordinates used are →azimuth and →altitude. The horizontal system is used, for example, in calculating the rising and setting times of celestial bodies. It is, however, not suitable for ephemerides in an almanac, because the azimuth and altitude of a body depend on the observing site, and constantly alter because of the rotation of the Earth.

Hour Angle: Difference between the local →sidereal time and the →right ascension of a star. The hour angle measures the sidereal time that has passed since the last →culmination.

J2000: A standard epoch, which is widely used in astronomy. It represents midday on 2000 January 1 (2000 Jan. 1.5 = JD 2451545.0). The letter 'J' indicates that it is a Julian standard epoch. Successive Julian epochs normally differ by a full Julian century, consisting of 36525 days (e.g. J1900 = JD 2415020.0). (→B1950)

Kepler problem: Another description for the →two-body problem.

Light-Time: During the time that light requires to cross the distance to the Earth (499 seconds per astronomical unit), a planet moves slightly farther along its orbit. The observed position therefore does not correspond to the actual point at which the planet was located at the time of observation.

Many-Body Problem: The task of calculating the motion of more than two bodies, taking into account their mutual gravitational attraction. Whereas the two-body allows for a closed-form solution, this is not generally the case with the many-body problem. Numerical methods are used to treat it, together with analytical approximations (series expansions).

Meridian: In astronomy, this is used to designate the great circle on the celestial sphere that passes through the observer's →zenith and intersects the →horizon at the north and south points (→culmination). In geography, the term 'meridian' means a great circle on the surface of the Earth that lies in the plane of the Earth's axis. All meridians pass through both of the Earth's poles and intersect the equator at a right angle. Meridians serve to define the geographical longitude of a point on the Earth.

Nutation: An oscillation of the Earth's axis about its mean position, which is superimposed on →precession. The period of nutation is 18.6 years, and is determined by the period for one rotation of the Moon's ascending node.

Parallax: The apparent alteration in the position of a celestial body relative to the background of fixed stars, and caused by changes in the observer's position. Parallax is greatest for the nearest celestial body, and is therefore most noticeable with the Moon and (to a far lesser extent) with comets and minor planets that pass close to the Earth (→topocentric coordinates).

Perifocal coordinates: Position of a celestial body relative to its orbital plane and the line of apsides.

Perigee: The closest point of the Moon's or a satellite's orbit to the Earth.

Perihelion: The closest point of a planetary or cometary orbit to the Sun.

Physical Ephemerides: Quantities describing the appearance of a planet as seen through the telescope, including e.g. the apparent diameter, the brightness, the location of the →central meridian and the →position angle of the rotation axis.

Planetographic Coordinates: Angles describing the location of a point on the surface of a planet, corresponding to →geographic coordinates on Earth.

Position Angle: The apparent angle on the celestial sphere between a reference direction and a given direction (e.g. pointing towards the Sun or the pole of a planet). Usually, the position angle is measured in degrees from North to East.

Prime Meridian: The reference meridian for determining longitude in the system of →planetographic coordinates.

Precession: The long-term shift in the positions of the ecliptic and the celestial equator. Because of perturbing forces exerted by the Sun and the Moon, the Earth's axis is not stationary in space. It sweeps out the circumference of a cone, centred on the mean pole of the ecliptic, with a period of 26000 years. As a result, the orientation of the celestial equator also changes. The forces exerted by the other planets also lead to a slow shift in the position of the ecliptic. The motion of the equator, →ecliptic, and →vernal equinox means that in giving the →equatorial and →ecliptic coordinates of a body the reference date must always be specified. Commonly the coordinates are referred to the equinox of date, the equinox of →B1950, or the equinox of →J2000.

Refraction: The deflection of rays of light in the Earth's atmosphere. A celestial body appears at a slightly greater altitude to an observer on the ground than it would without the effect of refraction. Close to the horizon, a ray of light has to pass through a particularly long portion of the atmosphere. Refraction is therefore most noticeable when a celestial body is rising or setting.

Right Ascension: One of the coordinates in the →equatorial system of coordinates. Right ascension is measured positive eastwards along the →equator from the →vernal equinox. The resulting angle is normally expressed as time ($15° \equiv 1^h$), subdivided into hours, minutes and seconds.

Semi-diurnal Arc: Half of a star's apparent arc above the →horizon. When measured in hours, the semi-diurnal arc gives the →sidereal time between rising and →culmination and between culmination and setting.

Sidereal Time: The right ascension of a star that appears exactly on the observer's meridian. One sidereal day corresponds to the time taken for the Earth to rotate once, and lasts approximately 23^h56^m Universal Time.

Stellar Aberration: The difference between the direction of a ray of light as seen by an observer stationary with respect to the Sun, and one moving with the Earth.

Terrestrial Time (TT): Terrestrial Time is the relativistic timescale that replaced →Ephemeris Time (ET) as the time reference for apparent geocentric ephemerides (→Dynamical Time). For practical purposes both time scales may be considered to be equivalent.

Topocentric Coordinates: Coordinates related to the position of the observer on the surface of the Earth. Topocentric and →geocentric coordinates differ by the amount of →parallax.

Two-Body Problem: The task of calculating the motion of two bodies under the influence of their mutual gravitational attraction. The two-body problem is a simplified representation of the motion of a planet, comet, or minor planet around the Sun, where perturbations caused by other bodies in the Solar System are neglected. The importance of the two-body problem for the calculation of ephemerides is that it is relatively simple to solve mathematically.

Universal Time: A measurement of time related to the rotation of the Earth, which forms the basis for civil time. Universal Time may be determined by the observation of the meridian transit of celestial bodies of known →right ascension. Because the rotation of the Earth is subject to irregular variations, Universal Time is not a uniform time scale (→Ephemeris Time, →Dynamical Time).

Vernal Equinox (First Point of Aries): The point of intersection of the →ecliptic and the →equator, at which the Sun crosses the equator from south to north, during its yearly passage along the ecliptic. This currently occurs about March 21 each year. The exact instant of this event defines the beginning of spring. The vernal equinox provides the reference point for the measurement of ecliptic longitude (→ecliptic coordinates) and →right ascension. Because the positions of the ecliptic and equator alter over the course of time through the effects of →precession, the date to which the coordinates refer must always be stated. Most frequently the equinox of date, the equinox of →B1950, and the equinox of →J2000 are employed.

Zenith: The imaginary point on the celestial sphere directly above the observer. The →altitude of the zenith therefore amounts to 90°.

Bibliography

The following listing contains a selection of basic and more comprehensive works on spherical astronomy and celestial mechanics, together with various items on mathematics and programming. They should serve to help readers explore more deeply various individual points that cannot be treated in detail here. Where applicable, supplementary references to world wide web sites are also provided. By their very nature, however, Internet addresses tend to change frequently. Readers are therefore advised to employ any of the available search engines (Altavista, Lycos, Yahoo, etc.), if they encounter an outdated reference.

General Works

[1] *Astronomical Almanac*; U. S. Government Printing Office, Her Majesty's Stationery Office; Washington, London.

The leading English-language almanac. The 1984 edition contains a summary of the numerical values and constants (sidereal time, precession and nutation, planetary masses and diameters, elements for physical ephemerides, etc.), that had just been recommended by the IAU for astronomical calculations.

[2] H. Bucerius, M. Schneider; *Himmelsmechanik I-II*; Bd. 143/144, Bibliographisches Institut; Mannheim (1966).

A modern description of celestial mechanics. Among other subjects discussed are general perturbation theory, lunar theory, and orbital determination.

[3] L. E. Doggett, G. H. Kaplan, P. K. Seidelmann; *Almanac for Computers for the Year 19××*; Nautical Almanac Office, United States Naval Observatory; Washington.

This yearly publication contains descriptions of the coordinates of the Sun, Moon and planets in terms of Chebyshev polynomials, which allow simple and accurate calculation of any required ephemerides, even on small computers. The introductory chapter gives advice, and simple formulae for calculation of astronomical phenomena.

[4] P. K. Seidelmann (ed.); *Explanatory Supplement to the Astronomical Almanac*; University Science Books (1992).

Supplements the yearly *Astronomical Almanac*, with details of the fundamental data and calculation methods used. It supersedes the *Explanatory Supplement to the Astronomical Ephemeris and the American Ephemeris and Nautical Almanac*, published by the U. S. Government Printing Office, Washington, and Her Majesty's Stationery Office, London, 1974, but lacks some of the material given there.

[5] R. M. Green; *Spherical Astronomy*; Cambridge University Press; Cambridge (1985).

The modern standard work on spherical astronomy.

[6] A. Guthmann; *Einführung in die Himmelsmechanik und Ephemeridenrechung*; Spektrum Akademie Verlag (ehem. BI Wissenschaftsverlag), Heidelberg (1994).

Introductory and practically oriented presentation of celestial mechanics and ephemeris calculation.

[7] D. McNally; *Positional Astronomy*; Muller Educational; London (1974).

Introduction to spherical astronomy.

[8] J. Meeus; *Astronomical Algorithms*; Willmann-Bell; Richmond, Virginia (1991).

Revised and greatly extended edition of *Astronomical Formulae for Calculators*. Practically orientated collection of interesting and unusual calculation procedures for amateur astronomers. Comprehensive examples, which also can be implemented on small computers. In addition floppy disks with Basic, Pascal or C programs are available.

[9] J. Meeus; *Astronomical Tables of the Sun, Moon and Planets*; Willmann-Bell; Richmond, Virginia (1983).

Comprehensive collection of interesting astronomical data and results include: phases of the Moon, solar and lunar eclipses, planetary conjunctions, occultations of planets by the Moon, etc.

[10] I. I. Mueller; *Spherical and practical astronomy*; Frederick Ungar Publishing Co.; New York (1969).

Description of spherical astronomy and its applications to geodesy. Among other topics covered are: the basis for the various definitions of time and their practical application; and the prediction and reduction of solar eclipses and lunar occultations.

[11] W. H. Press, S. A. Teukolsky, W. T. Vetterling, B. P. Flannery; *Numerical Recipes in C*; Cambridge University Press; Cambridge, 2nd ed. (1993).

A comprehensive collection of important procedures covering all fields of mathematical computation. Editions for Fortran and Pascal are also available. Source codes are available in a variety of media formats.

[12] A. E. Roy; *Orbital Motion*; Adam Hilger Ltd.; Bristol, 2nd ed. (1982).

General introduction to celestial mechanics.

[13] H. R. Schwarz; *Numerische Mathematik*; B. G. Teubner Verlag; Stuttgart, 2. Aufl. (1988).

Practically orientated description of important algorithms in mathematical computation.

[14] K. Stumpff; *Himmelsmechanik I-III*; VEB Deutscher Verlag der Wissenschaften; Berlin (1959, 1965, 1974).

Comprehensive discussion of celestial mechanics.

[15] B. Stroustrup; *The C++ Programming Language*; Addison Wesley; 3rd edition (1997).

Comprehensive text book written by the developer of C++. Special consideration is given to the current language standard and the Standard Template Library.

Chapter 2 (Coordinate Systems)

[16] J. H. Lieske, T. Lederle, W. Fricke, B. Morando; *Expressions for the Precession Quantities Based upon the IAU (1976) System of Astronomical Constants*; Astronomy and Astrophysics, vol. 58, pp. 1-16 (1977).

A list of the various auxiliary angles required to describe precession in ecliptic and equatorial coordinates. The numerical values are based upon the IAU's 1976 system of astronomical constants.

[17] G. Moyer; *The Origin of the Julian Day System*; Sky and Telescope, vol. 61, pp. 311-313 (April 1982).

Article on the history of the Julian Date, with formulae to convert between the running date number and the ordinary calendar.

Chapter 3 (Rising and Setting Times)

[18] L. D. Schmadel, G. Zech; *Empirical transformations from UT to ET for the period 1800-1988*; Astronomische Nachrichten, vol. 309, pp. 219-221 (1988).

Representation of the observed difference between Universal Time and Ephemeris Time by a 12th-order polynomial, with validity over a period of almost 200 years.

Algorithms for the calculation of rising and setting of objects and for handling refraction are given in *Almanac for Computers* [3].

Chapter 4 (Cometary Orbits)

[19] J. M. A. Danby, T. M. Burkhardt; *The Solution of Kepler's Equation I-II*; Celestial Mechanics, vol. 31, pp. 95-107, pp. 317-328 (1983);

Discussion of various iterative procedures for solving the Kepler equation and the corresponding starting values.

[20] B. Marsden; *Catalogue of Cometary Orbits*; Smithsonian Astrophysical Observatory; Cambridge, Mass.;

Regularly revised listing of osculating elements for all known comets.

[21] *Ephemerides of Minor Planets*; Institute for theoretical Astronomy; St. Petersburg.

Orbital elements and ephemerides for the known minor planets. Appears yearly.

[22] Yu. V. Batrakov, V. A. Shor; *Catalogue of Orbital elements and Photometric parameters of 7316 Minor Planets numbered by 25 November, 1996*; Institute for Theoretical Astronomy; St. Petersburg, Russia (1997).

Current orbital elements of the most important catalogued minor planets. Available from the Centre de Données astronomiques des Strasbourg under
`ftp://cdsarc.u-strasbg.fr/cats/I/245`

Methods for handling near-parabolic orbits after Stumpff are described in *Himmelsmechanik I* [14]. A comprehensive database of minor planets orbital elements is provided by the Lowell observatory under `ftp://ftp.lowell.edu/pub/elgb/astorb.html`.

Chapter 5 (Perturbations)

[23] L. F. Shampine, M. K. Gordon; *Computer Solution of Ordinary Differential Equations*; Freeman and Comp., San Fransisco (1975).

Introduction to the theory of multistep methods for the numerical solution of ordinary differential equations and description of the variable order variable step size Adams method DE.

[24] X. X. Newhall, E. M. Standish Jr., J. G. Williams; *DE102: a numerically integrated ephemeris of the moon and planets spanning forty-four centuries*; Astronomy and Astrophysics, vol. 125, pp. 150-167 (1983);

The description of what was then the best ephemeris over a period of several thousand years. The described algorithms apply likewise for later releases of the JPL Development Ephemerides. Current versions (DE200, DE405, etc.) are available from
`http://ssd.jpl.nasa.gov/eph_info.html`

The most recent orbital elements of minor planets are published annually in the *Ephemerides of Minor Planets* ([21], see also [22]) of the St. Petersburg Institute of Theoretical Astronomy as well as in various astronomical yearbooks.

Chapter 6 (Planetary Orbits)

[25] P. Bretagnon, J.-L. Simon; *Tables for the motion of the sun and the five bright planets from -4000 to +2800; Tables for the motion of Uranus and Neptune from +1600 to +2800*; Willmann-Bell; Richmond, Virginia (1986).

Series expansions and programs, with which the motion of the inner planets during historical times can be calculated with sufficient accuracy. The coordinates of the outer planets are represented by polynomials.

[26] T. C. van Flandern, K. F. Pulkkinen; *Low precision formulae for planetary positions*; Astrophysical Journal Supplement Series, vol. 41, p. 391; (1979).

Series expansions for the calculation of low-accuracy (ca. $1'$) heliocentric and geocentric planetary coordinates.

[27] E. Goffin, J. Meeus, C. Steyaert; *An accurate representation of the motion of Pluto*; Astronomy and Astrophysics, vol. 155, p. 323; (1986).

Representation of Pluto's orbit by a trigonometric series, which was obtained by fitting a numerically integrated ephemeris.

[28] G. W. Hill; *Tables of Jupiter, Tables of Saturn*; Astronomical Papers of the American Ephemeris, vol. VII, part 1-2; Washington (1898).

Analytical series expansions of the motions of Jupiter and Saturn.

[29] M. P. Jarnagin, jr.; *Expansions in elliptic motion*; Astronomical Papers of the American Ephemeris, vol. XVIII; Washington(1965).

Series expansions for the equation of the centre and radius, as well as other coordinates in the non-perturbed Kepler problem to high orders of eccentricity.

[30] S. Newcomb; *Tables of the motion of the Earth, Tables of the heliocentric motion of Mercury, Tables of the heliocentric motion of Venus, Tables of the heliocentric motion of Mars*; Astronomical Papers of the American Ephemeris, vol. VI, part 1-4; Washington (1898).

Analytical series expansions for the motions of the inner planets.

[31] S. Newcomb; *Tables of the heliocentric motion of Uranus, Tables of the heliocentric motion of Neptune*; Astronomical Papers of the American Ephemeris, vol. VII, part 3-4; Washington (1898).

Series expansions of Uranus and Neptune coordinates.

[32] F. E. Ross; *New Elements of Mars*; Astronomical Papers of the American Ephemeris, vol. IX, part 2; Washington (1917).

Improvement of the orbital elements in Newcomb's tables of the martian orbit.

Chapter 7 (Physical Ephemerides)

[33] M. E. Davies, V. K. Abalakin, M. Bursa, J. H. Lieske, B. Morando, D. Morrison, P. K. Seidelmann, A. T. Sinclair, B. Yallop, Y. S. Tjuflin; *Report of the IAU/IAG/COSPAR Working Group on Cartographic Coordinates and Rotational Elements of the Planets and Satellites: 1994*; Celestial Mechanics and Dynamical Astronomy vol. 63, 127-148 (1996).

Describes the rotational elements of the major planets currently used in the *Astronomical Almanac* [1]. Recent results of the working group on cartographic coordinates of International Astronomical Union are published approximately every three years. (Celestial Mechanics vol. 22, pp. 205-230 (1980), vol. 29, pp. 309-321 (1983), vol. 39, pp. 103-113 (1986), vol. 46, pp. 187-204 (1989), vol. 53, pp. 377-397 (1992)).

A detailed description of the computation of physical ephemerides is additionally given in the *Explanatory Supplement* [4].

Chapter 8 (The Orbit of the Moon)

[34] E. W. Brown; *An introductory treatise on the Lunar Theory*; Cambridge University Press (1896), Dover Publications (1960).

Description of the various perturbation-theory methods of analytically handling the motion of the Moon.

[35] M. Chapront-Touzé, J. Chapront; *ELP 2000-85: a semi-analytical lunar ephemeris adequate for historical times*; Astronomy and Astrophysics, vol. 190, p. 342; (1988).

Mean orbital elements and perturbation terms describing the orbit of the Moon over long periods of time.

[36] M. C. Gutzwiller, D. S. Schmidt; *The motion of the moon as computed by the method of Hill, Brown and Eckert*; Astronomical Papers of the American Ephemeris, vol. XXIII, part 1; Washington (1986).

Analytical series expansions of the main problem relating to the motion of the Moon (without taking planetary perturbations into account).

[37] *Improved Lunar Ephemeris 1952-1959*; Nautical Almanac Office; Washington, 1954.

Revised version of Brown's lunar theory of 1954 with all necessary perturbation terms and mean arguments.

A good introduction to the theory of the lunar orbit and the basic perturbations can be found in the volume *Himmelsmechanik II* by Bucerius and Schneider [2]. The method of approximating functions by Chebyshev polynomials is explained in textbooks on numerical mathematics (see [11] and [13]).

Chapter 9 (Solar Eclipses)

[38] F. Espanak; *Fifty Year Canon of Solar Eclipses*; NASA Reference Publication 1178 Revised; NASA (1987).

Elements and maps of solar eclipses in the 20th and 21st century with detailed data for eclipses in the years 1986 – 2035.

[39] H. Mucke, J. Meeus; *Canon of Solar Eclipses, -2003 to 2526*; Astronomisches Büro, Hasenwartgasse 32, Wien (1984).

Elements for 10774 solar eclipses for about 4500 years on the basis of Newcomb's theory of the solar orbit, and Brown's lunar theory. Simple diagrammatic charts showing the approximate path of the centre line.

[40] T. Oppolzer; *Canon der Finsternisse*; Denkschriften der Math.-Naturw. Classe der Kaiserlichen Akademie der Wissenschaften Bd. 52, Wien (1883); Dover Publications, Inc., New York (1962).

Elements for solar and lunar eclipses for the period -1207 to 2163. Maps of the paths of the centre line for total solar eclipses.

The book *Astronomical Algorithms* by J. Meeus [8] describes various methods, which are particularly suitable for rapid prediction and assessment of possible eclipse dates. Ready to use tables for all kind of solar eclipses are given in [9]. The classical method of calculating solar eclipses using Besselian elements is described in the *Explanatory Supplement* [4] and in Mueller [10]. Current predictions of solar eclipses are provided by the web site
`http://sunearth.gsfc.nasa.gov/eclipse/eclipse.html` of Fred Espanak.

Chapter 10 (Stellar Occultations)

[41] J. Robertson; *Catalog of 3539 Zodiacal Stars for the Equinox 1950.0*; Astronomical Papers of the American Ephemeris, vol. X, part 2; Washington (1917).

Reference catalogue for the prediction and assessment of stellar occultations. Contains stars near the ecliptic down to about ninth magnitude that may be occulted by the Moon.

The prediction of stellar occultations is described in various textbooks on spherical astronomy (for example, Green [5]) and in the *Explanatory Supplement* [4]. More detailed treatment of the orbit of the Moon and the rotation of the Earth is discussed in *Spherical and practical astronomy* [10]. Updated positions can be obtained from the PPM catalogue [47].

Chapter 11 (Orbit Determination)

[42] C. F. Gauss; *Theory of the Motion of the Heavenly Bodies Moving about the Sun in Conic Sections*; Dover Publications; New York;

English translation of *Theoria motus corporum coelestium*, in which Gauss describes his method of orbital determination. The book is mainly of historical interest, but it is less suitable as a textbook on orbital determination.

[43] H. Bucerius; *Bahnbestimmung als Randwertproblem I-V*; Astronomische Nachrichten, vol. 278, 280, 281, 282 (1950-1955);

A series of five papers, in which Gauss' simplified and full methods of orbital determination are examined, amongst other subjects. Numerous examples illustrate possible problems that may occur in determining orbits.

[44] D. L. Boulet; *Methods of orbit determination for the micro computer*; Willmann-Bell; Richmond, Virginia (1991).

Principles and methods of ephemeris calculation and orbit determination, including special perturbations and the orbit determination methods of Laplace, Olbers and Gauss. The listed Basic programs are also available on floppy disk.

An up-to-date introduction to orbital determination is also given by the corresponding chapter in *Himmelsmechanik I* by Bucerius and Schneider [2].

Chapter 12 (Astrometry)

[45] *SAO Star Catalog*; Smithsonian Astrophysical Observatory; Cambridge, Massachusetts (1966);

Four-volume star catalogue with positions, proper motions and magnitudes of 258 997 stars. Epoch and equinox 1950.0. The accuracy of the positions is around 1 ″.

[46] W. Dieckvoss; *AGK3 Star Catalogue of Positions and Proper Motions North of -2°.5 Declination*; Hamburg-Bergedorf (1975).

Comprehensive catalogue of positions for the northern hemisphere for astrometric work. Based on photographic surveys carried out by the Bonn and Hamburg-Bergedorf Observatories. Equinox 1950.0.

[47] S. Röser, U. Bastian; *Catalogue of Positions and Proper Motions*; Astron. Astrophys. Suppl. Ser. 74, 449 (1988).
S. Röser, U. Bastian; *PPM Star Catalogue*; Spektrum Akademischer Verlag, Heidelberg (1991).
U. Bastian, S. Röser; *Catalogue of Positions and Proper Motions - South* Astronomisches Rechen-Institut, Heidelberg (1993).
U. Bastian, S. Röser; *The Bright Stars Supplement to the PPM and PPM South Catalogue, Revised Edition* Astronomisches Rechen-Institut, Heidelberg (1993).

S. Röser, U. Bastian, A. V. Kuzmin; *The 90000 Stars Supplement to the PPM Star Catalogue*; Astron. Astrophys. Suppl. Ser. 105, 301 (1994).

Successor of the SAO Catalog with positions and proper motions of approximately 470 000 stars of the northern and southern hemisphere referred to the equinox J2000 (FK5). The various parts of the catalogue are available from the Centre de Données astronomiques des Strasbourg under
`ftp://cdsarc.u-strasbg.fr/cats/I/146/`
`ftp://cdsarc.u-strasbg.fr/cats/I/193/`
`ftp://cdsarc.u-strasbg.fr/cats/I/206/`
`ftp://cdsarc.u-strasbg.fr/cats/I/208/`
The printed version of the PPM catalogue is published by Spektrum Akademie Verlag, Heidelberg.

[48] W. Tirion, B. Rappaport, G. Lovi; *Uranometria 2000.0*; Willmann-Bell; Richmond, Virginia (1987).

Atlas of the whole northern and southern skies with 473 charts at a scale of 1.85 cm=1 ° for equinox 2000. Contains stars down to $9\overset{m}{.}5$ based on the Bonner Durchmusterung, the southern Bonner Durchmusterung, and the Cordoba Durchmusterung.

The calculation of standard coordinates is treated in many textbooks on spherical astronomy (cf. [5]). Details of regression fitting can be found in books on numerical mathematics, such as [11] and [13].

Index

www.ingramcontent.com/pod-product-compliance
Ingram Content Group UK Ltd.
Pitfield, Milton Keynes, MK11 3LW, UK
UKHW041309051126
16364UKWH00003B/4
9783642034350